Thermoforming

Scrivener Publishing
100 Cummings Center, Suite 541J
Beverly, MA 01915-6106

Publishers at Scrivener
Martin Scrivener (martin@scrivenerpublishing.com)
Phillip Carmical (pcarmical@scrivenerpublishing.com)

Thermoforming

Processing and Technology

Muralisrinivasan Natamai Subramanian

Scrivener Publishing

WILEY

This edition first published 2024 by John Wiley & Sons, Inc., 111 River Street, Hoboken, NJ 07030, USA and Scrivener Publishing LLC, 100 Cummings Center, Suite 541J, Beverly, MA 01915, USA

For more information about Scrivener publications please visit www.scrivenerpublishing.com.

Wiley Global Headquarters
111 River Street, Hoboken, NJ 07030, USA

For details of our global editorial offices, customer services, and more information about Wiley products visit us at www.wiley.com.

Library of Congress Cataloging-in-Publication Data

ISBN 978-1-119-55586-5

Cover image: Pixabay.Com
Cover design by Russell Richardson

Set in size of 11pt and Minion Pro by Manila Typesetting Company, Makati, Philippines

Printed in the USA

10 9 8 7 6 5 4 3 2 1

Contents

Preface

The field of thermoforming is experiencing rapid development driven by commercial factors. Today, millions of tons of polymers are manufactured for use in various applications, both as commodity and specialty polymers.

The first edition of the book *Update on Troubleshooting in Thermoforming* was published nearly a decade ago by iSmithers Rapra, UK. The second edition, titled *Thermoforming – Processing and Technology*, aims to incorporate new and revised material on recent developments in thermoforming. This edition includes new, fully revised chapters and updated information on materials and processes.

The book is designed to provide practitioners and students with essential information on processing and technology in a concise and portable format. It is a valuable resource for polymer processors, engineers, technologists, and students, offering a comprehensive update on thermoforming. The book caters to both engineers and experts in physics and chemistry, providing introductory aspects, background information, and an overview of thermoforming processing and technology. The troubleshooting section includes flowcharts to assist in correcting thermoforming processes.

Thermoforming – Processing and Technology offers a complete account of thermoplastics, covering properties and forming, with chapters providing perspective on the technologies involved. The book is practical and mechanistic in its approach, making it useful for both industry professionals and academics. It serves as a self-contained reference manual and is engaging, accessible, and attractively presented. It would be a valuable addition to any research group with an interest in polymers, particularly in the field of thermoforming.

I am deeply thankful to the people who have made significant sacrifices of their personal time to prepare the book. This includes my parents who raised me, my wife Himachalaganga, and my children Venkatasubramanian, Amrutha and Sailesh. I am also grateful to those who have supported me in my education and career, including my professors and teachers. Above all Lord Natarajar, who brought me to this world, and my Guru,

Thiruchendur Subramanya Swamy, who provided his knowledge to write this book. Furthermore, I would like to express my gratitude to Mr. Martin Scrivener and his team at Scrivener Publishing for their invaluable assistance in bringing this book to fruition.

Dr. Muralisrinivasan
Madurai, India

1
Introduction

Thermoforming was one of the earliest methods for fabricating thermoplastic materials. This involves shaping thermoplastic sheets into distinct parts. The United States has contributed to plastic processing methods. However, most of these techniques originated in Europe. It was not until the mid-twentieth century that a comprehensive understanding of its true nature was achieved [1].

Thermoforming has undergone significant growth and development with the introduction of cellulosic sheet materials, followed by acrylic and vinyl. In the 1890s, baby rattles and teething rings were thermoformed, but growth was slow until the 1930s, when roll-fed machines were developed in Europe. This process made significant progress in the 1950s using high-impact polystyrene for manufacturing containers and lids in the dairy industry. In addition, it has been used in the production of signage, exhibits, toys, and packaging materials. Initially, the thermoforming industry was limited by the lack of suitable sheet materials and forming equipment. However, in recent years, the process has rapidly advanced due to its advantages such as low machine and mold costs, low temperature and pressure requirements, ease of forming large parts, and fast mold cycles [2].

The emergence of plastics, synthetic materials that showcase human creativity and ingenuity, coincided with this understanding. Among the various methods utilized for processing plastics, thermoforming is a widely used technique for shaping extruded sheets into desired forms. This process is commonly used to produce large parts with low production volumes. It is a more cost-effective alternative to injection molding, which incurs higher expenses due to substantial fixed costs.

Thermoforming is a straightforward processing technology [2]. This involves modifying polymeric materials or systems to enhance their usefulness. Thermoforming is a cost-effective alternative to other plastic molding and forming methods. The process discussed here is often associated with the production of packaging materials such as blister packs and disposable

Muralisrinivasan Natamai Subramanian. Thermoforming: Processing and Technology, (1–4)

coffee cup lids. However, it is important to note that the cost and time benefits of this process can be extended to a wide range of products in various industries.

Thermoplastics dominate the global polymer market and account for 85% of all polymers [3]. Most thermoplastic production focuses on large-volume, low-cost commodity resins, such as polyethylene, polypropylene, polystyrene, and polyvinyl chloride. Engineering plastics, such as acrylics, acrylonitrile-butadiene-styrene (ABS), and high-impact polystyrene (HIPS), are chosen for their performance and cost-effectiveness. However, there is an increasing demand for high-performance materials, which has led to the utilization of engineering plastics such as polyacetal (POM), polyamides (PAs), and polycarbonate (PC), and polyesters such as polyethylene terephthalate (PET), polybutylene terephthalate (PBT), and polypropylene oxide, as well as their blends. Polymers such as liquid crystalline polymers, polyetheretherketone (PEEK), and fluoropolymers are commonly used in high-temperature applications [4].

Thermoforming presents numerous advantages over alternative thermoplastic fabrication techniques. These advantages encompass several aspects, such as the ability to fabricate components with a large surface area, reduced expenses for molds and equipment due to its low-pressure requirement, the capability to produce extremely thin-walled parts, increased production rates for high-volume thin-walled products, and the ability to manufacture low-volume heavy-gauge products at lower tooling costs [5].

This realization coincided with the advancement of plastics, which are synthetic materials that showcase human ingenuity and resourcefulness. The process involves shaping the thermoplastic sheets into precise components. Thermoforming has gained considerable importance as a method for processing plastic sheets, particularly for producing larger components [6].

Thermoplastics, which have gained significant popularity in the global polymer market, are exceptionally suitable for thermoforming processes because of their ability to be produced in high volumes at lower costs. This versatile procedure can be used for any thermoplastic sheet material, which demonstrates the necessary dimensional stability and impact resistance. Thermoforming presents numerous advantages over alternative thermoplastic fabrication techniques.

The market for thermoforming is experiencing growth due to its increasing use in the production of complex shapes and a wider range of materials. Thermoforming is widely regarded as a viable and economically advantageous alternative to various plastic molding and forming methods.

The development of applications has largely been brought about by the introduction of new manufacturing methods. New methods have encouraged the development of new forming machines and techniques. This has brought the thermoforming industry to a new level in terms of processing technology.

2
Polymers

2.1 Introduction

Polymers have experienced steady market growth and are widely acknowledged as a significant category of materials. These materials possess characteristics such as affordability, ease of manipulation, and versatility, rendering them suitable for a diverse array of applications. These applications encompass a wide range of household items, packaging materials, advanced fibers, medical devices, and wearable electronics. The application of polymers has greatly enhanced our standard of living and sparked revolutionary progress in various sectors [1, 2].

The prevalence of synthetic polymers can be attributed to the fact that 90% of these polymers are derived from finite fossil feedstock. The incorporation of these materials into diverse products is driven by their exceptional stability, processability, versatile mechanical properties, and durability. Although the terms plastics and polymers are often used interchangeably, it is crucial to acknowledge that there is a subtle distinction between them [3, 4].

Polymer engineering encompasses the technological processes used for the complete synthesis of meticulously regulated macromolecules. The objective of polymer engineering is to attain mastery over the physical characteristics of macromolecules, encompassing factors such as molecular weight, molecular weight distribution, end functionality, tacticity, stereochemistry, block sequence, and block topology. Polymers find applications in a wide range of consumer products such as carpets, furniture, glues, and clothing. In addition, they play a crucial role in advanced engineering, particularly in the development of materials used in the aerospace industry. Therefore, the utilization of polymers and their constituent monomers is important in our daily lives [5].

Muralisrinivasan Natamai Subramanian. Thermoforming: Processing and Technology, (5–26)

2.2 Physics and Chemistry of Polymers

The study of polymer materials is a complex and intriguing subject in physics and chemistry. The structure of a polymer is determined by the number and type of repeating units it contains. Polymer science investigates the characteristics, composition, and behavior of these substances at the atomic, molecular, and macroscopic levels. This multidisciplinary field combines concepts from biology, engineering, chemistry, and physics to understand and manipulate material properties for specific purposes. Many polymeric materials exhibit emergent properties, which are unique characteristics arising from the interaction of their constituent monomers.

One particularly fascinating aspect of chemistry and polymer science is the ability to transition materials between different phases, particularly in relation to the monomers. These transitions allow the manipulation of materials for real-world applications. This process is not only scientifically intriguing, but also essential for the development of new and innovative materials with a wide range of uses in modern technology and everyday life. This is particularly true for monomers that form polymers [6, 7].

Polymer molecules consist of a sequence of monomers that are joined by chemical bonds. The orientations of these bonds between successive monomers are correlated with those of adjacent monomers and crossing energy barriers is necessary for a bond to change its orientation. The arrangement of monomers within a polymer significantly affects its characteristics. Describing the behavior of materials at different length scales, from atomic interactions to macroscopic properties, is challenging. The development of polymer materials with specific properties is crucial for numerous technological advancements.

Even small variations in the arrangement of monomers can lead to significant differences in the material behavior. Polymers are highly versatile due to their flexible design, structure, and chemical composition, allowing them to exhibit a wide range of properties that can be tailored to meet specific requirements [8].

2.3 Natural Polymers

Natural polymers, which are associated with biopolymers, are formed through metabolic processes in living organisms [9]. These polymers consist of monomeric units that are linked by covalent bonds. Examples of natural polymers include proteins such as collagen and silk fibroin as well as polysaccharides such as chitosan, alginate, hyaluronic acid, and cellulose [10]. These polymers play important roles in nature, including the preservation and transmission of genetic information and the storage of cellular energy. One of their key advantages is their ability to biodegrade, with the released CO_2 rapidly absorbed by agricultural crops and soil. Among polysaccharide biopolymers, cellulose is particularly abundant and is present in approximately 33% of all plant components [11, 12]. Other notable natural polymers include chitin/chitosan, starch, and lignin. Chitosan, alginate, cellulose, lignocellulose, starch, and PVA are among the most promising and frequently studied natural polymers, either as standalone materials or in combination with other advanced materials [13].

2.4 Synthetic Polymers

Synthetic polymers belong to a distinct class of polymers derived from crude oil, petrochemicals, natural gas, or biomass. These polymers possess a wide range of desirable characteristics such as low density, high durability, and resistance to deterioration. In addition to their advantages in terms of weight and cost, polymer products offer enhanced durability that can help prevent damage during transportation. Furthermore, many polymers exhibit corrosion resistance, making them well suited for use in harsh environments, such as chemical manufacturing facilities [14].

Polymerization, which involves chemically bonding monomers to form long chains, is utilized to produce synthetic polymers, which are subsequently used in various manufacturing processes. Synthetic polymeric networks consist of repeatable inert units and are generally superior to natural polymers in terms of mechanical properties and immunogenic responses [15].

Synthetic polymers offer tailored structures and properties through the appropriate design of their functional groups. These advantages ensure predictable, reproducible, and adjustable properties, which can vary according to specific applications. For example, the degradation rate of synthetic polymers can be altered by manipulating their chemical compositions, crystallinities, and molecular weights.

The ability to process and shape polymers enables the efficient mass production of a diverse range of items. Techniques, such as injection molding, extrusion, blow molding, and thermoforming, facilitate the creation of complex designs and ensure reliable quality. The lightweight nature of plastic products contributes to reducing shipping expenses and energy consumption. Moreover, manufacturing products from plastics is often more cost-effective than manufacturing conventional materials [16].

2.5 Polymerization Methods

During the polymerization process, monomers undergo chemical bonding to form extensive chains or networks, resulting in the development of unique polymer properties. This chemical reaction, known as a polymerization reaction, leads to the formation of high-molecular-weight molecules from the monomers [17]. The structural features of polymers, such as linear, branched, or network configurations, are determined by the arrangement of monomers and the types of chemical bonds that connect them. The ability to manipulate the structure and composition of polymers is crucial in various industries and applications. Two primary approaches are utilized in the production of polymers: addition and condensation polymerization. Both mechanisms can be used in the polymerization of the same monomer or different monomers can be used to create the same polymer through both approaches, provided that suitable functional groups are available for each individual polymerization. Addition and condensation polymerizations are both essential in the creation of diverse polymers, fibers, rubbers, and other materials that have extensive applications in everyday life and industry. The selection of specific monomers and the desired characteristics of the final polymer product determine the appropriate polymerization method [18, 19].

2.5.1 Addition Polymerization

In this process, the reaction between monomers containing double or triple bonds leads to the formation of polymer chains. This involves the breaking of the double or triple bonds and the subsequent connection of the monomers without the production of any byproducts. The double bonds of the monomers react without releasing any molecules. These reactions occur *via* the addition of monomer molecules *via* unsaturated (double) bonds [20].

The initiation of the reaction can be achieved using chemical molecules, such as azo compounds or peroxides, or through physical sources, such as heat or electromagnetic radiation. These initiators create radicals, anions, or cations in the monomer. Depending on the type of initiation, addition polymerization can be classified as radical, anionic, cationic, or coordination polymerization. All addition polymerizations involve three stages: initiation, propagation, and termination [18, 21].

In the case of addition polymerization, a reactive center such as a radical, an anion, or a cation must first be created on a molecule that contains a double bond. New monomer molecules are then successively added to this active molecule, creating a new active center for further addition. This process continues until the reaction is terminated by other reactions or events. Termination results in the formation of "dead" polymer, which is no longer capable of further reaction. During these reactions, addition must be initiated and typically involves a termination reaction. No small molecules, such as water or alcohol, are released during this process [22].

The most common type of addition polymerization is free-radical polymerization. Most radical polymerizations require an initiator to generate the initial active radical and initiate a chain of addition reactions. The most common initiation reaction is the thermal decomposition of molecules containing weak bonds such as peroxides (-OO-) or azo compounds (-N=N-). Addition polymerization is generally rapid and can be moderately or highly exothermic. Additionally, polymerization of an individual chain is completed quickly, resulting in the formation of high-molecular-weight products even in the early stages. The concentration of monomers in the medium slowly decreased over time [23]. The most significant group of addition polymerizations involves the combination of monomers, such as ethylene, propylene, styrene, and vinyl chloride, which give rise to polyethylene, polypropylene, polystyrene, and polyvinylchloride, respectively.

2.5.2 Condensation Polymerization

The homogeneous two-component polycondensation reaction holds significant historical importance in the field of polymer science. Achieving a consistently high degree of polymerization (DP) is a crucial consideration in polymer synthesis, as the desired physical properties often arise from high molecular weights. Condensation polymerization involves the reaction between two distinct functional groups present in monomers. This process entails the elimination of small molecules, such as water, alcohol, or ammonia, while simultaneously forming a polymer chain from monomers that possess two or more functional groups, such as -OH and -COOH. The reaction occurs in multiple phases, with each phase releasing a small molecule, hence earning the designation of a "condensation" reaction [24] and presented in Equation (2.1).

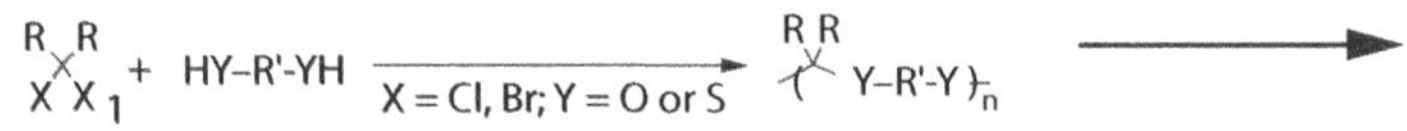

(2.1)

It is applicable to monomers that possess functional groups, such as -COOH, -COOR, -COOOC-, -COCl, -OH, -NH2, -CHO, -NCO, and epoxy. In condensation polymerization, the reactive groups located at the ends of each monomer react with one another. Consequently, a growing chain is formed with reactive groups at both ends, and when two chains combine, the length of the chain immediately increases.

However, the increase in the molecular weight of the product in condensation polymerization is slow due to the addition of growing chains to each other, leading to a depletion of monomers at the initial stages. Condensation polymerization is typically characterized by its slow rate, limited by equilibrium, and slightly exothermic nature [25].

In condensation polymerization, two functional groups form a bond by releasing a small molecule such as H_2O. While there are some condensation reactions where no molecule release occurs, such as polyurethane polymerization, nylon 6 is an example of a polymer with repeating units $-NH(CH_2)_5CO-$. It can be synthesized through either condensation of 6-aminocaproic acid or addition polymerization of caprolactam [26].

Furthermore, there have been recent advancements in polymerization techniques, including click polymerization, atom transfer radical polymerization (ATRP), and reversible addition-fragmentation chain transfer polymerization (RAFT). These techniques employ specialized catalysts and agents to produce polymers with controlled molecular weights, offering new possibilities for polymer design [27–29].

2.6 Polymer Molecules in Thermoforming

The behavior of polymeric materials is influenced by the chemical composition and physical structure of both the penetrant molecule and polymer itself. These factors play a crucial role in determining the mobility of the chain segments, presence of defects, and interactions that govern the extent of sorption and molecular mobility of the penetrant within the polymer.

Anisotropy, or directional dependence, occurs in polymers when the links of the polymeric chain align in the direction of stretching. In thermoforming, anisotropy is associated with the temperature at which a transition between two or more phases occurs as well as the temperature at which a change of state, such as melting, occurs. The molecular structure of the polymer contributes to these transitions, which depend on the monomer structure, degree of orientation, and extent of stretching at specific temperatures. It is important to note that a given transition in polymers can arise from either the amorphous or the crystalline phase.

Generally, polymers exhibit significant changes in their physical properties as a function of temperature. The theory of rubber elasticity states that strained elastomers with a low degree of crosslinking exhibit a combination of orientation and segmental motion. However, in stretched amorphous polymers, segmental motion is absent because it can only exist below the glass transition temperature. This resulted in a frozen-in orientation. By subjecting amorphous linear polymers to strain at temperatures above the glass transition temperature, they can be brought into an anisotropic state. Once cooled below the glass transition temperature, the polymer remains in the stretched state even when unloaded.

In mechanical applications, the amorphous regions between crystallites largely determine the response to stress, giving polymers desirable qualities such as ductility and toughness. In electronic applications, it is believed that tie molecules spanning adjacent crystalline lamellae can improve field-effect charge carrier mobility [30].

2.7 Classification

Polymers are classified on the basis of their unique properties, characteristics, or attributes. The objective of this categorization is to methodically arrange these elements and their connections, promoting comprehension and enabling a deeper investigation within each specific category.

2.8 Primary Classification

In the field of polymer science, plastics can be classified into two fundamental groups: homopolymers and copolymers, which are distinguished by their distinct molecular structures.

2.8.1 Homopolymers

Homopolymers possess a consistent and uniform structure throughout their composition because they are composed of identical repeating units of a single monomer. Despite their uniformity, the properties of homopolymers can still be modified by adjusting various factors, such as the processing parameters, molecular weight, or monomer structure [31].

The chemical structure of homopolymers is relatively straightforward, allowing relatively easy synthesis. Depending on the specific monomer and polymerization conditions, homopolymers can exhibit varying degrees of crystallinities. Higher levels of crystallinity can enhance mechanical properties but may reduce transparency. The homogeneous structure of homopolymers generally makes them more suitable for processing, making them applicable to a range of production techniques, including injection molding, extrusion, blow molding, and thermoforming. Additionally, homopolymers typically possess high thermal stability, enabling them to withstand high temperatures without significant degradation. This thermal resistance makes them valuable for applications that require heat resistance [32].

Homopolymers are known for their inherent rigidity, which is attributed to their high elastic modulus. This characteristic renders them particularly advantageous in scenarios where maintaining structural integrity is of utmost importance. Additionally, many homopolymers exhibit exceptional optical clarity, making them highly suitable for applications that necessitate transparency and visual acuity, such as display screens, optical lenses, and transparent packaging materials. Moreover, homopolymers generally offer a more cost-effective solution compared to copolymers or more intricate polymers due to their straightforward chemical composition and ease of production. However, it is crucial to acknowledge that homopolymers possess a limited range of properties, which can be perceived as a drawback [33].

2.8.2 Copolymers

The blending of polymeric substances is a cost-effective and straightforward method to create new materials with enhanced mechanical properties. It is anticipated that copolymers will exhibit a wider melting temperature range and a sigmoidal relationship between their crystallinity and temperature. When two polymer surfaces that are not soluble in each other come into contact, it is expected that their surface functionality will arrange or structure themselves in a way that minimizes free energy. Understanding the structure of polymer/polymer interfaces is crucial in various areas such as adhesion, polymer blends, and nanocomposites [34].

These new multiphase materials can be obtained as blends, block copolymers, and graft copolymers, all of which typically consist of two or more polymeric phases in the solid state. It is important to differentiate these materials from composite materials. Most homopolymer phases are insoluble in each other, resulting in materials with low strength due to the lack of interfacial adhesion [35].

Copolymers have a narrower range of compositions than blends, but they offer several advantages. The different segments in the copolymers are covalently bonded, eliminating the interface problem [36]. The molecular architecture of copolymers can be precisely controlled to produce novel materials. Copolymers can also strengthen blends of immiscible polymers by acting as emulsifiers and facilitating physical connections between the phases. This improves the interfacial adhesion and ability to transfer loads between components [37].

In an ordered copolymer, where structural units of a specific type are arranged in long sequences, crystallinity should disappear at a temperature slightly below the melting temperature of the pure polymer. The phase rule cannot be applied to the crystal-amorphous transformation in polymers because the free energy per unit amount of the amorphous "phase" is not a unique function of the composition [38].

The presence of copolymers at the interface between immiscible polymers has been widely recognized for their significant influence on interfacial properties, provided that an appropriate copolymer structure is utilized. The degree of toughening varies depending on the copolymer architecture and the interactions between the copolymer and polymer [39, 40].

2.8.2.1 Block Copolymers

Block copolymers are polymers that consist of two or more distinct sequences joined together by covalent bonds and have different chemical compositions. Blending different homopolymers together to form a block copolymer structure offers unique properties that cannot be achieved by simply blending individual homopolymer components. This is due to the ability of the block copolymer to segregate its chains, resulting in distinct characteristics. There are two types of block copolymers: coil–coil and rod–coil, with the latter having a rigid and inflexible rod-like segment [41].

Unlike homopolymers, block copolymers consist of distinct blocks of different monomers, rather than repeating units of the same monomer. Block copolymers are a fascinating group of polymeric materials that are of great interest in both the scientific and technological fields. These consist of sequences of different homopolymers within the same molecule. These block copolymers had distinct hard and soft blocks with substantially different compositions. In the strong segregation limit, they exhibit highly ordered long-range organization, which is a result of the interplay between the thermodynamics of polymer interactions, configuration of domain-forming blocks, and minimization of free energy with the morphology of microphase-separated domains [42].

These blocks can vary in size, and their arrangement can result in different morphologies and properties, as depicted in Figure 2.1. Within a block copolymer, blocks can form distinct and specialized regions. This phase separation led to the formation of microdomains with unique characteristics. This property makes block copolymers important in industries, such as nanotechnology and the production of self-assembling materials.

These copolymers tend to separate into different phases on the nanoscale due to the repulsive interaction between their repeating units. This results in the formation of well-organized morphologies, such as alternating lamellae, gyroids, hexagonally packed cylinders, and body-centered cubic spheres. The equilibrium morphology and domain size of block copolymers depend on the balance between the enthalpic contribution associated with short-range segmental interactions and the entropic contribution associated with chain packing and distribution within the microdomains [43].

At equilibrium, a dense collection of monodisperse diblock copolymer chains is arranged in minimum free energy configurations (ordered). Lowering the temperature, i.e., increasing the energy parameter, favors a reduction in the A–B monomer contacts (disordered) [44]. To manipulate the phase behavior of block copolymers, factors such as the volume fraction of components, degree of polymerization, and segmental interaction parameters are considered. Another approach to modify the phase behavior of block copolymers is to blend them with other polymers such as homopolymers or other block copolymers. This blending introduces an additional degree of freedom into the system, resulting in a more complex and diverse phase behavior than that of a neat block copolymer. This enhanced behavior includes microphase separation, macrophase separation, order–order transition, and order–disorder transition. For instance, studies have been conducted on binary block copolymer blends where two

Figure 2.1 Schematic representation of block copolymer.

block copolymers have the same repeating units but different molecular weights [45].

Commercially available block copolymers are more cost-effective than the other sequestering phases. These materials are designed to possess thermomechanical properties that meet engineering requirements, such as those required for automobile parts. However, there is limited knowledge about the structure-property relationships that govern their interactions with small molecules, which ultimately determine their performance in various applications. Notably, exceptional mechanical properties can be achieved by carefully selecting the type and molecular weight of the block copolymer.

Block copolymers have frequently been used as compatibilizers in polymer blends that undergo phase separation. The ability of block copolymers to act at interfaces is attributed to their preferential location at the interface and the extension of each block into the homopolymer phase, where their energetic interactions are favorable [46]. However, block copolymers have certain drawbacks when used as interfacial agents: they are relatively challenging to synthesize and tend to form micelles, which significantly reduces their ability to segregate at the interfaces [47, 48].

In the context of film crystallization, the aggregation of block copolymers significantly affects the optical and electronic properties of materials. This is because interchain contacts alter the absorption characteristics, quench emission, and affect conduction behavior. Block copolymer architectures have been successfully used to control the orientation and morphology of optically and electronically active materials that can be processed. The flexible segments in these block copolymers primarily serve to increase the overall solubility, reducing intermolecular contact and aggregation of the functional segments.

Block copolymers can enhance the bonding between substances with different surface characteristics. Block copolymers are utilized in the fabrication of nanoscale patterns and templates, which play crucial roles in nanolithography and nanodevice production. These materials have applications in various fields. Some block copolymers are suitable drug delivery systems because they can self-assemble into vesicles (polymersomes) or spherical micelles in solution. The mechanical characteristics and toughness of regular polymers can be significantly improved by incorporating small amounts of block copolymers. Due to their unique permeability characteristics, block copolymers are used in membrane technologies for water purification and gas separation [49].

2.8.2.2 Diblock Copolymers

The most basic forms of copolymers are diblocks, which consist of a sequence of A monomers connected to a sequence of B monomers. From a thermodynamic perspective, the impact of diblock copolymers on interfaces can be described as follows: copolymers exhibit amphiphilic behavior in a mixture of homopolymers [33]. As the concentration of the copolymers increases, the interfacial tension decreases continuously until a third phase, rich in copolymers, is formed, terminating this process. Block copolymers, which are partially compatible with both the A-rich and B-rich phases, tend to aggregate at interfaces, reducing the number of direct contacts between the A and B homopolymers, thus reducing the interfacial tension [50]. Simple AB-block copolymer melts undergo microphase separation on a single characteristic length scale. However, when more than two types of monomers are involved, microphase separation often occurs at different scales [7, 51, 52].

These blocks are composed of two separate units, A and B, which are alternately connected to each other, such as A-B-A-B-A-B. The blocks tend to separate into various domains and do not mix easily, resulting in a variety of microstructures, including spherical, cylindrical, and lamellar structures [53, 54].

2.8.2.3 Multiblock Copolymers

In the field of polymer chemistry, multiblock copolymers are commonly composed of consecutive blocks of the same type of monomers. Among these copolymers, the simplest and most interesting are diblock copolymers, which consist of two homopolymers connected end-to-end, with the first polymer composed of type A monomers and the second polymer composed of type B monomers [55]. At low temperatures, the A and B monomers undergo local demixing due to the irreversible chemical links between the blocks, but this demixing does not occur globally. Depending on the average composition of the chain and the temperature, various microdomain textures spontaneously appear at thermal equilibrium. By manipulating the displacement of different volumes by the monomers or by making one block significantly stiffer than the other, the boundaries between the different phases can be shifted [56].

The development of block copolymers has recently attracted significant attention due to their intriguing characteristics and potential applications in cutting-edge technologies. Phase segregation and microphase segregation in block copolymers provide them with distinct qualities. The two primary forms of block copolymers are multiblock and diblock copolymers. These materials display more complicated phase-separated microstructures and feature more than two types of monomer blocks in their structures, such as A-B-C or A-B-A-C-B-A [57].

2.8.3 Alternating Copolymers

The distribution of monomers within the chain of an alternating copolymer follows a predictable sequence. This type of copolymer is characterized by the composition of two or more distinct monomer units. In alternating copolymers, monomer units are arranged alternately along the polymer chain. Creating, a regular pattern in the polymer backbone, with each monomer unit followed by a different monomer unit [58].

The repetitive arrangement of monomers in alternating copolymers leads to distinct characteristics. Typically, these copolymers exhibit higher melting temperatures and increased stiffness compare to random copolymers [59].

For example, a basic alternating copolymer of type AB can be represented as A-B-A-B-A-B-A-B-A-B-... as shown in Figure 2.2.

In this structure, "A" and "B" denote different monomer units that are repeated in an alternating manner. The arrangement of monomers in alternating copolymers is often determined by the reactivity ratio of the monomers during polymerization. If the reactivity ratios are balanced, the monomers tend to undergo alternate integration. These copolymers often possess unique characteristics compared to homopolymers or other types of copolymers [60]. The regularity in their structure can lead to changes in their mechanical, thermal, and optical properties, resulting in more orderly packing in the solid state. Furthermore, the chemical reactivity exhibited by alternating copolymers makes them highly valuable for various applications, including optics, electronics, and specialty materials. Different polymerization processes, such as step-growth polymerization and chain-growth polymerization, can be used to produce alternating copolymers. Well-known examples of alternating copolymers include poly(styrene-alt-maleic) and poly(ethylene-alt propylene) [61].

Figure 2.2 Representation of alternating copolymer.

Alternating copolymers constitute a distinct category of copolymers wherein the two comonomers polymerize in a regular alternating sequence along the polymer chain. The intriguing physical and chemical properties of these copolymers, as well as the underlying mechanisms governing their formation, have garnered significant attention from both the academic and industrial sectors. In an alternating copolymer, the two comonomers arrange themselves in a consistent alternating sequence [50, 62]

2.8.4 Random Copolymers

Random copolymers are considered viable alternatives to block copolymers because of their cost-effectiveness and ability to offer a high degree of design flexibility. A simple representation of the random copolymer is shown in Figure 2.3. When added to an immiscible polymer blend, random copolymers can effectively lower the cloud-point temperature. Additionally, the interfacial activity of random copolymers allows for the desired reduction in interfacial tension when incorporated into homopolymer blends. For instance, the inclusion of styrene-methyl methacrylate random copolymers has been found to effectively reinforce the interface between polystyrene and polymethyl methacrylate, although the presence of the random copolymer may affect fracture energy [63].

The effectiveness of a long random copolymer in strengthening the interface between immiscible homopolymers surpasses that of a short, symmetric block copolymer. While random copolymers have a relatively weak impact on interfacial strengthening, both block and alternating copolymers show promise as interfacial modifiers [61, 64].

2.8.5 Graft Copolymers

Graft copolymers possess characteristics of a combination of physical blends. The presence of chemical linkages within these copolymers significantly affects their physical properties. The number of initiating groups along the backbone can vary, resulting in copolymers with different grafting densities and distinct phase-separation behaviors. The hydrodynamic volume of a graft copolymer changes as its molecular weight increases,

Figure 2.3 Representation of random copolymer.

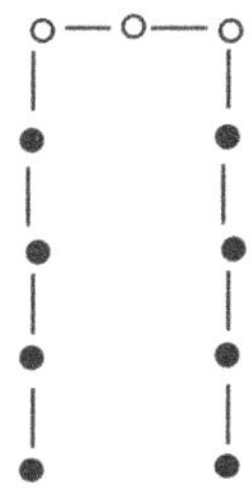

Figure 2.4 Graft copolymer with backbone of polymer "o" to which a number of "●" sequences.

which is responsible for its unique attributes. Grafting is widely recognized as an important technique for modifying the physical and chemical properties of polymers [65, 66].

All graft copolymers comprise two general structural features: a backbone of polymer "o" to which a number of "●" sequences are grafted, as shown in Figure 2.4.

Graft copolymers, known as A-g-B copolymers, are large molecular structures that resemble combs consisting of a linear backbone A with attached side B blocks. Typically, these two blocks are incompatible with each other. These copolymers have significant technological importance, particularly in their ability to make polymer blends compatible and create thermoplastic elastomers. In fact, they are the main agents used for compatibilization in commercial applications. The specific aggregation structures of these copolymers depend on various factors, including the level and distribution of grafting along the backbone, molecular weights of the backbone and grafts, and nature of the pendant grafts. In commercial applications, polydisperse graft copolymers with randomly distributed grafting points are commonly used because of their ease of production and cost-effectiveness [61].

2.8.6 Impact Copolymers

Impact copolymers, also known as elastomeric copolymers or impact modifiers, are a type of polymer blend that combine two distinct monomers to provide a unique set of properties. These copolymers are specifically designed to maintain a balance between stiffness and impact resistance, particularly at low temperatures where ordinary polymers may become brittle and prone to fracturing [67].

One of the key characteristics of impact copolymers is their ability to enhance the resilience of materials to impact and mechanical stress.

This is achieved through the inclusion of elastomeric segments, which can efficiently absorb and distribute energy, reducing the likelihood of fracture or shattering upon impact. Additionally, the formulation of impact copolymers allows for the adjustment of stiffness to suit the intended purpose, resulting in improved performance under different conditions [68].

Unlike regular polymers, which may shatter under cold conditions, impact copolymers exhibit ductility and resilience. This exceptional impact resistance makes them commonly used to enhance the durability and toughness of materials, such as plastics and rubbers. Furthermore, impact copolymers are often more processable than conventional elastomers, making them easier to produce and shape into desired forms [69].

The versatility of impact copolymers is evident in their utilization across various sectors, including consumer products, electronics, automotive, construction, and buildings. They are frequently used to enhance the properties of materials, such as polystyrene, polypropylene, and polyvinyl chloride (PVC). By adjusting the ratio and content of the two monomers, the characteristics of the impact copolymers can be tailored to meet the specific requirements for different applications. The flexibility of impact copolymers, even at low temperatures, is a significant advantage that makes them essential materials for many technical and industrial applications [70].

2.9 Secondary Classification

In materials science, plastics can be classified into two distinct categories: thermoplastics and thermosets. These groups are distinguished by their unique responses to changes in temperature and processing properties.

2.9.1 Thermoplastics

Thermoplastics are polymeric materials that can be reprocessed and fully recycled. In contrast to thermosets, thermoplastics consist of linear polymer chains that are not chemically linked. When heated, the intermolecular forces between the polymer chains weaken, allowing the material to reshape without any chemical bonding. This unique property enables the melting and reshaping of thermoplastics multiple times without significant degradation of their properties.

The classification of thermoplastics as either amorphous or semicrystalline based on their transition temperatures is a valuable categorization scheme in the field of materials science and engineering. The transition temperature provides insights into the molecular arrangement and behavior of polymer chains at different temperatures. It also aids in predicting the mechanical properties, heat resistance, and other relevant material characteristics for various applications.

Thermoplastics possess several key characteristics that make them widely used in various industries. They have a defined melting point, which is the temperature at which they transition from a solid to molten state. This property allows them to be melted, molded into various shapes, and then cooled and solidified again, thus enabling convenient processing and recycling. Thermoplastics are also more ductile than thermosetting plastics, meaning that they can be stretched and elongated without breaking.

Furthermore, thermoplastics exhibit a wide range of properties including flexibility, rigidity, transparency, impact resistance, chemical resistance, and electrical insulation. This versatility makes them suitable for various applications. Their mechanical properties can vary widely, with some thermoplastics being rigid and strong, whereas others are more flexible. Various processing techniques such as injection molding, extrusion, blow molding, thermoforming, and 3D printing can be used to manufacture thermoplastic products with intricate shapes and sizes.

Thermoplastics have applications in numerous industries including packaging, automotive components, household products, medical devices, toys, pipes, and cables. Common examples include high-density polyethylene (HDPE), low-density polyethylene (LDPE), polypropylene (PP), polyvinyl chloride (PVC), polystyrene (PS), polyethylene terephthalate (PET), poly(methyl methacrylate) (PMMA), and polycarbonate (PC). Their lightweight nature makes them useful in industries where weight is a critical factor. Additionally, they often exhibit good chemical resistance and weatherability and can be easily colored and molded into complex shapes. The recyclability of thermoplastics helps reduce waste and minimize environmental impacts [71].

Thermoplastics generally exhibit greater ductility than thermosets, although they can become brittle under certain conditions such as high strain rates, triaxial stress states, and low temperatures. The tensile moduli of most thermoplastics range from 1 to 4 GPa at room temperature, whereas thermosets have moduli ranging from 2 to 10 GPa. Due to their relatively low elastic moduli, pure resins are not commonly used in structural applications.

Overall, thermoplastics play a crucial role in modern manufacturing processes and have widespread applications in automotive, aerospace, packaging, electronics, construction, and consumer goods. Their ease of processing, recyclability, and wide range of properties continue to make them a preferred choice for various applications.

2.9.2 Thermosetting Plastics

Thermosets, which are referred to as highly crosslinked covalent network polymers, possess exceptional mechanical properties, chemical and heat resistance, and dimensional stability. Nevertheless, the irreversible nature of their chemical bonds renders thermosets incapable of reprocessing or recycling once they fail. Furthermore, the occurrence of shape alteration due to reversible bond-exchange reactions, commonly known as "creep," is regarded as a drawback of these polymers. Consequently, the irreversibility and absence of recyclability in thermoset networks are deliberate design characteristics [72, 73].

Thermosets are a class of polymers that undergo a chemical reaction known as curing or crosslinking during their curing process. This reaction causes them to transition from the liquid or soft solid state to the hardened or solid state. Unlike thermoplastics, which can be melted and reshaped multiple times, thermosetting plastics cannot be reprocessed once they have undergone curing. This characteristic distinguishes them from other thermoplastics.

The curing process of thermosetting polymers involves the formation of strong covalent bonds between polymer chains, resulting in the creation of a three-dimensional network structure. These crosslinked networks provide thermosets with superior mechanical properties, including high strength, stiffness, and resistance to heat and chemicals [74].

Thermosetting plastics are used in various industries, including aerospace, automotive, electrical insulation, adhesives, and composite materials for structural components. Their main advantage lies in their ability to maintain their shape and structural integrity at high temperatures, which makes them suitable for applications that require high heat resistance.

Common examples of thermosetting plastics include epoxy resins, phenolic resins, melamine-formaldehyde, and urea-formaldehyde. Each type of thermosetting plastic has specific properties and applications.

2.10 Distinction Between Thermoplastics and Thermosetting

The primary difference between thermoplastics and thermosetting plastics is their response to heat. Thermoplastics can be melted and reformed multiple times, whereas thermosetting plastics undergo irreversible chemical changes and become permanently hardened after their initial formation. The disparity in recyclability between these two types of plastics is a crucial factor to consider when evaluating their environmental impact and potential for integration into a circular economy. In a circular economy, materials are continuously reused and recycled to minimize waste and resource consumption. It is important to note that the distinction between thermoplastics and thermosetting plastics is vital when selecting materials for specific applications because their properties and behavior under different conditions can vary significantly.

2.11 General Classification

Plastics are a category of man-made materials derived from polymers, which consist of elongated chains of molecules comprising recurring units known as monomers. The primary distinction between commodity plastics and engineering plastics is their respective performance attributes and costs. Commodity plastics are more affordable and are commonly used for everyday objects, whereas engineering plastics offer superior properties, albeit at a higher price. The selection between these two options depends on the specific requirements of the application and desired level of performance [75]. Engineering plastics are favored when elevated strength, temperature resistance, chemical resistance, or other advanced characteristics are imperative, whereas commodity plastics are suitable for less-demanding applications. These are classified as follows:

2.11.1 Commodity Plastics

Plastic materials have undoubtedly brought about a significant transformation in numerous industries and everyday life, due to their exceptional properties and versatile applications. These materials offer several advantages that have resulted in their widespread adoption over traditional materials, such as metals, glass, and wood. Plastics often exhibit superior performance in specific applications compared with conventional materials. They can be engineered to possess a wide range of physical, chemical, and biological properties that render them suitable for diverse applications. They can be flexible or rigid, transparent or opaque, lightweight, or heavy-duty, and can even be electrically conductive or insulating [76].

Commodity plastics, also referred to as bulk or general-purpose plastics such as polyethylene, polypropylene, polystyrene, polyvinylchloride, and polyethylene terephthalate, constitute a group of widely utilized and economically viable polymers. They are typically produced in large quantities and can be applied in various industries due to their low cost and versatility [1].

2.11.2 Engineering Plastics

Recently, there has been a growing preference for the use of plastics in industrial structural components. This preference can be attributed to the advantageous characteristics of plastics such as their capacity to decrease weight, facilitate the creation of intricate shapes, and reduce expenses. Consequently, the engineering sector has witnessed a notable surge in the adoption of plastics in various applications. Presently, there is a prevailing inclination towards the production of substantial load-bearing components through molding techniques [77].

Engineering plastics, also referred to as high-performance plastics or advanced plastics, encompass a category of polymers, such as polyamide, polyacetal, polycarbonate, polyphenylene sulfide, and polyetheretherketone, which exhibit exceptional mechanical, thermal, and chemical characteristics. These materials are purposefully engineered to fulfill stringent performance criteria and are widely used in industries where conventional commodity plastics may prove inadequate [78].

In conclusion, despite the challenges encountered, ongoing research on polymers continues to make significant advancements, facilitating progress in diverse industries, including electronics, healthcare, energy, and transportation. The study of materials will remain indispensable in addressing global issues and constructing a more sustainable future, even as technology continues to evolve [76].

Polymers are important materials that have various characteristics and applications. They are affordable, easy to manipulate, and versatile. Synthetic polymers are popular because of their stability, processability, mechanical properties, and durability. Polymer engineering involves the synthesis of controlled macromolecules to manipulate their physical properties. Polymers are used in consumer products, such as carpets, furniture, glues, and clothing, as well as in advanced engineering, particularly in the aerospace industry. The use of polymers and their constituent monomers is important in daily life [79].

The continual development and modification of polymers have yielded a wide array of options, enabling the identification of suitable materials for nearly any application, ranging from basic consumer goods to intricate industrial components and medical devices. As technology and materials science continue to advance, it is anticipated that more advanced plastic materials will emerge in the field of thermoforming.

3
Thermoplastics

3.1 Introduction

Although synthetic plastics have only been around for 115 years and the advent of their widespread use typically began in the 1950s, affordable, comparably light, and highly versatile materials have become omnipresent in our daily lives [1–3]. Plastic packaging allows extended food preservation. In medical applications, as observed during the recent pandemic, plastic products allow for increased hygiene; in automotive applications, they reduce weight and thus increase fuel efficiency; in countless other applications, it is difficult or impossible to forgo plastics [4].

Thermoforming is a complex process, particularly for the production of intricate parts. The selection of plastics for thermoforming is influenced by several factors, including the desired properties of the final product, its intended use, cost considerations, and regulatory requirements. Thermoplastics are a class of polymers that are used for thermoforming.

3.2 Polyolefins

Polyolefins, especially polyethylenes (LLDPE, LDPE, and HDPE) and polypropylene (PP), contribute significantly to the growing amounts of plastics with a combined production share of almost 50% by mass. Polyolefins have been used in the packaging industry due to its low cost and density, attractive properties, and low energy consumption during processing [5]. The explosion in the dynamic development of PE and PP, which is still in progress.

They include large-volume materials such as polyethylene (PE) and polypropylene (PP), and specialty materials such as ethylene–propylene elastomers [EPR and EPDM] and polybutene-1 (PB-1). Even considering the universality of materials, POs are unique in their ability to perfectly match the basic prerequisites for the successful growth of a material [6]:

Muralisrinivasan Natamai Subramanian. Thermoforming: Processing and Technology, (27–68)

1) cost/performance balance and
2) Economic production, considering environmental factors.

The basis of their dynamic development and still tremendous potential lies in

- Versatility with respect to physical and mechanical properties and applications.
- Nontoxicity and bioacceptability.
- Energy savings during production and use in comparison with other materials.
- Low-cost and easily available raw materials.
- Economic, versatile, and nonpolluting production.

Polyolefins are synthetic polymers created through the polymerization of hydrocarbons containing a carbon–carbon double bond (C=C), known as olefins. These are high-molecular-weight hydrocarbons. The most important types are polyethylene and polypropylene [7]. Polyolefins have both crystalline and amorphous structures.

3.3 Polyethylene

Polyethylene is the simplest and most widely used polymer in the world in terms of volume. It consists a repeating unit of $-CH_2-CH_2-$. Polymer production has steadily increased globally, from 1.5 million tons per year in 1950 to 288 million tons in 2012. The use of polyethylene resins was approximately 100 million metric tons valued at $164 billion in 2019 [8]. Polyethylene is a semi-crystalline material that has good chemical resistance, good fatigue properties and wear resistance, and its properties vary over a wide range due to the difference in the length of the polymer chain. They can be distinguished from other plastics by their buoyancy on the water surface. Polyethylene has a good resistance to organic solvents and electrolysis and has a higher mechanical strength than polypropylene, but its operating temperature and tensile strength are lower, its weight is relatively small, and it has a low degree of absorption [9].

Linear polyethylene with a density of 978 kg/cm^3 at 23 °C was used to cool the material held at 160 °C for 5 min to room temperature at a cooling rate of 10 °C/min. Differential scanning calorimetry (DSC) showed a total heat of fusion of 212 J/g and a heat of perfect crystalline polyethylene of 293 J/g. Figure 3.1 shows the DSC cooling trace of the linear PE.

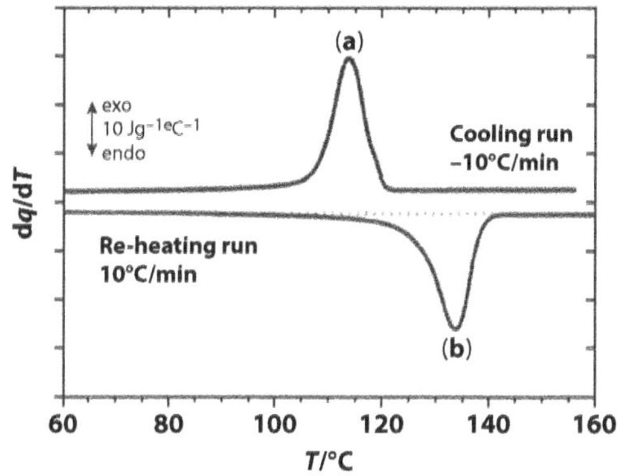

Figure 3.1 DSC traces for linear PE (a) cooled from the molten state at a cooling rate of 10 °C/min and (b) reheating run at a heating rate of 10 °C/min. The dotted line indicates the heat of fusion. [Reprint from [10]. Copyright © 2015, American Chemical Society. All rights reserved].

As shown in Figure 3.1, the heat of fusion changes are necessarily accompanied by changes in the free energies of both the amorphous and crystalline phases, and melting must be affected by drawing. The increase in the heat of fusion caused by drawing is attributed to the heat of relaxation of the amorphous orientation at the melting point. The combination of high melt strength and the right balance between crystalline and amorphous regions makes PE a preferred material for thermoforming applications [10].

The melting of polyethylene is a very complicated process involving both crystalline and amorphous phases. The melting point is the complete disappearance of crystallinity, although fusion in semicrystalline polymers occurs over a range of temperatures. In thermoforming, drawing causes structural changes, such as molecular orientation, size, and distortion of crystals, and the state of aggregation of amorphous chains.

3.3.1 Low Density Polyethylene (LDPE)

Low-density polyethylene (LDPE) is commonly used in various applications that require characteristics such as clarity, inertness, ease of processing, strength, flexibility, and moisture resistance. Low density polyethylene made in high-pressure plants is branched, with both long and short chain branching, and gives rise to polymers in the density range of 0.916–0.930 gm/cm^3 [11].

LDPE exhibits long-chain branching, resulting in melt strain hardening, which is an important characteristic of thermoforming. LDPE sheet material defects are predominantly caused by a higher melt elasticity and a higher level of long-chain branching during thermoforming. The ease of drawing during thermoforming depends on the elasticity of the melt. With a higher melt flow index, the drawdown performance of the ability of the sheet to be drawn down to the thing gauge is sufficient for thermoforming [12].

The optical properties of the sheet can be applied to the ease with which the LDPE melt can be drawn down to provide a thin gauge during thermoforming. LDPE showing less lateral reduction exhibits poor extensibility, i.e., a low level of maximum plug speed, and vice versa. Table 3.1 shows the physical properties of the LDPE. Typical LDPE contains short-chain branches (1–3 per 100 backbone carbon atoms) and long-chain branches (0.1–0.3 per 100 backbone carbon atoms) [13]. LDPE with many long-chain branches and a broad MWD, especially with a tail in a significantly high molecular weight region, is preferred for thermoforming.

Low-density polyethylene has a wider melting range and lower initial melting temperature due to the wide distribution of the crystallite regions. When thermoforming LDPE, it is important to consider its low melting point, heat it gradually to avoid warping, cool it down slowly to prevent deformation, control the pressure applied by the mold, use specific mold designs to support the material, and consider the thickness of the LDPE sheet. LDPE is commonly roll-fed during thermoforming to create products such as trays, lids, and cups.

3.3.2 High-Density Polyethylene (HDPE)

High-density polyethylene (HDPE) is one of the most widely used polymers for plastic packaging due to its excellent chemical resistance, ductility, tensile strength, and hardness [18]. HDPE has a high density due to its molecular structure with little branching. This results in stronger intermolecular forces, leading to a higher tensile strength and stiffness. HDPE exhibits exceptional moisture resistance. Despite its strength, HDPE is relatively light in weight, making it convenient to handle and transport [19, 20].

The versatility of polyethylene arises from the differences in its structure and molecular weight. Polyethylene is also known for its flexibility, chemical resistance, toughness even at low temperatures, and affordability. The advantages of HDPE in thermoforming include its ability to maintain

Table 3.1 Properties of low-density polyethylene (LDPE).

Parameter	Unit	Value	Reference
Density	kg/m^3	910–930	[14]
Glass Transition Temperature	°C	–120	[15]
Melt Temperature	°C	111	[16]
Heat Deflection Temperature	°C	49	[15]
Heat of fusion	J/g	112.6	[16]
Heat of crystallization	J/g	94.9	[16]
Crystalline temperature	°C	100	[16]
Tensile Strength	MPa	4–16	[15]
Elongation at break	%	300	[15]
Izod Impact Strength	J/m	>400	[15]
Crystallinity $\varkappa_c$	%	39.2	[16]
Shaping temperature	°C	90–135	[17]
Ultimate rupture strength	MPa	6.8–14.7	

toughness at temperatures as low as –118 °C, withstand steam sterilization at 121 °C, and exhibit excellent resistance to many chemicals. During thermoforming, because of tensile deformation, the original lamellar microstructure of PE is transformed into a fibrillar structure, where polymeric chains are preferentially oriented along the stretching direction. Under severe conditions of high pressure, low temperature, and high deformation speed, they become brittle [21–23].

Figure 3.2 shows that the microstructure of HDPE is the simplest polymer, with fine (a) and coarse (b) spherulites forming normally in the forming window. Mechanical stability is essential for thermoforming. During thermoforming, polymer chains are preferentially oriented along the stretching direction. For higher-density materials, crystallization detects can be significant and contribute considerably to the total haze during thermoforming [25–27].

High-density polyethylene (HDPE) also exhibits gas permeability and resistance to chemicals, moisture, and breakage. This hydrocarbon polymer consists of linear chains that allow for dense packing, resulting in a

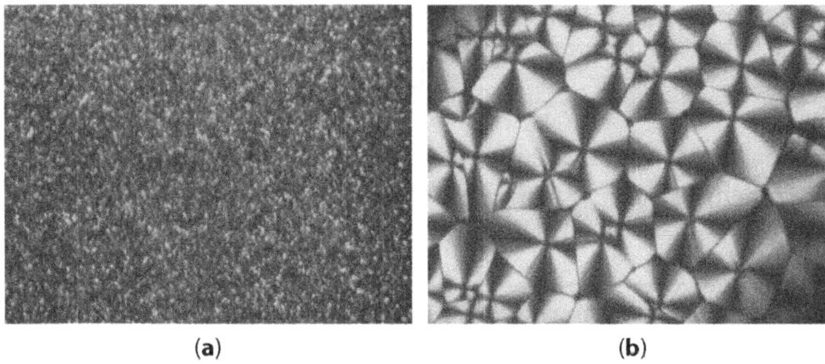

Figure 3.2 Typical microstructures of HDPE. (a) with fine spherulites (spherulite diameter ≈5 μm). (b) coarse spherulites (spherulite diameter ≈30 μm). [Reprint with permission from [24] Copyright © 2009, Society for Experimental Mechanics All rights reserved].

density ranging from 0.94 to 0.96 or higher. The properties of HDPE are listed in Table 3.2. Despite its softness, relatively low melting point, high elongation, and poor gas barrier properties, HDPE is still used for various applications [29–31].

High-density polyethylene (HDPE) has excellent dielectric properties and, therefore, a wide variety of industrial applications, including electrical applications. Due to their low dielectric loss and good heat resistance, they have been widely used as electrical insulators [32, 33].

Important factors to consider during HDPE processing include controlled heating to reach the forming temperature. The material exhibited a relatively high shrinkage rate during cooling. The benefits of slow and controlled cooling reduce the stress in the final formed part. Depending on the specific application and requirements, additional processing, such as trimming, punching, or surface finishing may be necessary for HDPE thermoformed parts [34, 35].

3.3.3 Linear-Low-Density Polyethylene (LLDPE)

Linear-low-density polyethylene (LLDPE), a copolymer of PE that typically contains octene or hexene, has numerous but very short side branches. LLDPE is known for its excellent clarity, high elongation, and low melting point among packaging polymers. They have major applications in heat-sealable coatings and films, stretch- and shrink-wraps, and retail and industrial bags [36].

Linear-low-density po1yethylene has a significantly higher level of short-side chain branching brought about by the copolymerization of

Table 3.2 Properties of high-density polyethylene (HDPE).

Parameter	Unit	Value	Reference
Density	kg/m^3	935–975	[28]
Glass Transition Temperature	°C	−120	[15]
Melt Temperature	°C	133.9	[28]
Heat Deflection Temperature	°C	82	[15]
Heat of fusion	J/g	168.2	[28]
Tensile Strength	MPa	19.7 ± 0.1	[28]
Elongation at break	%	700	[15]
Izod impact Strength	J/m	40	[15]
Crystallinity $\varkappa_c$	%	57.4	[28]
Shrinkage	%	2.0–3.5	Shrinkage normally used 2.5%
Shaping temperature	°C	120–135	[17]

α-olefins such as butene, octene, etc., at levels ranging from approximately 8%–10% for low-density and around 1%–2% for high-density copolymers [37]. The ability to use a range of comonomers, such as butene, octene, and hexene provides the same flexibility of product properties [38].

Linear-low-density polyethylene(LLDPE) is a thermoplastic; its properties are presented in Table 3.3. It has better stiffness, tensile strength, elongation, puncture resistance, environmental stress-crack resistance, and excellent chemical, mechanical, and thermal stability. However, higher energy requirements are required for extrusion. It can also be used to fabricate thinner films with better environmental stress-cracking resistance [39, 40].

3.4 Polypropylene

PP is one of the most popular thermoplastic materials and is extensively used for thermoforming due to its high tensile strength, stiffness, chemical resistance, and increased significance as a packaging material. PP has a high softening point, good chemical resistance, a strong melting point,

Table 3.3 Properties of linear-low-density polyethylene (LLDPE).

Parameter	Unit	Value	Reference
Density	kg/m^3	870–918	[28, 41]
Glass Transition Temperature	°C	-52.1	[41]
Melt Temperature	°C	125–169	[28, 41]
Heat of fusion	J/g	97.4	[28, 42]
Tensile Strength	MPa	12.3 ± 0.5	[28, 42]
Crystallinity $\varkappa_c$	%	33.2–44.5	[28, 42]

and low heat distortion [43, 44]. Figure 3.3 shows the polypropylene from melting to cooling to the crystalline state dynamically, and Figure 3.4, shows the spherulite morphology from extruded polypropylene.

Although solid-phase pressure forming (SPPF) below the melting temperature between 155 and 165 °C is possible for PP, it is not as popular as melt-phase forming because the high pressure required for forming results in the presence of residual stresses in the formed components. Only a narrow temperature window above the melting temperature is available for thermoforming because a steep drop in the modulus above the melting temperature poses a challenge due to its crystallinity towards cooling for sustaining the stresses induced during biaxial deformation during thermoforming. Furthermore, a low melt strength results in excessive sagging of

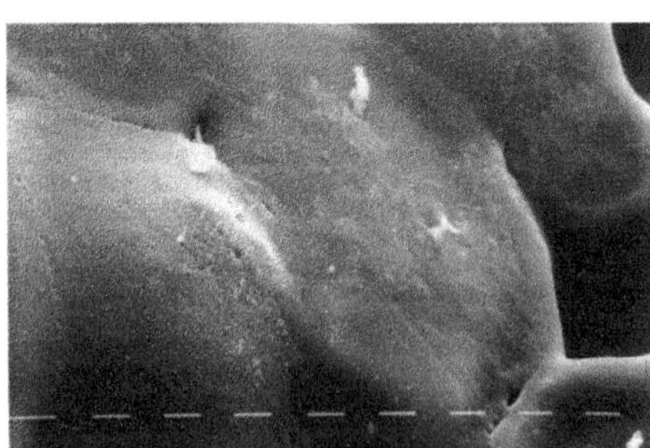

Figure 3.3 Scanning electron micrograph (SEM) of polypropylene. The DSC crystallized specimen was dynamically cooled from the melt at a scan rate of 10 °C/min. Magnification ×1,000. [Reprint with permission from [45]. Copyright © 2001, Kluwer Academic Publishers. All rights reserved].

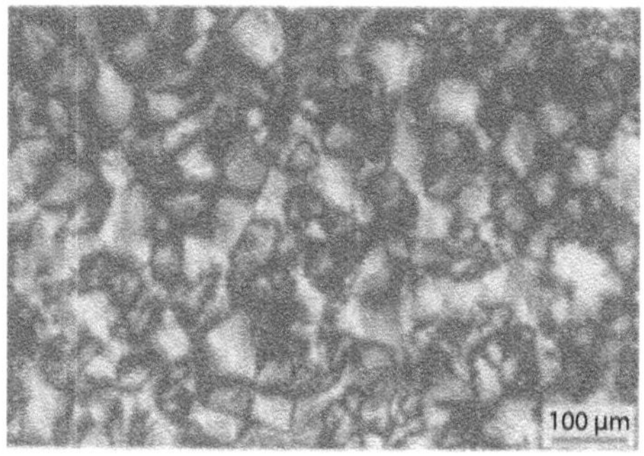

Figure 3.4 Spherulite morphology in the extruded polypropylene [Reprint with permission from [46].Copyright © 2004, Chapman & Hall. All rights reserved].

the sheet prior to forming, which leads to uneven deformation and uneven thickness distribution in the product.

Other properties of PP are presented in Table 3.4. PP has superior live-hinge properties, and PP continues to expand its existing properties with new diversified features, making it ideal for replacing traditional materials in rigid packaging applications. With excellent food contact properties, such as resistance to contributing off-flavors of foods, PP is safe for

Table 3.4 Properties of polypropylene (PP).

Parameter	Unit	Value	Reference
Density	kg/m^3	900–905	[48–50]
Glass Transition Temperature	°C	10–11 –20**	[51] **semi-crystalline
Melt Temperature	°C	185*	[52] *Crystalline
Heat Deflection Temperature	°C	110	[15]
Tensile Strength	MPa	18–33	[15, 53]
Elongation at break	%	415	[15]
Izod Impact Strength	J/m	20	[15]
Crystallinity $\varkappa_c$	%	56	[15]
Shaping temperature	°C	150–200	[17]
Ultimate rupture strength	MPa	29–40	

microwave cooking because of its temperature resistance and is attractive due to its transparency. PP has a poor melt strength, and hence tends to sag. Compared to conventional PP, high-melt-strength PP has greater shear thinning behavior and a lower tendency to sag during thermoforming. Branched PP is often added to conventional PP to increase its melting strength. The long-branched chains of branched PP act as reinforcing linkages over the matrix chain entanglements and prolong the sag time [47].

In the melting region, PP suddenly changes from a semicrystalline solid to a melt, and there is no appreciable rubbery plateau during heating. Therefore, thermoforming of PP is difficult because of the excessive sagging of the heated sheet, low melt strength, low elasticity, narrow processing window, and poor thickness distribution in the final part [54]. PP can be made more amenable to thermoforming by enhancing its melting strength and elasticity.

There are numerous methods to increase the melt strength of PP. They are:

- increasing the weight average molecular weight,
- broadening the molecular weight distribution particularly by creating bimodal resins containing considerable fraction of high molecular weight chains,
- blending PP with polymers such as low-density polyethylene (LDPE), and
- introducing long-chain branching (LCB); on the backbone of PP.

Increasing the molecular weight affects the processability. Polypropylene, like other crystalline materials, has a limited range of temperatures where it can be formed. Once the material becomes molten, it quickly begins to excessively sag [55–57].

The elastic modulus of PP is higher in the drawing direction, with a more pronounced effect for a higher draw ratio. It is well known that biaxially constrained stretching during the thermoforming process yields a nonuniform chain orientation along the drawing direction. Polypropylene, the elastic modulus is higher in the drawing direction, with a more pronounced effect for the higher draw-ratio. The drawing direction and stress–strain behavior should be considered representative of the average behavior of the material due to the nonuniform chain orientation.

There is a close relationship between the processability of PP in thermoforming, the strain-hardening behavior of its melt in uniaxial elongational flow, and the number of long chain branches on its chain. The use of clarifiers and nucleators in PP improves the clarity, stiffness, heat deflection temperature, and process rates [58].

3.5 Polystyrene (PS)

Polystyrene (PS), is a well-known commodity polymer with good processing characteristics and excellent mechanical properties (elastic modulus and tensile strength) [59]. Polystyrene (PS) is a commonly used thermoforming material due to its ease of processing. It possesses excellent thermoforming properties and is characterized by its hardness, clarity, and resistance to water, alkali chemicals, acids, and detergents. Oriented polystyrene forms a good base sheet for thermoforming. Polystyrene has proven to be the ideal material for these breakthrough technologies due to its rigidity and ease of processing. Figure 3.5 shows an SEM micrograph of pure PS which shows a clear semi-crystalline morphology.

Polystyrene is a semicrystalline thermoplastic that forms repeating units with a high degree of ordered tacticity in an alternating configuration. These thermoplastics have advantageous properties, such as exceptional heat and chemical resistance, desirable electrical properties, dimensional stability, and resistance to moisture.

When thermoforming polystyrene, there are several important factors to consider, including clamping, fixing the material, and the cooling process. Polystyrene has a relatively low melting point compared with other thermoplastics. Polystyrene absorbs heat evenly throughout the material

Figure 3.5 An SEM micrograph of pure PS sample. [Reprint with permission from [60] Copyright © 2016, Springer-Verlag Berlin Heidelberg. All rights reserved].

and stretching during thermoforming contributes to the production of parts with increased toughness and high deep-drawn ratios [61].

It is mostly amorphous and lacks regular crystalline structure. Polystyrene can soften or deform when exposed to high temperatures and is sensitive to tensile stress, which can cause it to break or lose its integrity. Contaminants or impurities in polystyrene can result in a loss of transparency in the final products. Polystyrene is relatively brittle and has limitations in terms of its resistance to certain solvents [62].

One of its notable qualities is its ability to maintain good impact resistance even at low temperatures, such as in freezing conditions. Its chemical resistance may be compromised under high temperatures or tensile stress, as it is susceptible to attack by aliphatic and aromatic hydrocarbons and their derivatives. However, it is inherently brittle unless modified, and has a relatively low melting point. They can be attacked by hydrocarbons and their derivatives.

It is important to consider these limitations, particularly in terms of solvent resistance, when choosing it for specific applications. It is transparent but can be colored by the introducing of colorants. It is heat resistant, lightweight, and exhibits good strength and durability, making it suitable for a variety of applications. Polystyrene (PS) is an easily processed resin used for thermoforming applications.

Table 3.5 lists the physical properties of polystyrene that are necessary for thermoforming. These properties are essential to thermoforming. One advantage of polystyrene is its uniform elongation, which allows for a wider processing window during thermoforming. This, along with its excellent draw ratios, makes polystyrene commonly used in various thermoforming applications. Polystyrene offers advantages in terms of energy efficiency, process control, flexibility, and the production of strong and durable parts.

Table 3.5 Properties of polystyrene (PS).

Parameter	Unit	Value	Reference
Density	kg/m^3	1,040–1,090	[63–66]
Glass Transition Temperature	°C	96	[63, 64, 66, 67]
Melt Temperature	°C	240	[63–65]
Specific Heat Capacity	J/kg/K	1,200	[68]
Thermal Conductivity	W/mK	0.17	

3.6 High-Impact Polystyrene (HIPS)

Polystyrene (PS) can be substantially improved by incorporating a dispersed rubber phase, leading to a well-known class of materials known as high-impact polystyrene (HIPS). This class of materials is known as a typical rubber-toughened polymeric material. HIPS is an amorphous polymer that is natually translucent white. Glassy polymers such as polystyrene (PS) can be greatly toughened by adding rubber particles to produce polymers known as high-impact polystyrene (HIPS) [69]. The HIPS offers extraordinary property combinations, thus making a vital contribution to everyday life. High-impact polystyrene is also used in many applications due to its excellent balance of properties and low cost. HIPS has good impact resistance, dimensional stability, aesthetic qualities, is easy to paint and glue, and can be manufactured at a low cost [70, 71].

The HIPS is the easiest material to deform. Solid rubber particles in HIPS are less able to accommodate large deformations without voids formed at the craze-rubber interface than particles containing many sub-inclusions, suggesting that the particle-size effect may be at least partially due to the different internal structure of the particles (as a function of size).

HIPS materials are composed of multicomponent and multiphase polymeric materials with glassy and rubbery phases; thus, end-use properties are dependent on many parameters. The rubber phase is a composite structure consisting of bulk rubber and PS sub-inclusions.

Table 3.6 Properties of high impact polystyrene (HIPS).

Parameter	Unit	Value	Reference
Density	kg/m^3	909–1,050	[72]
Glass Transition Temperature	°C	100	
Heat Capacity	J/kg K	2,000	
Heat transfer	J/s·m·K	0.13	
Linear Thermal Expansion	K^{-1}	$6–8 \times 10^{-5}$	
Izod Impact Strength	J/m	80–150	[59]
Shaping temperature	°C	110–150	[17]
Ultimate rupture strength	MPa	40–46	

Polybutadiene is used as the rubber, although styrene–butadiene and ethylene–propylene-copolymers, are also used. The volume fraction of the rubber phase, its chemical composition, degree of crosslinking, particle morphology, level of adhesion to the matrix, type of rubber, and most importantly, the rubber particle size and its size distribution depend on the percentage of rubber content. Advances in polymer chemistry have allowed the manufacture of rubber-toughened materials with different rubber morphologies [73, 74].

The properties of HIPS are listed in Table 3.6. The presence of rubber in HIPS makes it more flexible and less brittle; however, it is less transparent and has a reduced clarity. It has limited resistance to high temperatures and lower heat resistance than other plastics. HIPS is highly malleable and exhibits viscoelastic behavior. Despite these limitations, HIPS remains a popular and practical choice for various applications due to its unique combination of properties and cost-effectiveness. It is also a versatile material that can be recycled and is suitable for thermoforming applications [75].

3.7 Polyvinylchloride

Polyvinyl chloride is the most versatile plastic because of its blending capability. Pure PVC is a hard and brittle material that is challenging to process. The chlorine content of PVC can be characterized by its K value. PVC with a K value of 65 is typically used in the sheet form for thermoformed applications. PVC has good clarity and chemical resistance (which is important for holding household detergents and other harsh materials).

Most commercial PVC materials, which are manufactured at temperatures between 40 °C and 70 °C, have low crystallinity. The properties of PVC can be found Table 3.7. The melting range is very broad, stretching from approximately 130 °C to 230 °C. Because PVC is thermally unstable, it must be processed below its melting point [76]. Most commercial PVC samples have 5%–20% crystallinity in the area by volume.

The increased use of resins with a wide range of molecular weights in rigid PVC compounds causes the flow activation energy to decrease with increasing molecular weight of the resin. The change in the activation energy occurs in the temperature region, which is associated with the melting point of PVC. The melting point increased with increasing molecular weight. Therefore, if the change in activation energy is associated with melting, the inflection temperature should increase with increasing molecular weight [77].

PVC has good resistance to chemical and solvent attacks. Its vinyl content provides good tensile strength, and it can be changed from rigid to flexible using additives and modifiers. Its resistance to fire is attributed to the release of chlorine. Furthermore, in the compounding process, many ingredients such as fillers, stabilizers, plasticizers, and lubricants, are added to PVC materials to optimize the properties of the product. Thus, in the processing of PVC materials, one faces a material with a very complex structure [78].

PVC can be broadly divided into rigid and flexible materials. It is also versatile and has clarity, ease of blending, strength, toughness, and resistance to grease, oil, and other chemicals. PVC can also be flexible, chemically resistant, and flame-retardant. To improve thermoforming, acrylic processing aids have been used for rigid PVC sheet materials. Acrylic processing aid also improves cell structure during foam extrusion [81].

3.8 Acrylonitrile–Butadiene–Styrene (ABS)

Acrylonitrile–butadiene–styrene (ABS) is a graft copolymer that is well-known for its attractive properties, such as impact toughness, solvent resistance, and good processability [82]. Rubber particles are grafted with styrene and acrylonitrile copolymer (SAN) to interact with the SAN matrix, which endows ABS with a high impact toughness [83]. Due to their various applications in the engineering field, technologies for the synthesis of ABS have been developed.

It is one of the most widely used rubber-toughened commercial plastic materials. This is a two-phase system [84]. The most significant advantage gained by the incorporation of rubber is the transformation of brittle organic gloss into a ductile thermoplastic with high impact strength. The impact strength of ABS resin and other toughness measures are proportional to the amount of rubber in the material [85, 86].

Heterogeneous polymeric systems and specifically rubber modified polymeric materials such as acrylonitrile–butadiene–styrene copolymers (ABS) are of particular interest since they possess excellent properties for consumer goods (toughness, durability, chemical and heat resistance, to name a few) and can be produced over a wide range of processing temperatures, especially in the thermoforming process [59].

ABS, the range of mechanical properties attainable extends from the properties of an SAN copolymer containing approximately 26% acrylonitrile to those of an ABS containing 25%–30% rubber, normally polybutadiene (PBD).

Table 3.7 Properties of polyvinylchloride (PVC).

Parameter	Unit	Value	Remarks	Reference
Density at 100 °C at 120 °C at 140 °C	kg/m^3	1.352		[79]
		1.338		
		1.352		
Glass Transition Temperature	°C	65		
	°C	71	DSC, 20 °C/min	
	°C	98	DSC, 32 °C/min	
	°C	71	Dilatometry	
Sub Glass transition temperature	°C	−50	(Transition) DMA	
Melt Temperature (Transition)	°C	212–310	(Calorimetric) Decomposition	
	°C	240	DSC	
Specific heat at constant pressure	mJ/mg °C	1.92	at 170 °C	[80]
	mJ/mg °C	1.899	at 190 °C DSC, 20 °C/min	
	mJ/mg °C	1.798	at 210 °C DSC, 32 °C/min	
Shaping temperature	°C	100–160		[17]
Ultimate rupture strength	MPa	13.7–18.6		

DSC, Differential scanning calorimetry
GPC, Gel permeation chromatography

Polybutadiene (BD) constitutes a rubbery phase and exists as spherical particles in a glassy matrix of poly (styrene-co-acrylonitrile). Usually, SAN is grafted to BD to increase the compatibility between the two phases and achieve good impact resistance at low temperatures [87, 88].

Medium-to-high-impact ABS and flame retardancy can be accomplished with only small effects on the physical properties. The melt viscosity typically decreases. Table 3.8 provides an overview of the properties of ABS. The ABS with larger amounts of hard rubber particles exhibited greater melt elasticity and better thermoforming performance. Strain hardening in ABS provides simple and reliable indicators of the thermoforming performance [89].

ABS demonstrates good resistance to sagging, meaning that it maintains its shape throughout the process. ABS has the highest viscosity and elastic modulus, which contribute to its overall stability during thermoforming. Strain hardening, which is the increase in material hardness due to deformation, can be used as an indicator of ABS's thermoforming performance. The presence of rubber particles in ABS affects its rheological properties such as melt elasticity, elongational viscosity, strain hardening, and strain softening. These rheological changes affect the thermoforming performance of ABS [90].

For these toughened polymers, significant stress-whitening occurs prior to fracture, and the elongation at break and notched Izod impact toughness are many times higher than those of the unmodified parent polymer (typical elongations at break are approximately 2% for SAN and 30% for ABS) [91]. ABS is used for thermoforming because of its wide processing window. It does not sag, particularly at low processing temperatures, due to its high melt stiffness [92].

Extruded or calendered ABS sheets are readily thermoformed, but the depth of the draw is restricted to approximately the width of the part. At temperatures above 107 °C, ABS is tacky and adheres to metal. Therefore, PTFE coating is used for plug assistance. The formed sheet is relatively free from residual stress if it is adequately plasticized in the heating cycle prior

Table 3.8 Properties of acrylonitrile–butadiene–styrene (ABS).

Parameter	Unit	Value	Reference
Density	kg/m^3	1,010–1,050	[67, 93]
Glass Transition Temperature	°C	100	[93]
Heat Deflection Temperature	°C	88–89	[94]
Tensile Strength	MPa	42.5–44.8	
Elongation at break	%	23–25	
Flexural modulus	MPa	2,250–2,280	
Izod Impact Strength, notched	J/m	246–249	

to forming. Electric plate heaters, infrared ovens, and electric or gas-fired ovens may be used for thermoforming. Before removal from the mold, they are cooled to approximately 65 °C. In thermoforming, one recommended technique is deep draw bubble forming [95].

3.9 Polyethylene Terephthalate (PET)

Poly(ethylene terephthalate) (PET) is a most widely used thermoplastic polymer. It is a high molecular weight polymer produced by the condensation reaction of ethylene glycol and terephthalic acid. The primary applications of PET are in films and fibers. PET is a condensation polymer characterized as a slowly crystallizing material [96].

Polyethylene terephthalate (PET) is a plastic material that consists of long-chain polymers belonging to the polyester family. PET possesses several desirable physical properties for thermoforming, including clarity, toughness, and notch sensitivity. It can exist in either an amorphous or highly crystalline state and exhibits good clarity and barrier properties. In its pure form, PET is amorphous and glass-like, but it can be extruded in the crystalline state and oriented in the amorphous state [97]. Crystallinity can be developed using additive modifiers or heat treatment of the polymer melt.

Poly(ethylene terephthalate) [PET] is often processed in the solid state at temperatures above its glass transition temperature using a variety of stretching processes (drawing, biaxial stretching, blow molding, and vacuum forming). In many instances, the initial PET state is an isotropic amorphous state obtained *via* rapid cooling from the melt. The stretching processes impart a preferential orientation to the macromolecular network, which in turn can result in strain-induced crystallization. Thus, the end product will be molecularly oriented and partly crystalline. The level of molecular alignment and the percent crystallinity that develop with strain strongly depend on the strain rate and temperature, and the texture of the molecular orientation and crystallinity also depend strongly on the mode of straining (e.g., uniaxial stretching vs. biaxial stretching) [98].

The mechanical and functional properties of these materials depend on their degree of crystallinity and texture. The crystallinity of polymers depends on their thermal and mechanical history, and its value influences processability and other properties [99].

The effect of large deformations in PET polymeric materials has great technological importance in thermoforming because it is related to the processing conditions that affect their morphology and, consequently, their physical and mechanical properties, as shown in Table 3.9 [100, 101].

Table 3.9 Properties of polyethylene terephthalate (PET).

Parameter	Unit	Value	Reference
Density	kg/m^3	1,410	[79]
Glass Transition Temperature	°C	67.8	[102]
Melt Temperature	°C	248.1	
Heat Deflection Temperature	°C	78	[103]
Heat of fusion	J/g	34	
Crystalline temperature	°C	203.9	[102]
Tensile Strength	MPa	56	[103]
Impact Strength	J/m	25	

The amorphous PET structure changes during stretching. The change starts with the formation of nematic and smectic phases [104]. An interesting polymer, PET, typically has glass transition temperature (Tg) higher than room temperature. PET exhibits a very large density change with crystallinity (approximately 0.0012 g/cm^3 or 0.09% change with each 1% change in crystallinity). PET has a very poor melt strength. The thermal properties of PET, which in turn depend on the degree and type of crystallinity, are key factors in its processability. PET has a low crystallization rate because of the short ethylene chain that separates the aromatic rings [105].

3.10 Acrylics

Acrylics is a type of plastic material that possesses several advantageous properties, making it suitable for various applications such as skylights, windows, and light diffusers. These properties include high clarity, brittleness, excellent resistance to weathering, and the ability to softened and hardened repeatedly without significant changes in their properties or degradation.

Acrylic sheets offer exceptional surface finish, uniform thickness, and optical quality, with advantages such as resistance to breakage during handling, transportation, installation, and service, good stiffness, desirable thermoforming characteristics, and enhanced resistance to solvent craze. Additionally, acrylic sheets exhibit weathering advantages, such as outstanding clarity, gloss retention, and resistance to yellowing.

Compared with other plastic sheet materials used in thermoforming, acrylic sheets have shorter thermoforming cycles, lower forming temperatures, and wider forming temperature windows. These unique characteristics have allowed acrylic sheets to penetrate various markets and applications successfully. They offer greater toughness and breakage resistance, excellent thermoforming definition and flexural modulus features for designers, economic attractiveness, ease of application, exceptional quality, durability, and appearance for end-users. Acrylic polymer is known for its high transparency, good optics, surface hardness, chemical resistance, and ability to withstand weathering. Acrylic polymers can be repeatedly softened and melted upon heating and hardened upon cooling, with insignificant property changes and without considerable degradation.

The acrylic sheet ensures exceptional surface finish, optical quality (colorless), and thickness uniformity. These advantages include breakage resistance (in-shop, in-transit, during installation, and in-service), excellent clarity and weathering properties, good stiffness, desirable thermoforming characteristics, and enhanced solvent craze resistance. It also exhibits weathering advantages, such as outstanding clarity and gloss retention properties, and an unyielding resistance to yellowing [106, 107].

3.10.1 Polymethylmethacrylate (PMMA)

Polymethylmethacrylate (PMMA) is the most important homopolymer in the acrylic family, and its important properties are presented in Table 3.10. PMMA is a transparent thermoplastic polymer material that has been widely used in various industries of national production, such as aircraft windshields and portholes, radar screens, optical lenses of telescopes, and instrument protection [108]. PMMA, as a typical glassy polymer, has many excellent mechanical properties. Nevertheless, the mechanical behavior of PMMA is significantly influenced by extreme conditions in engineering applications, such as temperature and strain rate. PMMA has good elastic memory. PMMA sheets can be formed using unmodified thermoforming equipment [109].

The strength of PMMA exhibited a strong strain rate dependence at room temperature. Significantly, at low strain rates, PMMA is ductile to large strains. However, at dynamic strain rates, the strength of PMMA increases dramatically and becomes brittle at room temperature. Figure 3.6 shows an SEM micrograph of pure PMMA which, shows some nucleation within the morphology. This nucleation helps form clear parts during drawing in thermoforming.

Table 3.10 Properties of polymethylmethacrylate (PMMA).

Parameter	Unit	Value	Reference
Density	kg/m^3	1,170–1,200	[79, 110]
Glass Transition Temperature	°C	120	DMA [110, 111]
		113	[110]
Decomposition temperature	°C	279	[111]
Shaping temperature	°C	120–200	[17]
Ultimate rupture strength	MPa	40–66	

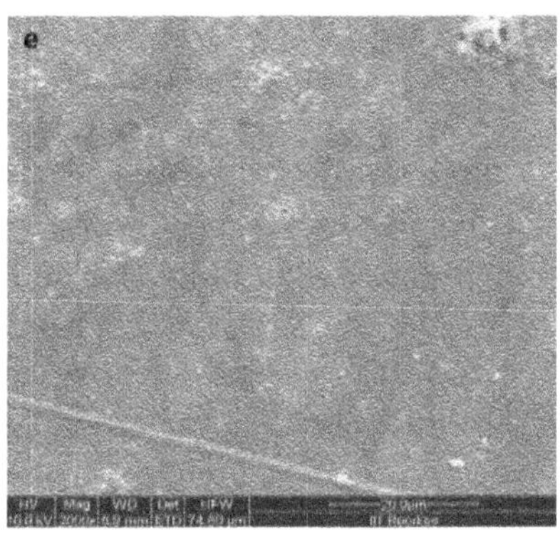

Figure 3.6 An SEM micrograph of pure PMMA. [Reprint with permission from [60]. Copyright © 2016, Springer-Verlag Berlin Heidelberg. All rights reserved].

The strength of PMMA also appears to lose strength at very high rates between 10^3 and 10^4 s^{-1}, but this is likely a result of a non-uniform material [112], and the yield strength is greatly affected by the strain rate, but less affected by the confining pressure. This is one important aspect of thermoforming PMMA. Its fracture toughness is in a wide range of temperatures, and it can be concluded that the change in fracture toughness can be reflected by the modulus change with temperature and rate [113].

In PMMA, the strain rate increases, and the compressive yield stress also shows a linear increasing trend over a wide range of temperatures and strain rates. The compressive mechanical behavior transitions from ductile to brittle. The tensile mechanical behavior of PMMA exhibited an obvious strain rate sensitivity. The yield strength and initial modulus of materials increase with increasing strain rate [114–116].

3.11 Nylon 6 (PA6)

Nylon 6 (PA6) is an important engineering thermoplastic that generally requires the addition of impact modifiers to enhance its performance. The properties of PA6 are presented in Table 3.11. PA 6 is a thermoplastic material, and its properties depend on its semicrystalline morphology. It has been found that solidification at temperatures above 150 °C leads to the formation of monoclinic alpha crystals [117, 118].

As shown in Figure 3.7, the optical micrograph shows the crystalline region of the transfer films for PA6. The transfer films are formed on the steel counterfaces using PA6. Nylon 6 is a semi-crystalline aliphatic polyamide that has a low friction coefficient, high fatigue strength, and high resistance to a wide range of chemicals, oils, and fuels. It is an important engineering polymer used in a wide range of applications such as automotive, construction, and transportations [120].

Nylon is used in many industrial applications, and there are many subtypes of nylon available in the market. Nylon 6 is a well-known member of the nylon family and has desirable properties that can be used in any mechanical assembly or component manufacturing. Although they have

Table 3.11 Properties of nylon 6 (PA6).

Parameter	Unit	Value	Reference
Density	kg/m^3	1,130–1,150	[121]
Melt Temperature	°C	221	[121]
Heat Deflection Temperature	°C	65	[122]
Tensile Strength	MPa	68.6	
Elongation at break	%	>100	
Tensile modulus	MPa	1,110	
Flexural strength	MPa	89.3	
Flexural modulus	MPa	1940	
Crystalline melt temperature	°C	225	[123, 124]

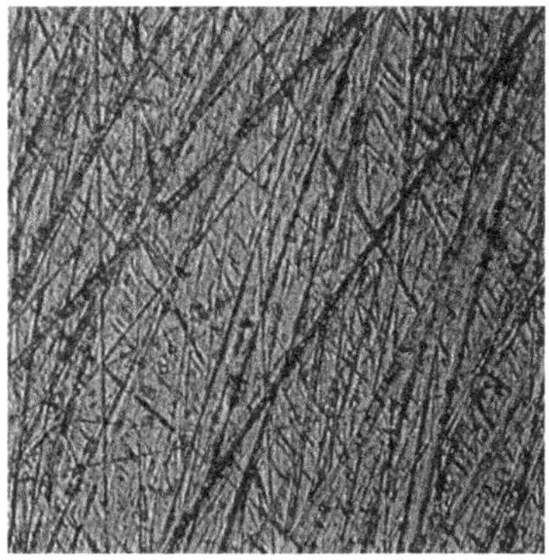

Figure 3.7 Optical micrograph of steady-state transfer films formed by PA6 [Reprint with permission from [119]. Copyright © 1986 Published by Elsevier B.V. All rights reserved].

good mechanical properties, they cannot be used where excessive load, friction, and wear are the main causes of failures due to their low tensile strength, low hardness, and high wear rate compared to metals [125].

Nylon 6 is a large-volume commercial polymer with several complementary attributes. The properties that account for its popularity include high crystalline melting point, good resistance to hydrocarbons, high strength, and ease of fabrication and processing [126, 127]. Polyamide has been reported to have a superior wear resistance when sliding against a steel counterface relative to other polymers. The coefficient of friction of polyamide is affected greatly by normal load, sliding speed, and temperature [128].

3.12 Nylon 66 (PA66)

Semicrystalline nylon 66 or poly(hexamethylene adipamide) have been widely used in various engineering applications. This is due to their excellent dimensional stability, mechanical strength at high temperatures, chemical resistance, and other advantages. Nylon 66 has high tensile strength, high thermal stability, and biocompatibility. The other properties are listed in Table 3.12. Moreover, the hydrophilicity of nylon 66 is higher than that of many other polymers due to the presence of amide groups in the polymer backbone. The properties of a semicrystalline polymer depend on the crystalline structure formed during processing. PA66 is a typical polyamide containing many amide bonds in its main chain and possesses excellent processing properties and lower melt viscosity. However, the limiting

Table 3.12 Properties of nylon (PA66).

Parameter	Unit	Value	Reference
Density	kg/m^3	1,069–1,220	[130]
Melt Temperature	°C	272	
Heat of fusion	J/g	118.28	
Tensile Strength	MPa	56	[131]
Elongation at break	%	60	
Flexural Modulus	MPa	2,882–3,157	
Crystallinity $\varkappa_c$ at 250 °C	%	56	[130]
Crystalline melt temperature	°C	265	[123, 124]

oxygen index (LOI) value of pure PA66 is approximately 22.5, indicating poor flame retardancy [129].

3.13 Polyoxymethylene (POM)

Polyoxymethylene (POM) is an important engineering polymer due to its exceptional properties and versatility [132]. It possesses a high degree of crystallinity exceeding 70%, which is attributed to its flexible main chain composed of repeating -CH_2-O- units. This molecular structure gives rise to an opaque white material with a relatively high density of approximately 1.41 g/cm^3 [133]. POM exhibits excellent short-term mechanical properties including tensile strength, toughness, and rigidity. It also demonstrated a low tendency for creep and fatigue, as well as a low linear coefficient of thermal expansion. Furthermore, POM displays remarkable stability, exhibited resistance to chemicals, organic solvents, and fuels at room temperature. It exhibits low water permeability and absorption, as well as favorable tribological properties, such as high hardness, glossiness, and smooth molded surfaces. Additionally, POM maintains its electrical and mechanical properties over a wide temperature range, from −40 to −140 °C, with long-term stability at 90 °C and short-term stability at 140 °C. These exceptional characteristics have led to the extensive utilization of POM in various engineering applications, including bearings, gears, conveyer-belt links, wear surfaces, creep-resistant housings, gas caps, chemical sprayers, soap dispersers, paint mixing paddles, and safety systems such as seatbelts [134].

POM, as a polymer, exhibits superior mechanical strength, dimensional stability, and resistance to chemical corrosion. In comparison to commonly used synthetic materials, such as polypropylene (PP) fiber, polyvinyl alcohol fiber (PVA), and polyethylene (PE), POM exhibits better dispersion and lower air entrainment. This is attributed to the large number of ether bonds present in its molecular structure, which enhance the interfacial bonding strength with inorganic materials. Moreover, POM demonstrates a higher corrosion resistance than steel, making it suitable for use in harsh environments. The exceptional overall performance of POM suggests its significant potential for a wide range of engineering applications [135, 136], and the properties listed in Table 3.13 indicates the suitability of its use in thermoforming.

The properties of thermoformable polyacetals, which are available in sheets and coils, include a high strength-to-weight ratio, high dimensional stability, scratch and wear resistance, and good chemical and solvent resistance. In addition, the material has a surface that readily accepts paint and ink for decorating and printing applications [137].

3.14 Polycarbonate (PC)

The polycarbonate of bisphenol-A (PC) is one of the most important engineering thermoplastics. In addition to its high toughness and excellent

Table 3.13 Properties of poly(oxymethylene) (POM).

Parameter	Unit	Value	Reference
Density	kg/m^3	1.491	[79]
Glass Transition Temperature	°C	–75	[79]
Melt Temperature	°C	165.4	[138]
Heat Deflection Temperature	°C	113	[139]
Enthalpy of melting	J/g	159.9	[136]
Enthalpy of crystallization	J/g	153.9	
Crystalline temperature	°C	180	[123, 124]

optical properties, high-temperature resistance is often required in many applications, such as lenses for automobile forward lighting and light-emitting diodes. The Tg of PC depends on its molecular weight, end groups, and additive package [140].

PC is an amorphous material with excellent toughness, thermal and dimensional stability, and transparency, making it one of the most widely used thermoplastics in application that demand high-performance properties. Polycarbonate is a highly durable material with excellent heat resistance and weathering properties. Polycarbonate sheets maintain their impact properties for a longer period, resist yellowing, and exhibit a high level of light transmittance over time [141].

Polycarbonates are very tough, have good heat resistance, and weathering properties, and are used in applications demanding a high degree of impact resistance, such as vandal-resistant glazing and sign faces in high-traffic, high-vandalism areas. Table 3.14 shows the physical properties that help in thermoforming. Polycarbonate sheets significantly preserve the impact properties, resist yellowing, and maintain a high degree of light transmittance over time [143].

The thermal conductivity of polycarbonate in Figure 3.8, increases slightly with temperature in the solid state and becomes almost unchanged with temperature in the molten state, except for a temperature around 220 °C. Polycarbonate requires special techniques for thermoforming, which differs from most thermoplastic sheets. PC loses rigidity very rapidly above 188 °C. It has a narrow high-temperature forming range. At higher

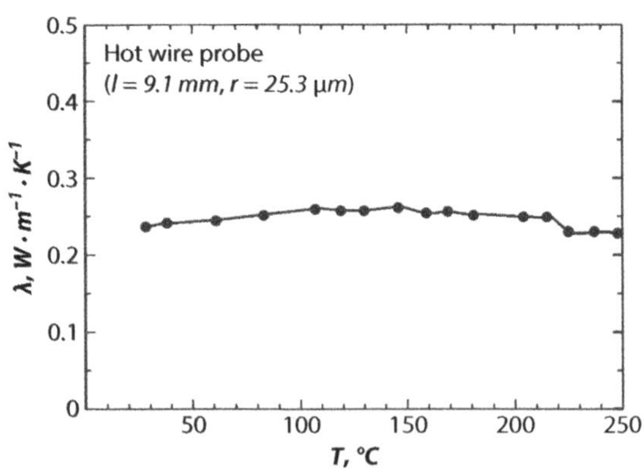

Figure 3.8 Thermal conductivity of polycarbonate. [Reprint with permission from[141]. Copyright © 2002, Plenum Publishing Corporation. All rights reserved].

Table 3.14 Properties of polycarbonate (PC).

Parameter	Unit	Value	Reference
Density	kg/m^3	1,200	[79]
Glass Transition Temperature	°C	150	[79]
Heat Deflection Temperature	°C	154	[139]
Tensile Strength	MPa	59–66	[142]
Yield strength	MPa	55–62	
Elongation	%	60–100	
Young Modulus	MPa	2,400	
Modulus of rigidity	MPa	800	

temperatures, entrained moisture can result in bubbling. In thin-sheet materials, water vapor typically escapes to the surface before it develops a bubble in the hot sheet. Thicker sheets must be dried in an oven to ensure that bubbling does not occur during thermoforming [144, 145].

3.15 Poly(ether-ether-ketone) (PEEK)

Poly(ether-ether-ketone) is a semi-crystalline polymer belonging to the polyaryl-ether-ketone family. It exhibits excellent mechanical properties, high thermal stability, and chemical resistance. Other properties of PEEK are listed in Table 3.15. Due to its large stiffness-to-weight ratio, the suitability of PEEK for the fabrication of thermoformed products is deemed particularly beneficial for high-end applications. PEEK for the application of thin-walled, thermoformed products is governed by its characteristically large stiffness to mass ratio (2.571 MNm/kg) or specific stiffness when compared to similar thermoplastics in this field, such as poly(ethylene terephthalate) (PET) and polypropylene (PP), which exhibit stiffness-to-mass ratios of 1.206 and 1.935 MNm/kg, respectively [146].

The forming temperature, Tf, is a significantly technological challenge. PEEK is a thermoplastic that exhibits the highest performance. Due to the advantageous thermal and mechanical properties [151], PEEK and its

Table 3.15 Properties of polyetheretherketone (PEEK).

Parameter	Unit	Value	Reference
Density Crystalline density Non-crystalline density	kg/m^3	1,400.6 1,262.6	[147]
Glass Transition Temperature	°C	144	[147]
Melt Temperature	°C	331.2	[148]
Heat Deflection Temperature	°C	203	[149]
Tensile Strength	MPa	84	[149]
Elongation at break	%	40–80	[149]
Young's Modulus	MPa	3,100 ± 0.04	[148]
Flexural modulus	MPa	2,600	[149]
Crystallinity $\varkappa_c$	%	32–35	[150]

composites are widely used and the thermal properties of PEEK. The superior mechanical properties of PEEK have led to its implementation in a wide spectrum of diverse applications in bulk form and as a matrix material for short fiber-reinforced composites. With PEEK exhibiting a profound dependence on both the temperature and strain rate, the yield stress decreased with increasing temperature, and the opposite is true for increasing strain rate. Ductility is positively correlated with temperature [133, 152].

3.16 Polyphenylene Oxide (PPO)

Polyphenylene oxide (PPO) is a thermoplastic material that exhibits flame resistance, excellent chemical resistance, and favorable electrical properties. It finds applications in various industries, including electrical and medical, where it is commonly used for manufacturing electrical and medical enclosures [153, 154]. Table 3.16 presents the selection of the physical properties essential for the thermoforming process. Polyphenylene oxide (PPO) is flame-resistant, chemically resistant, and has good electrical properties and applications [155, 156].

Table 3.16 Properties of polyphenylene oxide (PPO).

Parameter	Unit	Value	Reference
Density	kg/m^3	1,050	[157]
Glass Transition Temperature	°C	479	[158]
Heat Deflection Temperature	°C	171	[139]
Tensile Strength	MPa	66	[159]
Tensile Modulus	MPa	2,248	
Young's Modulus	MPa	2,344	[139]
Notched Izod Impact Strength	J/m	96	[159]

3.17 Polybutylene Terephthalate (PBT)

Polybutylene terephthalate (PBT) is a thermoplastic polymer with excellent dimensional stability, is very rigid, and offers good thermal resistance, electric isolation, and low dielectric losses. The increasing use of this material is related mainly to the electronics sector, particularly with the incorporation of new functionalities into the automotive sector by the integration of connectors and sensors because of its good performance in micro-molding processes and 3D printing technology with good thermal and mechanical properties [160, 161]. Poly(butylene terephthalate) (PBT) is a semi-crystalline, high-molecular-weight polymer with an excellent balance of properties and processing characteristics.

Polybutylene terephthalate (PBT) is in a rubbery state where the polymer chains have sufficient energy to untangle and become mobile, leaving the crystalline phase to provide mechanical strength [162]. PBT has higher crystallinity than PET. PBT has a lower tensile strength due to its longer butylene chain, making it more flexible. PBT exhibited the longest elongation at break. The entanglement of the long chain of PBT when a load at a low strain rate is applied becomes flexible with its butylene chain within PBT, enabling it to absorb the impact energy

PBT exhibits higher tensile, flexural, and dielectric strengths and faster and more economical forming characteristics. The properties listed in Table 3.17 indicate the suitability of the material for thermoforming. PBT exhibits an excellent resistance to a broad range of chemicals at room temperature. Due to its rapid crystallization, the required cycle time can be lower than that required for many engineering plastics. This thermoplastic

Table 3.17 Properties of polybutylene terephthalate (PBT).

Parameter	Unit	Value	Reference
Density amorphous Crystalline	kg/m³	1.280 1.396	[120]
Glass Transition Temperature	°C	42	[165]
Melt Temperature	°C	221	
Crystalline temperature	°C	185	
Storage modulus	MPa	1,795	[102]
Flexural modulus	MPa	1,777	[102]
Crystallinity $\varkappa_c$	%	33.4	[102]

polyester has very good dimensional stability. They also exhibit high heat resistance, chemical resistance, and good electrical properties.

PBT offers extremely good versatility when mixed with other materials, which has enabled additives to be incorporated, such as flame retardants, and particularly more environmentally friendly, non-halogen additives to improve their properties against fire.

PBT is an ideal candidate for the growing number of industrial components in the transport sector, due to its excellent properties, such as electric insulators. These components could be subjected to high-temperature environments for long periods of time, and the stiffness and energy absorption are key factors due to the dimensional stability and durability requirements [163, 164].

PBT has a faster crystallization rate than PET, which makes it the preferred material for industrial-scale molding. PBT is the preferred material when high strength, good dimensional stability, resistance to various chemicals, and good insulation are required, as well as when bearing and wear characteristics play a role in the material selection. PBT is a semicrystalline feed polymer because of its better thermal stability, low moisture absorption, and higher mechanical strength than polyamides.

By annealing PBT for a long time, the crystallinity of the material allows for a shift in the glass transition and melting temperatures to higher values, which helps in thermoforming. Furthermore, annealing can increase the crystallinity of PBT. PBT is one of the toughest thermoplastics, and the major drawbacks of using PBT are its rapid crystallization and limited processing duration [165].

3.18 Liquid Crystalline Polymers

Thermotropic liquid crystal polymers (LCPs) are aromatic copolyesters that exhibit densely packed rod-like polymer chains and exhibit distinctive mechanical properties in a single direction. LCPs demonstrate exceptional resistance to radiation, hydrolysis, and chemicals and possess inherent flame retardancy with minimal smoke production and a low expansion coefficient.

Liquid crystalline polymers (LCPs) possess various interesting properties and unique processing opportunities. Traditional methods for processing LCPs, such as injection molding, extrusion, and fiber spinning, have been extensively studied and utilized. Due to the long relaxation times of LCPs, the end product retains a high degree of molecular orientation induced during processing, resulting in distinct anisotropy in structure and properties. This anisotropy can be advantageous in certain situations but may also present challenges to others. For instance, thermoforming, which involves the biaxial deformation of an extruded sheet, is negatively affected by the anisotropy of the LCP sheets, affecting both the thermoforming process and the resulting products.

The process of thermoforming LCP-reinforced thermoplastics involved heating them at temperatures above the melting point of LCP (284 °C), which allowed for additional orientation in the LCP phase. However, annealing the LCP at this temperature may lead to a loss of orientation [166].

3.19 Cyclic Olefin Copolymer (COC)

Cyclic olefin copolymer (COC) are amorphous thermoplastics with excellent moisture barrier, high-temperature stability, and stiffness, along with other properties listed in Table 3.18. COC have excellent transparency, good stiffness, excellent thermoforming characteristics, short cycle time, low machine direction (MD) and cross-direction (CD) shrinkage and can typically be run on existing equipment without modifications, low density, and low environmental risk. As with any type of plastic polymer, its chemical structure makes the resin well suited for some applications and poorly suited for others [167].

Table 3.18 Properties of cyclic olefin copolymer (COC).

Parameter	Unit	Value	Reference
Density	kg/m^3	1,020	[79]
Glass Transition Temperature	°C	85	[167]
Heat Deflection Temperature	°C	75	[167]
Forming temperature	°C	110–130	[79]
Tensile Strength	MPa	66	[167]
Tensile Modulus	MPa	2.6	

Cyclic olefin copolymer (COC) blends can be extruded into thermoformable sheets. COC can provide film and packaging converters with the opportunity to create thermoforming films. Mono- and multi-layer examples of LLDPE-based forming films, compared to commercially available products, clearly demonstrate how well the addition of COC improves the physical properties, thermoforming, and packaging performance. COC improved the material distribution of the LLDPE-formed trays. These improvements enable the formed tray to withstand a greater crushing force. Enhanced performance permits down gauging [168].

Cyclic olefin copolymer (COC) blends can be extruded into sheets suitable for thermoforming. COC, being an amorphous thermoplastic, possesses advantageous characteristics such as exceptional moisture barrier, high-temperature stability, and stiffness [169].

3.20 Plastic Foams

The thermoforming of foamed plastics has the benefits of low cost, good formability, uncomplicated molds, and the capability to build large or thin structures. However, viscoelastic phenomena have hindered the overall success of foam-plastic thermoforming. The temperature of the foam sheet needs to be raised well above the glass transition temperature for amorphous plastic foam materials.

Amorphous materials have broad thermoforming temperature ranges and large draw ratios, whereas equivalent low-density closed-cell foams have very restricted ranges and very low draw ratios. During heating of the low-density foam sheet, expansion occurs due to the increasing cell gas pressure and decreasing cell wall strength. Excessive heating rates can lead to catastrophic cell collapse [170].

During thermoforming, the sheet must be heated to the glass transition temperature (4). The stiffness of the polymer is significantly reduced while the gas expands. Overall, the foam expands, especially in the thickness direction, resulting in a low internal pressure and little residual anisotropy.

PLA foams are considered to be one of the most feasible alternatives to traditional PS foams. PLA-based low-density foam has been commercialized as a compostable packaging tray for meat, fish, fruits, and vegetables [171, 172].

Figure 3.9 illustrates the SEM micrographs of PLA foam sheets as-extruded (a), after thermoforming using heating temperatures ranging from 170 °C to 185 °C (b–e), and thermoformed bowl (f). The thickness and density of the as-extruded foam sheets before thermoforming (a) consist of both closed and open cells. The morphology reveals that the melt strength of PLA due to the chain extender induces a long-chain branching reaction. SEM (a) also indicates that the melt temperature during foam extrusion was not completely optimized. The sheet was not optimized; therefore, it led to cell rupture. (b)–(e) SEM images after foam sheet thermoforming at

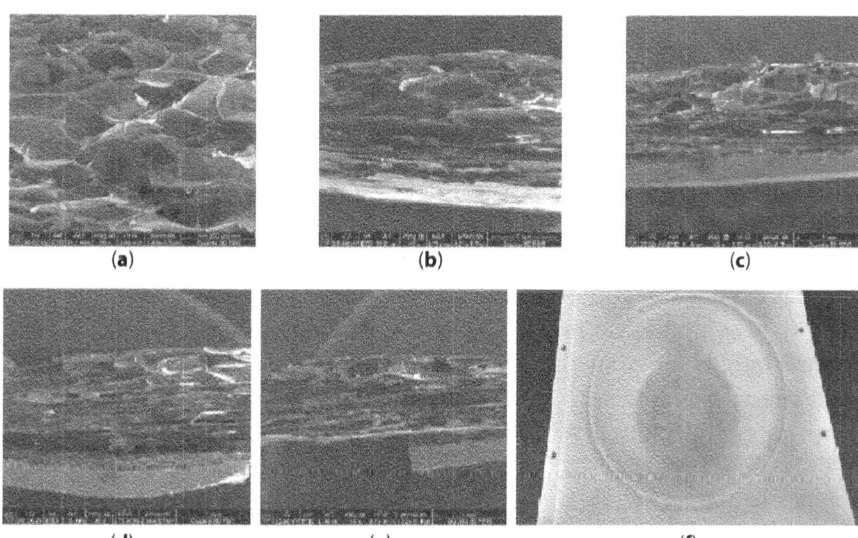

Figure 3.9 SEM micrographs of thermoformed PLA foam sheets (a) before thermoforming, at (b) 170 °C, (c) 175 °C, (d) 180 °C, (e) 185 °C, and (f) thermoformed PLA bowl [Reprint with permission from [173]. Copyright © 2016, American Chemical Society. All rights reserved].

different temperatures. The SEM images are from the center of the thermoformed bowls along the radial direction. The cells elongated in the straining direction after thermoforming. Higher temperature thermoforming suppresses stiffness; hence, the foam sheet deforms and elongates during the shaping process [173].

3.21 Thermoplastic Elastomers

Thermoplastic elastomers (TPEs) exhibit excellent physicomechanical properties, elasticity, processability, recyclability, and resistance to heat, chemicals, and ultraviolet radiation. These lightweight materials have a pleasant touch, luxurious feeling, and excellent price performance ratio, all of which make them ideal candidates for enormous applications [174].

TPEs are special types of polymers that exhibit the hybrid behavior of thermoplastics and rubbers. They can be processed into thermoplastics. However, they behave like rubber over a broad temperature range. When thermoforming TPE, the biphasic morphology of the hard plastic phase softens and flows, which solidifies reversibly upon cooling. The hard phase provides thermoplasticity at the sheet temperature during processing [175].

3.21.1 Thermoplastic Olefin (TPO)

Thermoplastic olefins (TPOs) are blends of polypropylene and rubber, with additional fillers and additives. Thermoplastic olefins (TPOs) are very tough and flexible, have outstanding resistance to low temperatures, low thermal expansion, and excellent UV resistance, and are used in applications that combine the properties of high-quality polyurethane with the processing efficiency of thermoplastics [176]. The Young's modulus and yield stress increased with strain rate, while the elongation decreases with strain rate. The influence of temperature on the mechanical behavior of TPO is significant.

TPO is gaining wide acceptance in automotive applications. An exterior application, is fascia. It should be referred to as an olefinic-based thermoplastic elastomer, which can be reactor TPO, physical blend TPO, or partially crosslinked TPO. TPO has emerged as one of the most viable substitutes in automotive interior trim applications, and this replacement continues to be a major objective in the automotive industry. The powerful drivers for the ever-increasing growth rate of olefinic materials in automotive applications are weight savings, economics, processability, expanded property profiles, weatherability, and low temperature ductility [177].

Thermoplastic polyolefins (TPO) based on polypropylene (PP) are desirable for cutsheet thermoforming applications because of their superior physical properties, weathering performance, and chemical resistance. However, their acceptance has been limited due to their poor processing performance [178–180].

3.21.2 Thermoplastic Urethane (TPU)

Thermoplastic urethane elastomers are linear block copolymers, and the polymer chain consists of a relatively long flexible polyester or polyether diol. Amorphous polyol blocks are usually termed soft segments because they impart elastomeric character to the polymer. The hard segment is formed by the reaction of aromatic diisocyanates with low molecular weight diols or triol chain extenders. Due to the polar nature of the urethane groups in the hard segments and their ability to form hydrogen bonds, these hard segments are capable of intermolecular association [181].

The thermally reversible network structure of these copolymers provides the elastomeric or apparently crosslinked nature of these polymers. Table 3.19 lists the properties of TPU. The dynamic mechanical properties of these polymers can be interpreted in terms of the corresponding modifications in either the hard or the soft segments. In thermoforming, the lower-temperature glass transition region for these elastomers appears to be affected. The degree of flexibility of the amorphous soft segments and their interactions with the hard segments and the higher-temperature relaxation region is controlled mainly by the extent of association of the rigid and hard segments. In addition, the relative concentrations of the hard and soft segments affect the plateau modulus at temperatures between the major relaxation regions [182, 183].

Table 3.19 Physical properties of Thermoplastic Urethane (TPU).

Parameter	Unit	Value	Reference
Melt flow index	g/10min	15.9	[184]
Glass transition	°C	−24.0	
Heat deflection temperature at 0.455 MPa	°C	74.0	
Tensile modulus	MPa	26.0	
Tensile stress at yield	MPa	8.6.0	
Tensile stress at break	MPa	39.0	
Elongation at yield	%	55.0	
Elongation at break	%	580.0	
Flexural strength	MPa	4.30	
Flexural modulus	MPa	78.7	

3.22 Thermoplastic Composites (TCs)

In the field of thermoforming, the behavior of thermoplastics with varying melting temperatures is influenced by the addition of fiber reinforcement. Thermoplastics are polymers that can be melted and re-melted without undergoing significant chemical changes. When fibers are incorporated into these thermoplastics, the resulting composites exhibit improved mechanical properties such as increased strength, stiffness, and impact resistance. Commonly used fiber materials for reinforcement include glass, carbon, and aramid.

Thermoplastic composites (TCs) are typically used in load-bearing applications. TCs consist of a combination of two or more types of material. Thermoplastic composites have several advantages and good prospects for replacing metal and thermoset composites in the market [185]. Thermoforming of these composites is aimed at applying them to complex-shaped parts for large-volume production [186].

The resulting products benefit from the advantageous properties of each component. In certain cases, composites are the only viable option for achieving desired properties. The aerospace industry primarily relies on high-performance composites, also known as "advanced composites," and rarely uses pure resin forms of plastics in applications that require high structural integrity [187]. Instead, fillers are added to plastics for various purposes, such as reducing costs, enhancing stiffness, minimizing shrinkage, improving impact resistance, enhancing performance at extreme temperatures, and increasing flame retardance [188].

The process of thermoforming composite sheet materials involves the correct positioning of fibers to prevent wrinkling or buckling. When the fibers are compressed during forming, they can wrinkle or buckle. To prevent this, it is important to ensure that tensile stresses are present throughout the sheet during the forming process, as this helps reduce fiber buckling [189].

The ability of thermoplastic composite materials to form in thermoforming is limited by the properties of the material. Without a proper understanding of the forming process, the development of a thermoformed component will involve a lengthy and expensive trial-and-error process. This includes various stages such as manufacturing the preform, forming the component, trimming if necessary, and assembling it into the overall structure.

When low-temperature melting thermoplastics are reinforced with fibers, they tend to exhibit enhanced formability during the thermoforming process. The lower melting point of these thermoplastics allows for easier shaping over the mold and requires less powerful airflow during the forming stage. This ease of formation is advantageous for achieving complex shapes and reducing manufacturing costs.

Conversely, high-temperature melting of thermoplastics reinforced with fibers may present challenges during thermoforming. Their higher melting points may necessitate higher temperatures and greater energy inputs for the forming process, resulting in increased production costs. Additionally, the higher concentrations of reinforcement in these composites can reduce formability, requiring more powerful airflow or additional processing steps to achieve the desired shape.

The selection of the thermoplastic material and fiber reinforcement concentration is contingent upon the specific requirements of the thermoforming process. Thermoplastics with low-temperature melting points, when combined with fiber reinforcement, are often favored due to their enhanced formability and reduced energy consumption. Consequently, these materials are well suited for a range of applications in industries such as automotive, packaging, and consumer goods. However, despite posing potential challenges in the thermoforming process, thermoplastics with high-temperature melting points may still be utilized in applications that require superior mechanical properties.

The forming stage can be completed quickly using preconsolidated thermoplastic composite laminates, as they only require heating to become easily moldable and then cooled to revert back to their strong and stiff state. Temperature conversion can be achieved relatively quickly, especially for materials with lower melting temperatures and thinner laminates [190].

Thermoplastic composites made from polypropylene (PP) and long sisal fibers can be thermoformed using various processing techniques. By applying small reductions in the wall thickness, these composites can be shaped into three-dimensional forms with minimal energy input, resulting in exceptional properties and a high-quality surface finish. Thermoforming techniques for thermoplastic composites have been well established because of their shorter processing time, good formability, and ease of handling [191].

The blank forming temperature is the dominant factor that influences the surface roughness and reprocessing melt temperature [192]. The choice of woven fabric reinforcement is imperative to overcome this instability. This should be performed to ensure that the fabric has a high deformation potential to accommodate larger complex deformations [193, 194].

The effects of the parameters on the incidence of deformation in bamboo-reinforced PLA composites indicate that hot tooling conditions provide good quality domes and the predominance of deformations in the wrap direction of composite domes. The main issues with forming are the occurrence of instabilities such as buckling and wrinkling on the part surface [195]. The fiber density and bland holding force are strongly damaging.

Processing composite materials with thermoplastic matrices that have high-temperature melting points (Tm >200 °C) poses a significant challenge due to the risk of thermal degradation of the fibers. The elevated processing temperatures necessary for these matrices can lead to deterioration of the fiber properties, adversely affecting the overall performance of the composite.

One potential solution to this issue is the use of polar matrices such as polyethylene terephthalate (PET). PET contains carbonyl groups in its structure that can form hydrogen bonds with the hydroxyl groups found in cellulose fibers. These hydrogen bonds enhance the interfacial adhesion between the fibers and matrix, resulting in improved mechanical properties and reduced susceptibility to thermal degradation during processing.

The incorporation of fiber reinforcements and thermoplastic matrices has led to significant advancements in various industries. The selection of matrix materials and fiber types depends on the specific application requirements, processing conditions, and desired mechanical properties of the composite. The emergence of polar matrices, such as PET, offers new opportunities for the development of composites with enhanced performance, particularly when utilizing natural fibers that possess abundant hydrogen bonding sites.

Composite materials have gained widespread usage across various industries due to their versatility and distinctive properties. Sectors such as the construction, aerospace, furniture, and automotive sectors extensively employ composite materials. Natural fibers, including wood flour, sawdust, and hemp bast fibers, possess a thermal stability that aligns well with the low processing temperatures of these thermoplastic matrices. The utilization of thermoplastic matrices with low melting points not only reduces the energy consumption but also lowers the processing temperature requirements, enhancing the cost-effectiveness of the manufacturing process [196].

3.23 Bioplastics

Bioplastics, such as poly(lactic acid) (PLA) are bio-based and biodegradable. They have a variety of applications, with the expectation of improved strength and stiffness [197]. Due to technological advancements and research, bioplastics can play a vital role by gradually replacing plastics in various sectors such as the biomedical, agriculture, and household utility industries [198, 199]. Bioplastics are less common than synthetic plastic materials because of various factors, such as complications in processability, higher per-unit cost due to low production, comparably low mechanical strength, and dissimilarities in resources, which restrict their widespread utilization. However, among bioplastics, PLA is preferred due to its adequate compatibility with comparable mechanical and physical properties [200].

3.23.1 Polylactic Acid (PLA)

Polylactic acid (PLA) has been recognized as a potential polymer candidate with attractive characteristics for applications in both the engineering and medical sectors. PLA is produced from lactic acid formed by the fermentation of carbohydrates (found in corn, which is a cheap source) by bacteria using the Lactobacillus genus (an optimized strain). This is economically feasible and can be achieved. PLA has several advantages, e.g., it is a source of agriculture-based products, helps fix a substantial amount of CO2 (a greenhouse gas), requires less energy in processing, is easily recycled into lactic acid without much effort, and has tailored physical–mechanical properties through material modifications [201].

Poly (lactic acid) (PLA) is a promising polymer with valuable properties (Table 3.20) and potential because of its sustainability, low carbon footprint, and superior bio-based polymer properties compared to other bioplastics. PLA is a bioresin that is rapidly expanding into a variety of applications. It requires a range of processing techniques including extrusion, calendering, blow molding, and thermoforming. The brittleness and low melt strength of the polymer pose challenges in processing and in the final product performance, and the use of additives can improve the processing, melt strength, and impact strength of PLA. It provides a lower coefficient of thermal expansion and increased heat deflection temperature [202].

Table 3.20 Properties of poly(lactic acid) (PLA).

Parameter	**Unit**	**Value**	**Reference**
Glass Transition Temperature	°C	55	[15]
Melt Temperature	°C	170–180	
Heat Deflection Temperature	°C	55	
Tensile Strength	MPa	66	
Elongation at break	%	4	
Izod Impact Strength	J/m	29	
Tensile modulus	MPa	2,346.5	[203]
Tensile stress at yield	MPa	49.5	
Tensile stress at break	MPa	45.6	
Elongation at yield	%	3.3	
Elongation at break	%	5.2	
Flexural strength	MPa	130	
Flexural modulus	MPa	3,150	

3.23.2 Poly(Butylene Succinate) (PBS)

New materials such as poly(butylene succinate) (PBS) and thermoforming require extensive and expensive qualifications. They must meet manufacturing challenges, including strength, stiffness, durability, inspection, and maintenance requirements, as well as emerging recycling-reuse targets, obligations, and directives. The properties listed in Table 3.21 helps predict the thermoformability of poly(butylene succinate).

Poly(butylene succinate) (PBS) synthesized from succinic acid and 1,4-butanediol (BDO) exhibits balanced performance in terms of thermal and mechanical properties, as well as thermoplastic processability, compared with other common plastics. The thermal stability of PBS materials is sensitive to the water content, content of residual carboxylic acid terminals, and molecular structure of the polymer chains. Before thermal processing, PBS pellets are dried with hot air to decrease the water content to less than 0.1% to avoid hydrolysis, which occurs with traces of water at high temperatures. When the processing temperature of PBS exceeds 200 °C, a considerable decrease in shear viscosity is observed [204].

The melt strength of PBS is considerably enhanced by long-chain branching, which meets the demands posed by extrusion and thermoforming. The industrialization of PBS and its copolymers has been successfully pursued in several countries. In the future, attempts will be devoted to improving the tear resistance, impact strength, and transparency of PBS

Table 3.21 Properties of poly(butylene succinate) (PBS).

Parameter	Unit	Value	Reference
Glass Transition Temperature	°C	−18	[15]
Melt Temperature	°C	112–115	
Heat of fusion	J/g	69.8	
Enthalpy of melting	J/g	67.4	
Tensile Strength	MPa	35 ± 1	
Elongation at break	%	275 ± 35	
Tensile strength at break	MPa	29 ± 2	
Izod Impact Strength	J/m	73 ± 28	
Crystallinity $\varkappa_c$	%	61.1	

and its copolymer films which will greatly expand the thermoforming of these biodegradable plastics. Additives compatible with biodegradable materials, such as nucleating agents and impact modifiers, are being developed to manufacture PBS articles that meet the requirements of various applications [205].

In summary, thermoforming processing can be customized to optimize the material processing parameters. Determining the desired characteristics across a range of thermal and physical parameters is a complex task in the development of thermoforming processes for thermoplastic materials. Additionally, the design process traditionally relies on experience and simplified performance; however, there is a need for design tools that can quickly evaluate new materials and processes. Computational modeling should be further developed to reduce the reliance on experimental methods in product development. Furthermore, materials often have limited physical properties or processing values that meet their quality control requirements.

4

Properties of Thermoplastic Sheet Materials

4.1 Introduction

Synthetic polymers have proven to be highly versatile and valuable materials for contemporary technology. Their affordability and ease of fabrication using automated machinery have made them extensively utilized in the production of consumer goods, particularly in the form of packaging materials for agriculture, forestry, dairy, and other consumer products.

Understanding the relationship between the mechanical and physical properties of various polymers, such as the modulus, strength, strain to failure, and thermal properties, is crucial for determining the resulting thermoformability of the materials. The desirable strength and durability of polymers are key factors for their widespread application, as their resistance to environmental degradation, including biodegradation, is a significant consideration [1].

4.2 Polymer Characteristics

The stability and fragmentation of polymeric materials are regulated by various properties of macromolecules, including their length, rigidity, and intermolecular interactions [2, 3]. Thermoformable thermoplastics used in many applications must meet stringent requirements such as exceptional resistance to chemicals, hydrolysis, radiation, and abrasion. These materials should also possess precise and stable dimensions that can be consistently reproduced. Furthermore, they should exhibit inherent flame retardancy, a high heat deflection temperature, and resistance to thermal cut-through. Other important characteristics include cryogenic stability, low dielectric constant, high dielectric breakdown strength, affinity for reinforcement, and the ability to allow color coding and printing [3].

Muralisrinivasan Natamai Subramanian. Thermoforming: Processing and Technology, (69–110)

The characteristics of polymers can vary significantly depending on factors such as the molecular weight, monomer type, and synthesis process. Understanding these characteristics is crucial for tailoring materials for specific applications and for improving their performance. The physical properties of polymers are affected by the length and composition of bonds within the polymer chains. These properties can be modified by incorporating different polymers and reinforcement materials as well as by adding additives and modifiers. Finally, the behavior of the polymers is influenced by temperature.

4.3 Polymer Morphology

Polymers are composed of repeating monomers and small molecules that are bonded together to form a long chain. These chains possess carbon–carbon bonds that are capable of rotation, allowing for the rearrangement of the molecular chains into countless conformations. This rearrangement is primarily influenced by factors such as deformation rate, stress, and temperature. The behavior of polymers differs significantly from that of metals.

A polymer molecule is a macro system composed of numerous atoms and can be effectively described using phenomenological concepts such as free energy and the chain friction coefficient. The importance of volume interactions in the physics of a polymer macromolecule stems from the fact that the natural conformation of a polymer chain is not linear but rather takes the form of a coiled structure entangled in space. This situation is the result of the chain's flexibility. From the perspective of solid-state polymers, available movements can be classified into two main categories: vibrations and internal rotations. The translation and rotation of the entire molecule are not considered because of the large sizes of the polymer molecules. The morphological structure of linear polymers plays a critical role in determining their potential movement. In the solid state, there are two distinct states of disorder and order: completely amorphous and highly crystalline [4]. Conversely, a highly crystalline polymer is typically observed when a molecule possesses long, uninterrupted sequences of identical repeating units. However, even in these materials, defects are more prevalent than in comparable monomeric solids [5]. It is important to note that every movement, including vibrations and internal rotations, contributed to the specific heat of the polymer.

Initially, synthetic polymers were developed to enhance durability and resistance to degradation. Polymer molecules can be described statistically, as they lack thickness and can move independently without external

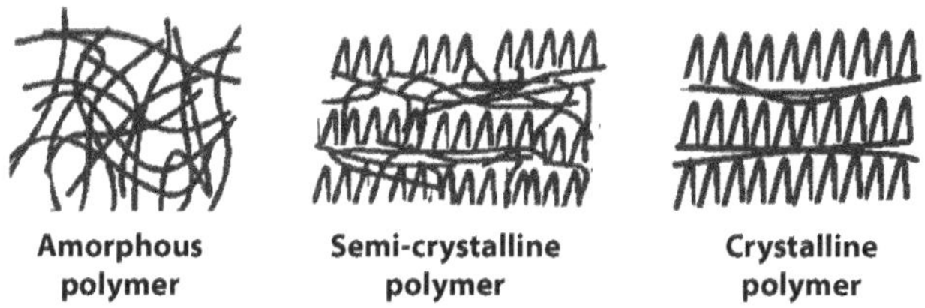

Figure 4.1 Schematic diagram of the molecular structure of polymers.

forces, provided their connectivity and topological integrity remain intact. This means that they cannot pass through themselves or through other polymers. Based on their molecular structures, polymers can be classified into three main groups: amorphous, semi-crystalline, and crystalline (Figure 4.1).

4.3.1 Amorphous Polymers

Amorphous materials undergo thermoforming within a specific temperature range, known as the rubbery region, which lies between the glass transition temperature and the point at which the material modulus reaches approximately 1 MPa. This temperature window offers a favorable combination of viscous and elastic properties, as well as a high flow activation energy and strain hardening, rendering the materials highly suitable for thermoforming processes [6].

The morphological structure of polymers is a significant consequence of deformation. The completely amorphous and crystalline polymers form the limiting cases of disorder and order in the solid state. The Semi-crystalline polymer showed a glass transition temperature as well as a melting transition at non-identical temperatures. In crystalline polymers, the molecules have long uninterrupted sequences of identical repeating units [7].

The idealized structure of an amorphous polymer is characterized by a random arrangement of long molecular chains that intersect, creating entanglement points. This leads to the formation of an isotropic network-like structure similar to that of an elastic material, but with physical entanglements replacing the chemical crosslinks. When the material reaches its yield point, there is resistance to the rotation of the chain segments, which is isotropic and occurs between entanglements at the rigid links. Intermolecular interactions between the physical entanglements and rigid links play a crucial role in this process. Once the material is allowed to deform, molecular alignment occurs, resulting in an anisotropic internal resistance to further inelastic deformation.

Amorphous polymers are characterized by a lack of distinct crystalline structures, resulting in a random or disordered arrangement of molecular chains. This arrangement gives the amorphous polymers a more uniform and translucent appearance. The molecular chains in amorphous polymers are assembled in a noncrystalline manner, leading to a more irregular chain structure. Consequently, amorphous polymers have lower softening and melting temperatures than crystalline polymers [8].

The stress response of amorphous polymers undergoes significant changes in the glass transition region, transitioning from a soft viscoelastic response to a stiff viscoplastic response. Furthermore, the mechanical properties of amorphous polymers under high deformation rates are primarily influenced by the forces between neighboring chain molecules. These properties can be characterized by bulk modulus, tensile fracture stress, and strain.

Amorphous polymers possess advantageous properties for applications that require transparency, such as optical lenses or clear packaging materials. This is primarily due to the random arrangement of molecular chains within these polymers, which contributes to their enhanced clarity and optical characteristics [9, 10].

4.3.2 Semi-Crystalline Polymers

The characterization of a semicrystalline polymer involves assessing its crystallinity percentage and the average dimensions of its crystallites. Semi-crystalline polymers consist of both ordered (crystalline) and disordered (amorphous) regions, with amorphous regions containing disordered chains, and crystalline domains consisting of polymer chains arranged in an organized manner. The mechanical properties of these polymers are primarily influenced by the molecular connections between the crystallites, such as chain entanglements and tie molecules [11].

The weak van der Waals or hydrogen interactions that hold crystallites together make them susceptible to slipping or splitting under stress, resulting in macroscopic brittleness. The presence of crystalline domains in semi-crystalline polymers contributes to their opacity due to light scattering, while the amorphous portions enhance toughness and impact resistance. However, crystalline sections contribute to more rigid and stiff structures [12].

When a load is applied to semi-crystalline materials, it is distributed among a limited number of intercrystalline molecular connections. These connections gradually lose their mechanical efficiency due to chain scission, the generation of chain radicals, or chain disentanglement through the crystalline component.

The presence of intercrystalline molecular connections in melt-crystallized semicrystalline polymers can be explained by two factors. The first is the existence of a critical molar weight at which chains become intertwined into a macromolecular network. As the molecular weight increased, the likelihood of generating tie molecules and chain entanglements also increased. This leads to reduced drawability, increased strain hardening, and a higher tensile strength. Consequently, the viscosity of the material increases significantly, limiting its technological application to weight-average molecular weights [13].

The thermoprocessability of semicrystalline polymers is influenced by factors such as thermal stability, viscosity, melt strength, and crystallization rate. In comparison to amorphous materials, semicrystalline thermoplastics generally have lower thermoformability due to a significant decrease in stiffness and viscosity upon crystalline melting. To prevent excessive sagging during the heating stage, the polymer must maintain sufficient stiffness over a wide temperature range. This poses a challenge in the vacuum thermoforming of semicrystalline polymers because they exhibit a sharp decline in modulus at the melting temperature and have a narrow range of sufficient stiffness compared to amorphous thermoplastics. Despite their favorable properties such as chemical stability and tribological properties, most semicrystalline thermoplastics are unsuitable for thermoforming processes. However, the stiffness of the heated sheet or film, as well as the viscosity and strain-hardening behavior, can be enhanced through techniques such as chain branching or by increasing the average molecular weight [14].

The thermal and mechanical properties of semi-crystalline polymers are significantly affected by various factors, including the molecular weight (MW), molecular weight distribution (MWD), branching content, and density [15]. By intentionally manipulating these structural parameters, a wide range of products with distinct thermal and mechanical properties can be fabricated. Commercially available shear-modified polymers are also available [16].

Semi-crystalline thermoplastics exhibit high service temperatures and excellent resistances to environmental factors and organic solvents. The morphology and degree of crystallinity of semicrystalline thermoplastics are highly dependent on the cooling rate. Therefore, it is crucial to control the cooling rate of these materials because crystallinity is influenced by it. The exceptional service temperature and resistance properties of semi-crystalline thermoplastics make them suitable for use in thermoformed products [17].

This organized molecular arrangement results in increased stiffness and strength, making these materials mechanically more robust than amorphous plastics. The crystalline regions act as reinforcing structures that resist deformation and enhance the load-bearing capabilities. In addition to their mechanical properties, semicrystalline plastics generally exhibit excellent heat resistance. The structured molecular arrangement enables stronger intermolecular forces, leading to higher melting points and improved thermal stabilities. This characteristic makes them suitable for applications in high-temperature environments. One of the benefits of semicrystalline thermoplastics is their ability to be customized for specific applications [18].

4.3.3 Crystalline Polymers

All crystalline polymers consist of a certain amount of amorphous material mixed with crystallites. The orientation of the crystalline structure of these polymers has a significant impact on their morphology and mechanical properties. In highly oriented crystalline polymers, shrinkage occurs simultaneously with melting. In systems with low to moderate draw ratios, shrinkage typically occurs between the glass transition temperature (Tg) and the melting temperature (Tm) [19].

The fusion process is expected to occur within a specific temperature range, resulting in the complete disappearance of the crystallinity at the melt temperature. When heating the crystalline polymers, two distinct heat absorptions are observed at different temperatures. These heat changes are associated with the disorientation of crystalline regions and the actual fusion of crystallites rather than the transformation or melting of polymorphs. In polymers with low draw ratios, there is a broad range of heat absorption. The melting point of each crystallite is significantly influenced by its surface energy, as it is believed that the crystallites in these polymers are very small. During heating, the highly crystalline polymers do not exhibit a significant rubbery plateau. The regularly repeating structural unit of a polymer plays a role in its crystalline state, which is analogous to the monomeric molecule in a crystal lattice. The unit cell of a given crystalline polymer is occupied by only a small number of structural units [20].

The inclusion of a crystalline phase in a material enhances its resistance to solvents while simultaneously reducing properties, such as the glass transition temperature (Tg). Increasing the level of crystallinity can improve properties such as the elastic modulus; however, it can also decrease the toughness of the material. Conversely, a decrease in crystallinity can affect the sensitivity of the material to environmental factors such as humidity and aggressive agents such as organic solvents.

In the case of polymer sheets, a higher degree of crystallinity can enhance the modulus, improving the strength and self-supporting properties of the sheets. This in turn minimizes sagging during the heating stages of thermoforming and enhances the strength and rigidity of the final products. Additionally, the presence of small spherulites in the sheet and thermoformed part can improve both the clarity and impact properties. It should be noted that different crystallization conditions within a part can result in variations in the optical, thermal, or mechanical properties. These factors directly influence the processing behavior in the subsequent steps of a process chain.

4.3.4 Amorphous Versus Semi-Crystalline Polymers

Semi-crystalline polymers possess a higher degree of molecular order because of the regular and repeating patterns of the polymer chains in the crystalline regions, in contrast to amorphous polymers. However, semi-crystalline polymer parts tend to be larger than amorphous polymer parts because of the greater shrinkage exhibited by crystalline polymers. This shrinkage is a result of the volumetric contraction that occurs during crystallization, which is generally more pronounced in crystalline polymers than in amorphous polymers.

It is important to recognize that both amorphous and semicrystalline polymers have distinct advantages and limitations. The selection of an appropriate material depends on the specific requirements of its application at hand. Factors such as the desired properties, processing methods, and environmental conditions should be carefully considered when choosing the most suitable polymer for a particular application. For a detailed comparison between amorphous and semi-crystalline polymers, refer to Table 4.1.

Moreover, semicrystalline polymers maintain rigidity. They exhibit sharp melting points and narrow thermoforming windows. The melt strength is typically low at higher thermoforming temperatures. Even above their Tg, they must be heated close to their melting temperature during thermoforming. This causes excessive sagging in several instances. On the other hand, amorphous polymers become sufficiently soft above their Tg in the rubbery plateau with a wide thermoforming window. Therefore, they are thermoformed by heating above their Tg. They are easier to thermoform than semi-crystalline polymers.

Table 4.1 Comparison of amorphous and semi-crystalline plastic materials.

	Amorphous	**Semi-crystalline**
Melting point	Not sharp/soften gradually	Relatively sharp
Orientation	Random chain orientation in both solid and melt phase	Ordered arrangement of chains of molecules. Regular occurrence of crystalline structure only in solid phase
Flowability	Do not flow easily	Flow easily above melting point
Transparency	Can give transparent part	Part is usually opaque
Reinforcement	Slightly improves deflection temperature	Increases load bearing capabilities

4.4 Molecular Structure

The concept of molecular structure encompasses a range of parameters that characterize the distribution of molecular weight, including number, weight, z-average molecular weight, polydispersity index, and degree of branching. An increase in polydispersity has the potential to enhance the melt strength of a resin by increasing its elasticity [21–23]. However, this phenomenon is not consistently observed in practical industrial applications because increasing polydispersity can decrease the melt strength of certain polymers [24]. Furthermore, the specific manufacturing technology used can significantly affect the polymer properties. Different polymerization techniques can produce polymers with distinct molecular architectures, leading to technology-dependent rheological and processing characteristics. Factors such as the rate of polymerization, solvent, and other external factors involved in each technique can influence the number and size of the resulting polymer molecules, resulting in variations in the molecular architecture and properties. Therefore, these differences are expected to affect the properties of the polymers. It is important to note that the melt strength parameter does not provide well-defined rheological properties because the strain and temperature are not uniform throughout the stretched polymer melt [25].

Hierarchical structure formation occurs in polymers due to van der Waals interactions, where a primary structure is accompanied by an entangled chain network. Phenomena such as shear yielding, necking, or cavitation are preceded by the breakdown of this primary structure during thermoforming [26, 27].

4.5 Molecular Weight

Higher-molecular-weight polymers are characterized by longer chains and increased entanglement between them. This results in stronger intermolecular forces, leading to an enhanced tensile strength, stiffness, and durability of the material. These polymers are typically solid and rigid, making them suitable for applications that require structural integrity. Conversely, polymers with lower molecular weights have shorter chains and fewer intermolecular interactions, which results in reduced strength and rigidity. Consequently, they exhibit softer and pliable characteristics. These lower-molecular-weight polymers are commonly utilized in applications that prioritize flexibility and ease of processing, such as soft plastics used for packaging or consumer products [28, 29].

Thermoplastic sheets with lower molecular weights are formed more easily due to their reduced melt strength and elastic memory. These materials demonstrate greater flowability, which facilitates the shaping of intricate details. However, it is important to acknowledge that materials with lower molecular weights are more susceptible to differential heating. This implies that certain areas of the sheet may heat up or cool down at a faster rate than others, potentially leading to the uneven formation or warping of the final product.

The molecular weight of thermoplastics plays a crucial role in their thermoforming. A higher molecular weight is known to contribute to improved heat strength during thermoforming. However, it can also lead to an increase in the melt strength and elastic memory, which can hinder the stretching process and cause the material to return to its original shape after deformation. This heightened melt strength can impede the flowability of the material and pose challenges to achieving the desired shape conformity to the mold during the forming process [30].

4.6 Molecular Weight Distribution

The molecular weight distribution (MWD) of a polymer is of the utmost importance in determining the final properties and performance of a material in various applications. The properties of polymers are greatly influenced by their molecular weight distributions (MWDs), which affect various aspects such as processability, mechanical strength, and morphological phase behavior. Polydispersity is commonly used to describe polymer MWDs and is defined as the ratio of the weight-average molar mass (Mw) to the number-average molar mass (Mn). The breadth and shape of the MWD in the final polymeric material plays a crucial role in determining its properties [31, 32].

The presence of a distribution of chain lengths in synthetic polymers allows the manipulation of rheological behavior through the incorporation of high-molecular-weight tails. This technique has been used in the industrial production of plastics to create materials that are easily processable and commercially viable. For instance, in the industrial processing of polymers, blending is used to improve material toughness by incorporating high-molecular-weight tails, whereas low-molecular-weight chains are added to lower viscosity and enhance processing. The addition of a small percentage of polymer can alter the shape of the MWD, increasing its stiffness. However, it should be noted that long-chain branching has a significant impact on the MWD of resins, affecting properties such as processability, melt strength, and film optical properties [33]. A larger fraction of high-molecular-weight chains in the MWD improve the thermal stability.

4.7 Melt Flow Index

The melt flow index can be considered an approximate measure of shear viscosities. The assessment of melt flow in thermoplastics is significant in the manufacturing industry due to its ability to provide valuable insights into the thermoformability of these materials during processing. Melt flow refers to the ease with which thermoplastics can be melted and manipulated. The melt flow index (MFI) serves as a quantitative measure of the rate at which a material flows over time and offers valuable information for various processing techniques. Both melt flow and thermoforming play crucial roles in thermoplastic manufacturing [34].

The processability of a material can be assessed by measuring melt flow. Thermoforming processes take advantage of the unique properties of thermoplastics to efficiently produce a wide range of products and hold significant importance in various manufacturing and shaping applications. However, it is important to acknowledge that although the Melt Flow Index (MFI) is a commonly used rheological property of polymers, it alone is insufficient to fully characterize the flow properties of commercial materials. The presence of a small number of high-molecular-weight tails or long-chain branching can significantly impact the rheological and final properties of these materials. Nevertheless, MFI remains an important property for plastics and can be utilized for the quality control of products [35]. The primary advantage of the MFI property lies in its correlation with other parameters that characterize plastic materials, such as viscosity and molecular weight.

Thermoforming can only be performed within a narrow temperature range near the melting point. The melt strength of the polymer decrease linearly as the MFI increases. Polymers with lower MFI possess longer molecular chains, resulting in a greater number of entanglements within the polymer structure. These entanglements contribute to a higher resistance to extensional deformation, leading to a higher melt strength in polymers with a lower MFI. The decrease in MFI is attributed to an increase in the molecular weight distribution and strain-hardening index. In addition, the presence of long-chain branching leads to a reduction in the strain at which the melt breaks. However, excessively high degrees of branching are not necessary for optimizing thermoformability [36].

4.8 Glass Transition Temperature

Polymers are non-liquid crystalline materials below their melting points or amorphous materials below their glass transition temperature. The glass transition temperature at which the specific heat or thermal expansion coefficient of the sheet material shows a sudden change due to relaxation effects. The completely amorphous structure is made up of randomly entangled chains. Heating above the glass transition temperature is necessary to establish a short-time equilibrium configuration. When cooling below the glass transition temperature, the chains are frozen in [4].

During the thermoforming process, the thermoplastic material is subjected to a range of dynamic multi-axial deformations at strain rates of 1–10 s^{-1} and higher while processing the polymer in its rubber-like state above and below its characteristic glass transition temperature (Tg) [37]. The glass transition temperature (Tg) is an important physical characteristic of polymers. This represents the critical temperature at which an amorphous polymer undergoes a transition from a rigid, glassy state to a more flexible, rubbery state upon heating. This property is of great importance in polymers.

At higher temperatures, the molecular chains could easily move in response to applied deformation. Conversely, at temperatures below the glass transition temperature, the material became more brittle due to a significant reduction in the mobility of the molecular chains. Consequently, the temperature attained during deformation plays a critical role and leads to significant variations in the final results of the manufactured part.

This transition involves the polymer becoming softer and more malleable while maintaining its amorphous structure. Semi-crystalline polymers, particularly those with a high defect content, exhibit both glass and melting transitions. However, these transitions occurr at different temperatures. Semi-crystalline high polymers can exist as either a two-phase system, comprising solid crystalline and amorphous phases or a one-phase system, consisting solely of an amorphous polymer [38].

Polymers display time-dependent thermomechanical behavior near their glass transition temperatures, with local stresses continuously relaxing until the local temperature falls well below the glass transition temperature. Consequently, the local strains caused by volumetric changes during solidification are minimal.

The glass transition is not a thermodynamic phase transition, but it is characterized by discontinuities in certain derivative thermodynamic quantities, such as expansivity. When the temperature is below the glass transition temperature (Tg), the polymeric material exists in a metastable state at thermodynamic equilibrium. At temperatures lower than Tg, the material is in an isoconfigurational state. In the glassy state, the temperature is close to Tg. The properties of amorphous polymers change over time at a slower rate, when the difference between Tg and the current temperature (T) is large [39].

This phenomenon involves softening and increasing the flexibility of the polymer, while maintaining its amorphous structure. This refers to the temperature at which the modulus decreases rapidly, and the polymer begins to exhibit liquid-like behavior. The glass transition temperature (Tg) is used to identify amorphous thermoplastics, which are typically processed at temperatures significantly higher than their Tg [40].

In the glassy state, the polymer material is brittle and can undergo only small deformations. However, in the rubbery state above Tg, the material becomes ductile and can undergo large deformation. When the thermoforming temperature exceeds Tg, the sheet material becomes adhesive and cannot easily slide on the plug surface [41].

The glass transition temperature plays an essential role in the stability and formability of plastic sheet materials during the thermoforming process. The thermomechanical properties of polymers undergo significant changes in the glass transition region. When a polymer in a rubbery state is cooled below its glass transition temperature (Tg), its local viscosity gradually increases, leading to vitrification and reduced molecular mobility. This hindered molecular motion eventually freezes, resulting in excess volume and enthalpy without reaching equilibrium. However, there is still residual mobility that allows for the gradual evolution of the polymer's microstructural state over time towards equilibrium. This process, known as physical aging or enthalpic relaxation, is associated with changes in the physical properties of the polymer, including its mechanical behavior and diffusion [42, 43].

During the thermoforming process, the plastic sheet is heated above the glass transition temperature (Tg) of the polymer. At temperatures above Tg, amorphous thermoplastic polymers are ready to deform and flow under loads. If the material undergoes flow, it can retain its permanent deformation even after the load is removed. The physical properties of polymers are significantly influenced by temperature, particularly by the glass transition temperature, Tg.

In the amorphous region of a fully crystallized or mechanically oriented semicrystalline polymer, the movement of the polymer chains is somewhat restricted because of their proximity to the crystal face. The glass transition, which represents the constrained motion of this amorphous region, exhibits a smaller magnitude than other transitions [44]. Once suitable process parameters are established, the restricted glass transition temperature (Tg) of a completely crystallized and sporadically aligned polymer becomes evident [45].

Thermoplastic systems typically have a more limited temperature range between the glassy state and rubbery plateau, resulting in relatively rapid changes in their physical properties. On the other hand, thermoset systems also have a glass transition temperature (Tg), but this transition may occur over a broader temperature range, leading to similar properties at different temperatures [46].

In amorphous polymers such as polystyrene and PMMA, the presence of large pendant groups and strongly interacting double-bond oxygen atoms contribute to their amorphous nature. The glass transition temperature is the point at which the kinetic units of the main polymer chain in an amorphous polymer exhibit some mobility. When polymers are deformed significantly below Tg, they exhibit a viscoplastic response characterized by yield and post-yield strain softening. The stress response below Tg also depends on the thermal history, as the yield strength of polymers increases with aging time [47].

Below the glass transition temperature, the mobility of the kinetic units in the main polymer chain is zero, although there may be some mobility in the pendant group units. The relaxation process in amorphous polymers involves the release of stored elastic energy within the intercrystalline (or "amorphous") regions of the material, which can lead to shrinkage [48, 49].

A fully amorphous polymer consists randomly intertwined chains of varying degrees. When the temperature exceeds the glass transition temperature (Tg), the polymer quickly establishes its equilibrium configuration after a disturbance. However, when the temperature fell below Tg, the configuration of the chains at that temperature becomes fixed.

In contrast, polymers display rigid solid behavior below their glass transition temperature (Tg), resulting in decreased extensivity and impact properties, which leads to brittleness. Thermoplastics, including polystyrene (PS), polyvinyl chloride (PVC), polymethyl methacrylate (PMMA), and polycarbonate (PC), have relatively high Tg values. On the other hand, polyolefins and rubber have low Tg values, resembling high-strength rubber or leather. Polymers with higher Tg and thermoforming temperatures would provide a wider range of temperatures for use [50].

4.9 Melt Temperature

The behavior of polymer melts is significantly influenced by their molecular structure, particularly the molecular weight, molecular weight distribution (MWD), and presence of long-chain branching (LCB). The management of the thermal conditions in thermoforming processes is of great significance because of the strong influence of the sheet temperature distribution on the thickness distribution. The variability of the melt temperature in thermoforming processes is dependent on the specific thermoplastic material used. To ensure optimal forming behavior and prevent issues, such as overheating or incomplete forming, precise temperature control is typically necessary [51, 52].

The melt temperature of thermoplastics can vary depending on the grade and formulation of the material. Furthermore, certain thermoplastics may have a wide range of processing temperatures, allowing adjustments to meet the specific requirements of the manufacturing process. Consequently, the melting temperature of thermoplastics is determined by the specific materials being used.

During the melting phase, the polymer transitioned from a solid to an amorphous state. Regardless of the processing technique used, the material must be heated to a very high temperature for a short duration to prevent thermal degradation, formed, and then cooled to the solid state. For most polymers, crystallization occurs relatively slowly in regions close to the melting temperature. Lower temperatures, which are relevant for polymer processing, cannot be achieved because crystallization begins during the cooling process [53].

4.10 Heat Deflection Temperature (HDT)

The heat deflection temperature (HDT) is a measure of a polymer's ability to withstand short-term exposure to high temperatures without deformation under an external load. Several factors influence the HDT of a polymer, including the specific composition of the polymer matrix, type and amount of filler material added, and thermal history of the sample. Different thermoplastics exhibit varying HDT values, and the choice of appropriate material depends on the desired forming temperature for a specific application. These high-performance polymers generally exhibit elevated glass transition temperatures and are moldable. The HDT values of these polymers range from 170 to 355 °C [2]. Figure 4.2 shows a graph illustrating the increase in HDT with filled semi-crystalline polymers.

The Heat Deflection Temperature (HDT) plays a significant role in ensuring the stability and shape retention of materials during the heating and forming stages. In the field of thermoforming, particularly plug-assisted

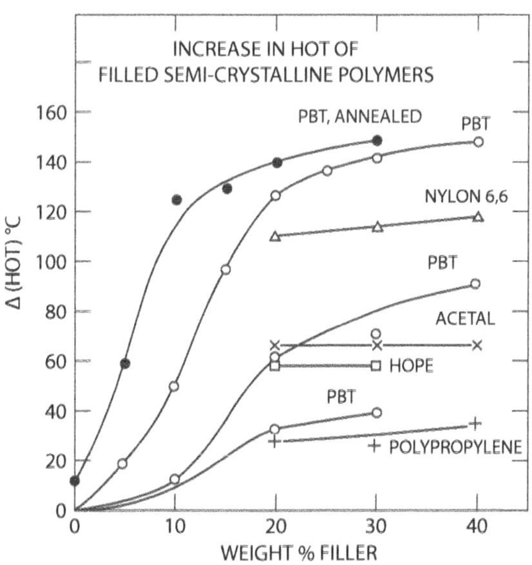

Figure 4.2 Increase in heat deflection temperature of some of the filled semi-crystalline polymers. [Reprint with permission from [54]. Copyright © 1979 Society of Plastics Engineers, Inc. All rights reserved].

thermoforming, HDT is a critical property [54]. It is essential to select a material with an appropriate HDT that aligns with the required forming temperature for a specific application. The impact of the thermal history on the HDT is attributed to its influence on the residual stresses [55] and free volume [56]. If the residual stress is eliminated, the HDT values approach a limiting value, that is, the glass transition temperature (Tg) [57].

4.11 Crystallization Temperature

The order of the higher forming temperatures resulted in an increase in the crystallization temperature. This can be attributed to the accumulation of thermal residual stress after the matrix underwent crystallization. It is reasonable to assume that the residual stress levels increase at higher crystallization temperatures and slower cooling rates. Slower cooling rates allow more time to alleviate internal stress, and the relaxation properties are influenced by the viscoelastic behavior. Consequently, at the crystallization temperature, the amorphous polymer undergoes a transition to a semicrystalline state [58].

Figure 4.3(a) illustrates the stretching of polymer chains and their impact on crystallization. The stretching of polymers induced by flow causes a transition from interpenetration to no interpenetration of neighboring chains. Optical micrographs are used to observe the influence of the initial melt state on the nucleation density of the PS crystals obtained

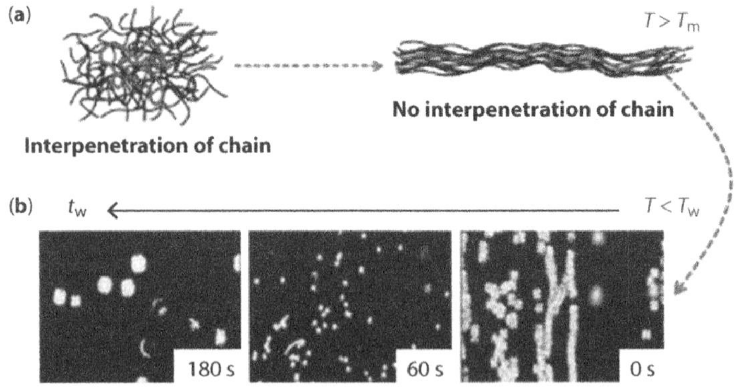

Figure 4.3 Stretching polymer chains and their consequences on crystallization: (a) flow-induced stretching of polymer, and (b) influence of the state of the initial melt on the nucleation density of PS crystals obtained at 180 °C. [Reprint with permission from [59]. All rights reserved. Copyright © 2008, American Chemical Society].

at a temperature of 180 °C. It is important to note that the nominal melting point of PS is 230 °C.

When extensional flow processing is applied during thermoforming, the polymer is stretched and aligned in the direction of the flow. Even after the flow stopped, a significant amount of time is required for the polymer to reach equilibrium. This long-lasting memory effect is commonly referred to as flow-induced memory. At temperatures below the melting point, the aligned chains facilitate nucleation and accelerate the crystallization process (Figure 4.3(b)). Crystals formed from flow-induced precursors exhibit morphologies that cannot be obtained from equilibrium melts. In cases where the sheet slips rapidly on the mold, such as during dewetting, flow-induced polymer alignment occurs at temperatures above the melting point. By controlling the arrangement of the side groups and flow conditions during forming, the viscoelastic response of the polymer can be harnessed. However, it is challenging to suppress unwanted nucleation and crystallization during cooling [60].

During isothermal crystallization, unexpected changes in morphology can occur, accompanied by variations in the growth rate and melting temperature of spherulites. These changes are attributed to local temperature variations and the final morphology of the spherulites, which are influenced by competition between the rate of crystallization and the rate of heat removal [62].

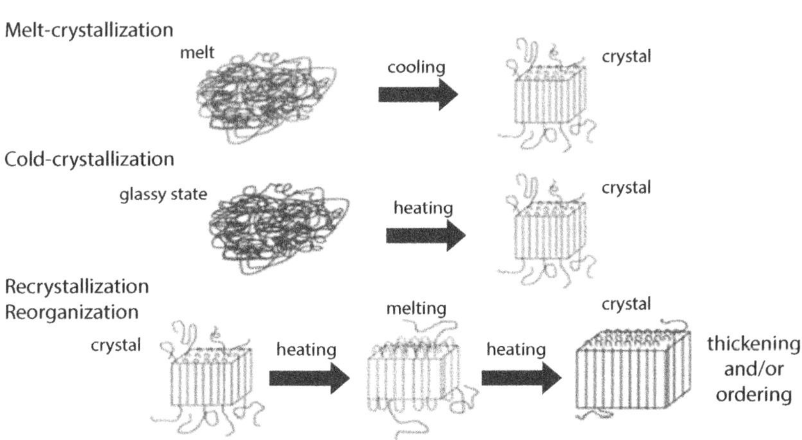

Figure 4.4 Schematic representation of the three types of non-isothermal crystallization of polymers. [Reprint with permission from [61], © 2018 Wiley Periodicals, Inc. All rights reserved].

The three distinct forms of non-isothermal crystallization processes in Figure 4.4 provide a visual representation of the non-isothermal crystallization of polymers. Melt crystallization occurs when the material cools during thermoforming. Cold crystallization, on the other hand, refers to the crystallization that takes place above the glass transition temperature while the material is heated from a glassy state. Finally, the relatively low heating rate used in thermoforming induces structural modifications in metastable crystals, leading to the formation of a more stable crystal structure during the heating process [61].

4.12 Melt Strength

Melt strength determines the tendency to sag and increases the deep draw capability. Therefore, the melt strength of polymers plays a role in determining the sagging as well as the depth of the draw. A higher melt-strength material reduces the tendency to sag and increases the deep drawing capability. The melt strength of the polymers determines the sagging and depth of the draw. A high melt strength is beneficial for thermoforming.

The melt strength of a polymer is an important processing parameter that plays a significant role in thermoforming which involves stretching and drawing. This refers to the polymer's ability to resist extensional deformation during the melting process. Insufficient melt strength can have a detrimental effect on the thermoforming behavior, leading to sagging. When the polymer is stretched, the molecular chains align themselves in the direction of the stretch, resulting in the formation of a regular structure in the hot sheet. This alignment enhances the plasticity of the melt. However, at low temperatures, the sheet may soften but not fully melt, resulting in an inaccurate replication of the mold's detail. Conversely, high temperatures can cause the sheet to lose its dimensional stability and excessively sag under gravity, leading to uneven wall thickness and potential tearing [63].

The challenges faced in melt-phase thermoforming with crystalline polymers are attributed to their sharp melting points and poor melt strengths. In contrast, amorphous polymers exhibit a wider temperature range, where they exhibit a rubber-like elastic state. Sagging, which involves downward movement of a polymer sheet under gravity, can be influenced by the melt strength. However, it is important to note that the melt strength does not represent a well-defined rheological property due to the nonuniform strain and temperature distribution in the stretched polymer melt. Various parameters, such as melt temperature, extrusion rate, and ambient temperature, can affect the melt strength of a polymer. The effect of these parameters on melt strength has been observed to increase sharply, which is believed to be caused by flow-induced crystallization. Crystalline structures can form under high stress, particularly when combined with high pressure and low temperatures. The formation of these melt structures primarily occurs because of polymer stretching rather than shear flow. The elongational flow of melts has been demonstrated to be highly effective in producing a high degree of macroscopic orientation [64].

The viscoelastic properties of the melt of sheet materials refer to their strength and ability to be shaped during heating. When a plastic sheet melts, the force required to shape it at a certain rate is directly related to its viscoelasticity. The sheet can be shaped, except under the influence of surface tension forces, which can cause sagging in thicker sections. However, if the sheet accumulates elastic energy during stretching, there is a point where the stored energy exceeds the amount required to create a free surface in front of a flaw or tear. This allows the tear to grow and ultimately leads to breaking of the sheet. The higher the amount of stored elastic energy per unit of deformation in the sheet, the lower the rate at which it will fail [65–67].

The melt strength of a polymer is a significant characteristic that plays a crucial role in its behavior during thermoforming procedures. This property governs the capacity of the polymer to withstand stretching and is closely associated with the entanglement of its molecular chains and their ability to endure strain without becoming disentangled. Insufficient melt strength can have detrimental consequences on the thermoforming process, such as causing the sheet to sag [68].

In contrast, the use of high-melt-strength polymers has the potential to improve the processing capabilities of thermoforming applications. The melt strength of a polymer is a crucial factor influencing the performance of the the material and the ultimate quality of the end product. This pertains to the ability of the polymer to withstand deformation when exposed to external stress, a characteristic of particular significance during thermoforming processes. This is because of the requirement for the molten plastic material to retain its shape and structural integrity while being stretched and molded over a form [69, 70].

Figure 4.5 presents the melt strength curves for various types of polyethylene, which are also applicable to thermoforming processes. Figure 4.5.(a) depicts the melt strength curve for an ethylene-α-olefin coordination copolymer, demonstrating that the force required for drawing remains relatively small and does not vary significantly with increasing drawing speed. In contrast, Figures 4.5.(b) and 4.5.(c) exhibit similar curves for a

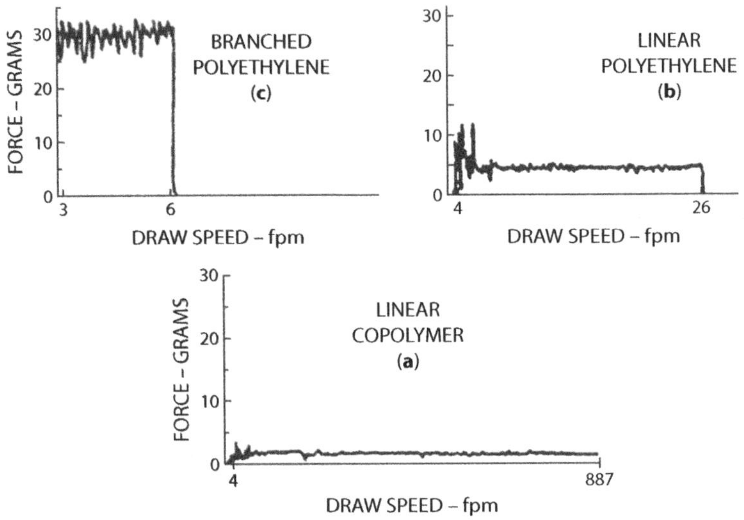

Figure 4.5 Melt strength of various polyethylene. [Reprint with permission from [71]. Copyright © 1967 John Wiley & Sons, Inc. All rights reserved].

linear polymer and branched (low-density) polymer, but with different drawing forces. The curves in Figure 4.5.(c) suggest some flow instability under these conditions, although the average of the peaks offers a reasonable measure of the necessary drawing force. At lower molecular weights, branched molecules exhibit a lower viscoelastic nature than linear molecules, although the opposite may hold true at higher molecular weights [71].

The melt strength of a polymer is a measure of its resistance to extensional deformations. The force at which the polymer melt breaks is called the "melt strength." The melt strength parameter does not provide a well-defined rheological property because neither the strain nor the temperature is uniform in the stretched polymer melt. The melt strength of a polymer is affected by several parameters, such as the melt temperature, extrusion rate, ambient temperature and the distance between the sheet material and heaters.

During the extrusion process, the material experiences shear forces during heating, which is also a process variable in thermoforming. The melt strength varies with the extrusion rate (Figure 4.6). The molecular orientation that occurs during extrusion undergoes deformation during the thermoforming process.

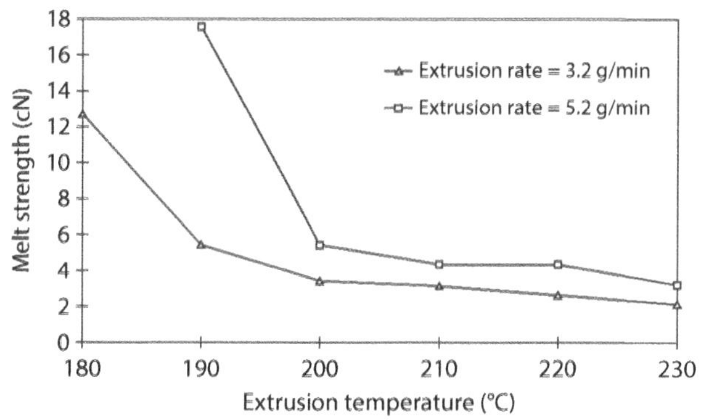

Figure 4.6 Extrusion temperature versus melt strength. [Reprint with permission from [53] Copyright © 1998 Society of Plastics Engineers. All rights reserved].

4.13 Rheological Properties

The rheological properties of polymers are influenced by various factors, with molecular structure being one of the most significant [72]. The rheological properties of polymers are influenced by three main types of deformations: isotropic compression, extension, and shear. The behavior of polymers during axial extension in thermoforming above the glass transition temperature is particularly significant in the manufacturing of various products made from polymeric materials.

4.14 Viscoelastic Behavior

The mechanical properties of polymers can vary depending on the loading conditions and temperature, causing a transition from brittle to viscoplastic behavior. Polymers can possess either an amorphous or semi-crystalline structure, with the latter being influenced by the degree of crystallinity and the size and distribution of the crystallites. In the case of amorphous polymers, the inelastic response is determined by factors such as molecular chain flexibility, entanglement, and structural variations among molecular chains. Furthermore, the mechanical behavior of polymeric materials is influenced by both time and temperature [73].

Viscoelasticity is a characteristic of materials that exhibit properties of both elastic solids and viscous fluids [74, 75]. The rheological properties of polymers in their molten state are significant in the field of polymer engineering and science, for several reasons. First, these properties are highly dependent on the chemical structure of the individual macromolecules, making them a subject of interest for polymer characterization. Second, understanding rheological properties is crucial for developing an accurate fluid dynamics model of polymer melts, which serves as the foundation for any theory related to polymer processing [76]. Finally, these properties play a role in the formation of large molecular orientations during the melting process, which are then preserved in the final products. Consequently, they contribute to the creation of inhomogeneous anisotropy in various physical quantities and can affect the technological properties of end-use products [77].

Challenges arise from the intricate and nonlinear rheological properties of polymer melts. These melts possess characteristics of highly viscous liquids, as well as being pseudoplastic (where viscosity decreases with increasing shear rate) and viscoelastic (where the mechanical response is dependent on time and frequency). Additionally, their viscous flow is often accompanied by significant elastic deformations that resemble rubber-like behavior. Moreover, polymer melts can undergo substantial elongation without experiencing rupture [78].

In the case of polymers, this behavior is attributed to their chain structure and low melting point [79]. The viscoelastic behavior of thermoplastics is particularly significant in the thermoforming process, as these materials possess the properties of both solids and fluids, which have notable implications in thermoforming. The temperature at which thermoplastics are processed is one of the factors that influence their viscoelastic behavior during thermoforming. Higher temperatures lead to increased mobility of the polymer chains, resulting in decreased viscosity and increased elasticity [80, 81].

The strain rate, or the rate at which a thermoplastic material is deformed, is a significant factor influencing the viscoelastic response. When subjected to higher strain rates, thermoplastics tend to exhibit a more rigid behavior because of the limited time available for the molecular chains to relax. Conversely, at lower strain rates, thermoplastics may exhibit more ductile behavior. This is particularly relevant during the thermoforming process, where thermoplastic materials undergo prolonged deformation [82].

The rheological behavior of polymer melts is a crucial aspect to be considered when evaluating their processing performance. It is influenced by various external parameters, such as temperature, time, and deformation rate, as well as material properties, such as the molecular structure of the polymer. Additionally, the mode of deformation differs depending on the specific processing operation used.

Because processing primarily occurs in the nonlinear regime, it is important to examine the rheological properties of elongation. This allows for a better understanding of the material's behavior during processing. Moreover, to make meaningful comparisons and assess processing properties, it is necessary to consider the time scales and ranges of elongation and elongational rates that occur during processing [83].

The rheological properties of the blends depend on the proportions of the different polymers in the blend. It has been observed that the viscosities of blends vary monotonically with composition [84–86].

Unusual rheological behavior has also been observed in blends of two homopolymers with the same structures but different molecular-weight distributions, as well as in blends of two homopolymers with different structures [87, 88]. Specifically, at a certain blending ratio, the melt viscosity of the blend reaches a minimum. Furthermore, the viscosity and elasticity of the individual components of the blends are found to influence the morphology of the blends.

Polymer blends typically exhibit phase separation in the molten state, which results in the formation of two or more incompatible phases. This behavior can be attributed to the low entropy of mixing [89]. The degree of phase separation can be categorized into two types of two-phase flows.

- The first type is a dispersed two-phase flow, where one component forms discrete droplets dispersed within the continuous phase formed by the other component.
- The second type is a stratified two-phase flow, where the components are separated from each other by a continuous boundary.

In the thermoforming process, the initial period is characterized by the development of a high-elastic deformation, which is of great importance. This high-elastic deformation is accompanied by an increase in viscoelasticity and modulus of high elasticity, as well as a delay in the relaxation processes in the region of longer relaxation times. At lower deformation rates, the increase in viscoelasticity and high-elasticity modulus, as well as the delay in relaxation processes, may be replaced by a decrease in elasticity and high-elasticity modulus and an acceleration of relaxation processes, leading to the attainment of stationary flow regimes. The transition from strain regimes with increasing viscoelasticity and a modulus of high elasticity to those with a decrease in these quantities corresponds to an increase in the rate of irreversible deformation. This transition is influenced by the orientation effect and destruction of the network of intermolecular bonds, which becomes evident during the attainment of stationary flow. The orientation effect is responsible for the delay in the relaxation processes, while the rupture of intermolecular network bonds accelerates the relaxation processes. Unlike shearing, the orientation effect is more important during extension. Therefore, in stationary flow regimes, the viscoelasticity may not only remain independent of the rate of strain but may even increase with it. In this case, the contribution of longer relaxation times to the relaxation spectrum increased with increasing stress in the stationary flow regime. It should be noted that the independence of longitudinal viscosity and modulus of high elasticity from stress in stationary flow regimes does not guarantee linearity of the mechanical properties of the polymer in the pre-stationary stage of deformation, where complex changes occur in its relaxation characteristics. At high deformation rates, the viscoelasticity and modulus of high elasticity continued to increase until rupture. For both semi-crystalline and amorphous polymers, thermoforming is always carried out in a semi-fluid/semi-solid state. The viscoelastic properties of polymers play an important role in their thermoforming.

4.15 Coefficient of Friction

The relationship between the friction experienced by the sheet and mold is influenced by temperature, particularly near the glass transition temperature. Manipulation of the thickness distribution of a thermoformed part can be achieved by controlling the mold temperature with the assistance of vacuum and/or pressure. Given the high temperatures of both the sheet and the mold, the friction force is particularly pronounced at the initial point of contact. Consequently, the thickness distribution of the thermoformed parts became nonuniform. During thermoforming, the coefficient of friction (COF) at temperatures below the glass transition temperature (Tg) generally ranges from 0.1 to 0.3. However, as the temperature increases above Tg, the COF undergoes a substantial increase [90].

The coefficient of friction (COF) exhibited a rapid increase when the temperature exceeded the glass transition temperature (Tg), resulting in a significant reduction in slip and an increase in uniaxial stretching at the clamp. This elevated temperature causes the sheet material to become sticky, impeding its ability to slide smoothly on the plug surface [91].

The influence of the friction coefficient on the thickness distribution of thermoformed parts is a significant factor to consider. The interaction between the sheet and mold friction is of utmost importance in determining the thickness distribution of thermoformed parts, especially in pre-stretching methods, such as plug-assist thermoforming [92]. It is important to achieve more consistency in the distribution of thickness within a range of friction coefficients from 0.05 to 0.2, a temperature range of 140–200 °C for the sheet, and a constant mold temperature of 90 °C. Additionally, the use of forced cooling equipment results in a more uniform thickness distribution [93].

4.16 Thermal Conductivity

The thermal conductivity (k) refers to the ability of a material to conduct heat. The thermal conductivity of polymers plays a crucial role in determining the processing parameters and understanding the melt flow and cooling behavior of polymeric materials during forming. This property is significant for assessing the suitability of a finished product for specific applications [94].

Thermal conductivity values play a crucial role in the selection of materials for specific applications because they determine the performance of a material at different temperature levels [95]. This property is particularly important in problems related to steady-state heat transfer and significantly affects the performance of materials in high-temperature applications [96].

To minimize heat loss, materials with low thermal conductivities are preferred. Conversely, when heat transfer from one location to another is desired, materials with higher thermal conductivities are used. The thermal conductivity of a material can be influenced by various factors, including the temperature, density, porosity, moisture content, degree of crystallinity, grain orientation, molecular size, and impurities [97].

Polymers generally exhibit low thermal conductivity and undergo exothermic crystallization, which leads to heat generation. As a result, the temperature distribution within a polymer sheet may not be uniform, even during seemingly isothermal crystallization, unless the sheet is very thin. Several factors, including crystallization conditions, heat generation during solidification, part geometry and size, and presence of secondary phases, can contribute to significant thermal gradients within a sample.

The study of heat conduction in polymers has focused on three main aspects of thermal conductivity: temperature dependence, crystallinity dependence, and orientation effect. The thermal conductivity of amorphous polymers typically falls within the range of 0.1–0.2 $Wm^{-1} K^{-1}$, while for semicrystalline polymers, it can reach 0.5 $Wm^{-1} K^{-1}$ [98]. Furthermore, the orientation of semicrystalline polymers results in a significant anisotropy in the heat propagation, particularly at higher temperatures [99].

When amorphous polymers undergo stretching, their thermal conductivity increases in the direction of the stretch and decreases perpendicular to it. This phenomenon can be attributed to the alignment of chain molecules in the stretched polymer, where the primary valences with low thermal resistance are predominantly oriented in the direction of stretching, whereas the van der Waals bonds with high thermal resistance are mainly aligned perpendicular to it.

The thermal conductivity behavior in partially crystalline polymers is not uniform and can be categorized into two groups. The first group, which included polyethylene and polyoxymethylene, demonstrated a decrease in the thermal conductivity as the temperature increased. Conversely, the second group of polymers exhibited an increase in the thermal conductivity with increasing temperature. Examples of polymers in this category include polyethylene terephthalate, isotactic polypropylene, polychlorotrifluoroethylene, and polytetrafluoroethylene [100].

Interestingly, the thermal conductivity of the crystalline phase of polyethylene terephthalate increases with temperature, a behavior similar to what is observed in the amorphous phase below the glass-transition temperature. In the crystalline region of the polymers, the atoms within the crystals are closely connected and undergo slight vibrations near their equilibrium positions, facilitating rapid heat transfer along the direction of the molecular chains. However, the level of crystallinity is not very high due to the random entanglement of the polymer chains and the large relative mass of the molecules. Additionally, the formation of complete crystals is hindered by the polydispersity of molecular weight.

The inclusion of fillers in polymer materials can potentially affect the thermal conductivity of the system. This effect is dependent on several factors, such as the size, shape, orientation, distribution, and thermal conductivity of the filler, as well as the characteristics of the interface between the matrix and filler [100, 101].

4.17 Thermal Diffusivity

Thermal diffusivity is a fundamental parameter that quantifies the rate at which heat spreads through a substance. This holds significant relevance in scenarios where the heat conduction is not in a steady state. Accurate knowledge of thermal conductivity and diffusivity, as well as their temperature dependence, is crucial in polymer processing operations involving both heating and cooling cycles. In cooling processes, fluctuations in thermal conductivity can lead to uneven cooling and shrinkage, ultimately

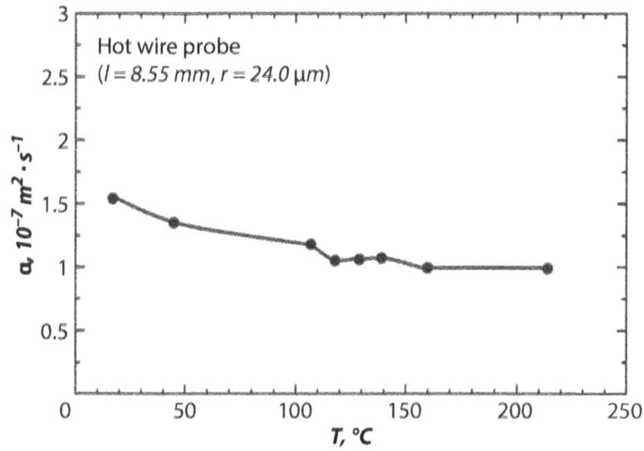

Figure 4.7 Measured thermal diffusivity of polyethylene. [Reprint with permission from [103]. Copyright © 2002, Plenum Publishing Corporation. All rights reserved].

resulting in the development of frozen-in stresses during thermoforming [102].

The thermal conductivity of polyethylene, semi-crystalline polymer, is primarily influenced by its degree of crystallinity in the solid state. As depicted in Figure 4.7, the thermal diffusivity of polyethylene also decreases as the temperature increases in its solid state, while it remains relatively constant in its molten state.

4.18 Specific Heat

Specific heat (c_p), which refers to the heat capacity per unit mass, is a fundamental thermodynamic parameter that quantifies the amount of heat required to raise the temperature of a unit mass of a substance by 1°. Anomalies in the specific heat at low temperatures, due to the existence of additional vibrational states, are a dominant feature of polymers. Consequently, it is closely linked to the energy consumption of the heating procedures. In the context of thermoforming, the value of c_p experiences a significant increase because of the phenomenon whereby a substantial amount of heat absorbed by the semi-crystalline material is utilized as the latent heat of fusion, rather than contributing to a change in temperature. The crystalline regions hinder the thermal degradation of the amorphous chains, with growing crystallinity that decreases the fragility of the amorphous phase [104, 105].

4.19 Stress

The stress experienced by a material can be influenced by the rate at which the strain is applied. The mechanical properties of polymer materials can vary depending on factors such as molecular weight and structure [106]. The plasticity of a solid is typically characterized by specific deformation characteristics, including:

a) A distinct inflection point in the stress-strain curve when subjected to a constant strain rate, resulting in significant strain increments for minor changes in stress,
b) A pronounced and nonlinear relationship between strain rate and stress increments at low levels of stress, and
c) Persistent deformations even after the removal of applied stresses.

Polymer sheets produced through extrusion can exhibit significant anisotropy due to their resulting molecular orientation. The thickness distribution in thermoformed products is significantly influenced by the viscoelastic characteristics of the polymer materials [107, 108].

The van der Waals stress behavior can be summarized as follows:

1) In the elastic region, the stress increases due to the decrease in density.
2) During plastic deformation, the stress decreases because of inter- and intrachain collisions resulting from chain sliding.
3) At higher strains, the stress decreases further due to the increasing significance of the loss of long-distance van der Waals interactions.

During heating, the polymer sheet (either in the biaxial or machine direction) loses its orientation and stress relaxation occurs. Again, the polymer chains become oriented due to stretching during thermoforming. Therefore, residual stress occurs during freezing due to the orientation of the polymer. Stress relaxation during thermoforming is closely associated with creep recovery. The residual stress retained during thermoforming limits the service temperatures of the parts. Polymers with faster stress relaxation, i.e., short stress relaxation times, can cause lower residual stresses in parts, becoming dimensionally stable [109].

4.20 Strain Hardening

Polymers typically undergo elastic deformation before transitioning into a plastic flow. During plastic flow, the true stress generally decreases as the strain increases, indicating the occurrence of strain softening. However, beyond this stage, further strain leads to a significant increase in stress before failure, which is known as strain hardening.

When the temperature of the sheet is consistent throughout, the extent of strain hardening and the behavior of failure during extension are the key parameters in controlling the kinetics of the process and the thickness profile. In situations where the sheet temperature is not uniform, the findings indicate that a higher degree of strain hardening has a greater impact on the dynamics of the process than relatively minor differences in sheet temperature [110].

Plastic deformation, which occurs after yielding, leads to a decrease in the growth rate of the bond, angle, and van der Waals (vdWs) stresses through interchain collisions. Additionally, the alignment of the system contributes to an increase in stress. Subsequently, the main-chain bonds and angles experience further stretching after strain softening, resulting in strain hardening [111].

Strain uniformity in thermoforming is governed by three types of variables: design variables (shape and curvature), process variables (plug speed, pressure, temperature, friction, and mold alignment), and material parameters (thickness, elongational viscosity, sheet orientation, and thermal stability). Strain hardening is observed primarily in long-chain branched polymers. This phenomenon is primarily attributed to bonded interactions, specifically bond and angle terms. As plastic deformation occurs, the vdW interactions that hold the polymer chains together weaken, allowing the chains to move freely during the strain-softening stage.

Eventually, the geometrical interchain entanglements become effective, initiating stretching of the entanglement network. Consequently, the bonds within the polymer chains experience higher stress as they are pulled by the entanglement points, leading to an accelerated increase in the bond stress. Simultaneously, the entanglement effect hinders the alignment of the polymer chains, resulting in somewhat saturated stress due to the system alignment. Additionally, the chain-sliding characteristic of plastic deformation continues to occur [112, 113]. Figure 4.8 shows the governance of strain uniformity.

Strain hardening has a positive impact on the uniformity of material deformation, indicating the influence of nonlinear viscoelastic behavior's

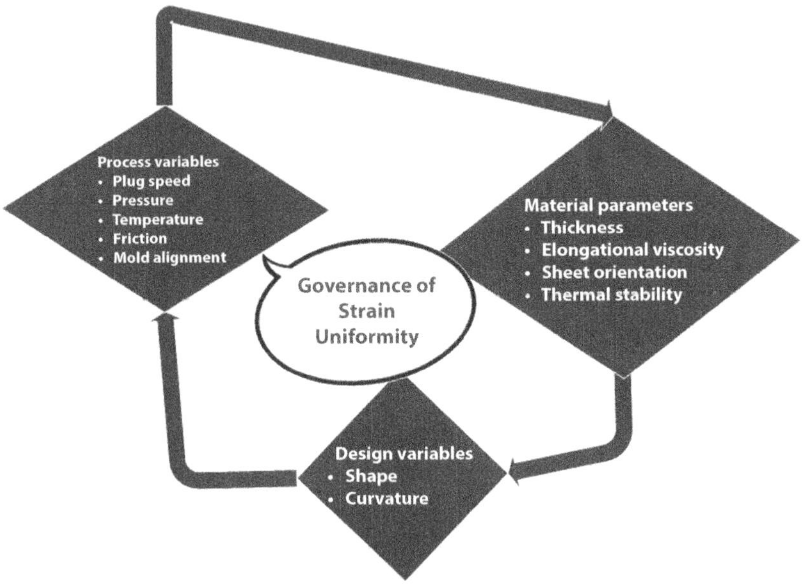

Figure 4.8 Governance of strain uniformity.

structural properties on elongation flow. Increasing strain rate results in increased strain hardening, indicating branching. Long-chain branched polyethylenes like LDPE and LLDPE are known to exhibit strain hardening [18, 19]. Strain hardening does not occur at high strain rates, but it increases with the strain rate. In LLDPE, strain hardening happens during phase separation in the heating stage of thermoforming [20].

In strain-hardening materials, homogeneity occurs in the material, which disappears as the elongation of the thinner cross-section area increases. This elongation leads to higher elongational viscosity. Furthermore, it follows that a more pronounced resistance to deformation occurs. For a non-strain-hardening material, the inhomogeneous material becomes more pronounced by stretching it further, and the material will finally break. Strain deformation causes molecular orientations, which is frozen by cooling. With a higher degree of orientation, the stiffness and strength increased to higher level. The long-chain branched polymer exhibited distinct strain hardening compared to the linear polymer. Materials with strain hardening in uniaxial deformation also exhibit biaxial deformation [114].

4.21 Plastic Strain

The plastic strain is unrecovered and permanent after the recovery of strain when the stress drops to zero. This caused the formed sheet to retain its shape. Polymers with higher plastic strains are expected to offer better replication of the contours of the mold. Controlling the process strain rate and temperature leads to better control of the thermally activated micromechanical processes during the thermoforming process.

The polymer exhibits slight deformations and has the ability to enhance its flexibility while retaining its linear elastic behavior [115]. The impact of heat transfer has been reduced. A form of plasticity is observed, which depends the rate of its progression. Faster evolution results in greater dissipation of energy than slower evolution.

The contours of the plastic shear strain rate of isotropic PMMA demonstrate an increase in intensity as deformation occurs within the band (Figure 4.9). The plastic shear strain rate within the band could be up to

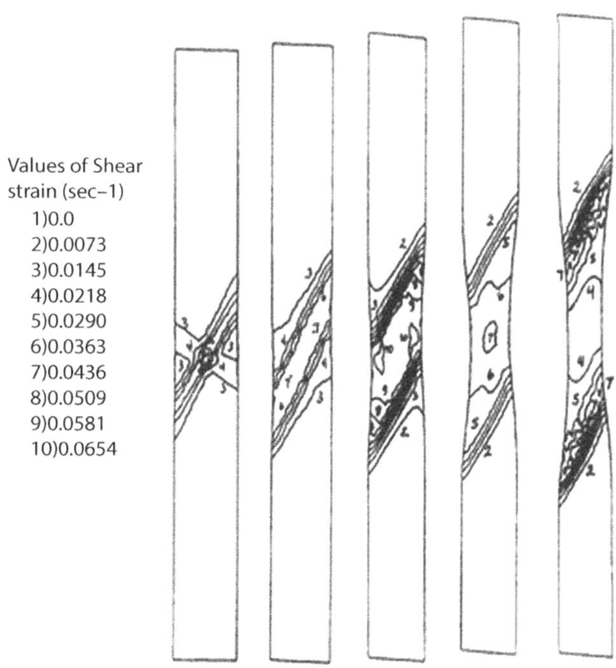

Figure 4.9 Contour plots of the plastic shear strain rate of isotropic PMMA as a function of deformation for the case of an initial draw ratio of 2.5 and an initial draw direction of 30°. [Reprint with permission from [116]. Copyright © 1989 Published by Elsevier Ltd. All rights reserved].

five times greater than the applied nominal tensile strain rate of $0.01 sec^{-1}$. When the inclination angle is 30°, the rate contours also reveal a widening of the band. This is accompanied by the movement of the region with the highest shear strain rate away from the center of the specimen. This suggests that the material near the center of the specimen, where the shear band initially formed, is effectively immobilized. Consequently, active plastic deformation spreads both upward and downward along the specimen, resulting in band broadening [117].

Rate-dependent dissipation in materials is supported by the evolution of dislocations at the microscopic level. This microscopic level of analysis prevents the material from undergoing excessive hardening and formation of localized regions of plastic strain. Consequently, problematic interactions between the concentrated plastic strain rates and thermal effects are avoided [118]. As the strain rate increased, the range of applicability of the material expanded to encompass higher strain values. This phenomenon is particularly evident in polymers with high molecular weights and viscoelastic properties, where the deformation within which the viscoelasticity increases with increasing strain becomes more pronounced.

Thermoforming is characterized by high deformation rates, with strain rates ranging from 0.1 to 10 s^{-1} and above. For semicrystalline polymers above Tg, the chain segments not involved in the crystalline domains can undergo deformation, the extent of which depends on the level of crystallinity. The mechanical behavior of polymeric materials is often influenced by the molecular orientations that develop because of stretching during the transformation process. This effect is particularly pronounced in semi-crystalline polymers due to potential changes in the orientations and microstructures of the crystalline domains, leading to significant anisotropy in the fracture properties. This behavior is observed in thermoformed polypropylene products, where the resistance to crack propagation along the drawing direction may be compromised. It is important to note that semi-crystalline polymers below the melting point of their crystalline content should be considered solids [119].

4.22 Tensile Strain

Tensile strain refers to the elongation or contraction of a material when it is subjected to pulling force. It is quantified by comparing the change in the length to the original length of the material. During thermoforming, plastic materials undergo tensile strain as they are molded and shaped into specific products [120].

During thermoforming, the polymeric material is strengthened by reorienting the lamellae and forming a fibrillar crystalline structure. However, the extent of the structural changes at high tensile strains is limited due to the restriction of polymeric molecules in the amorphous phase. The integrity of the crystalline skeleton is compromised during the tensile drawing process, resulting of small residual strains. Semi-crystalline polymers have a low sensitivity to sudden changes in strain rates. The viscoelastic response of these materials is governed by the interaction of multiple elastoplastic domains, which is influenced by the uneven density and isotropic distribution of the orientation of the crystalline lamellae [121–125].

4.23 Tensile Yield Stress

The yield stress of a polymer is influenced by various factors, including temperature, strain rate, and the presence of additives. When considering a specific temperature and strain rate, the inclusion of additives led to an increase in the yield stress. This can be attributed to the impact of the additive on the stretching rate, resulting in a decrease in temperature.

The load–elongation curves, shown in Figure 4.10, represent the general results of the polymer sheet material at normal temperatures. The figure shows that the Young's modulus and yield stress in the necking region start from the yield point and extend to the low-orientation region. As the load increases, the elastic region also increases with the orientation. At low

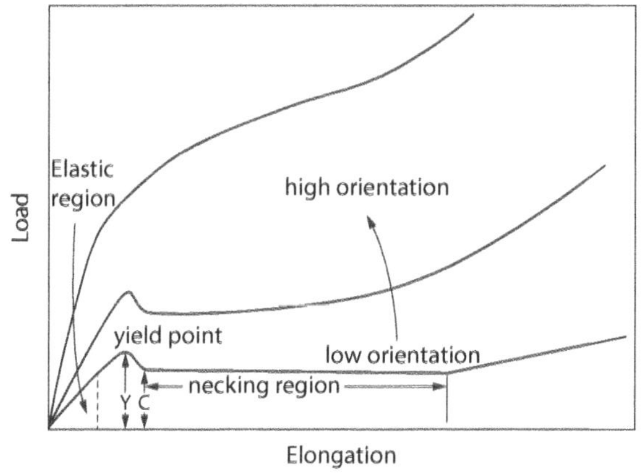

Figure 4.10 Load–elongation curve. [Reprint with permission from [126]. Copyright © 1993 John Wiley & Sons, Inc. All rights reserved].

loads, the orientation is also low. The general result of the tensile test of the oriented PET samples at room temperature (approximately 25 °C) is the measurement of Young's modulus and yield stress. Additionally the ratio of the stresses immediately before and after the occurrence of necking is used a representation of the extent of necking.

4.24 Deformation

Polymers tend to return to their original form after forces have been applied to them or removed, or they have the capability of resisting; deformation up to a certain limit. Forces such as air, vacuum, or both can cause deformation. The magnitude and duration of the forces required for the sheet to eventually deform depend on the stress applied. Polymers of all degrees of rigidity are available, and there is no sharp boundary between the softened sheet material and the deformation process. There are two stages in thermoforming: heating the sheet material near its melting point and forming it to obtain the desired shape. Deformation kinetics becomes important because of the elongational behavior of the sheet material during the forming stage. This influences material selection and processing parameters.

The behavior of polymer chains under high extension is characterized by approaching limiting values due to the application of strong stretching forces. This leads to the deformation of the local structure, which is influenced by factors such as bond lengths, valence and dihedral angles, and the potential for internal rotation. To achieve steady flow, it is necessary to maintain low deformation rates. However, an increase in the strain rate results in large draw ratios, which can only be attained through the development of elastic (reversible) deformations in the polymer. Consequently, the flow (irreversible deformation) is effectively suppressed [127].

Figure 4.11 illustrates the behavior of the ABS material under stretching conditions, with the maximum strain serving as an indicator of its strength. The central portion of the diagram depicts the transition from flow-induced deformation to rubber-like behavior of the molten material. On the right side of the diagram, there is a gradual shift towards glassy-like behavior at extremely high deformation rates.

During the initial stages of polymer deformation, the longitudinal viscoelasticity exhibited a linear relationship with strain. This relationship is particularly applicable to the viscoelastic behavior of a material in response to irreversible strain, and the law of linear increase in viscosity with increasing strain holds true for various strain rates.

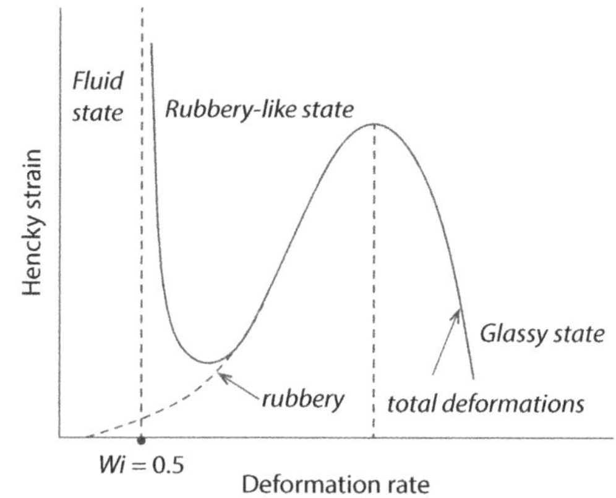

Figure 4.11 Diagram of limiting deformations as a function of the deformation rate. [Reprint with permission from [128]. Copyright © 2014 Elsevier Ltd. All rights reserved].

Figure 4.11 effectively showcases the correlation between the maximum Hencky strain, deformation rate, and division of this strain into viscous and elastic (rubbery) components. In the case of ABS and similar materials subjected to low deformation rates (as depicted on the left side of the diagram), significant elongation can be observed in the thermoforming processes. The central section, delineated by two vertical dotted lines, is of particular importance. Within this range, ABS deformation increases in tandem with the deformation rate, ultimately accounting for nearly the entire strain, whereas the contribution of viscous flow becomes negligible.

Thermoforming involves the formation of a deformation band when a polymer yields. The characteristics of these deformation bands, such as their geometry and optical properties, are influenced by the stress concentration between the plug and sheet materials. The propagation of the plug across the material leads to the appearance of nucleation and propagation stages, which is a direct consequence of these stress concentrations [129].

The deformation processes in amorphous polymers are typically explained in terms of conformational changes in the chain molecules. However, in crystalline polymers, the more ordered regions, such as crosslinks or inert fillers, have a significantly higher modulus than the amorphous matrix [130, 131].

Plastic deformation of polymers often occurs through a non-uniform mechanism. This process can be categorized into four groups based on morphology: circles, shoulders, slip bands, and kink bands. Slip and kink bands are commonly observed during the plastic deformation of oriented polymers. During thermoforming, slip bands occur primarily during the elongation of oriented polymers at small angles to the axis of the primary orientation. Kink bands or kink-zones, on the other hand, are formed during compression along the direction of orientation or in cases of uniaxial elongation at large angles close to or equal to 90° to the axis of the initial orientation, as well as in displacement and bending processes [132].

4.25 Stress Deformation

Polymers undergo a substantial increase in stress when exposed to extensive deformation, primarily due to strain hardening. This phenomenon arises from the alignment of the entangled network within the polymer as the strain intensifies. During thermoforming, the stress induced by significant deformation can be separated into two parallel constituents. The first constituent corresponds to a constant frictional force comparable to the yield stress. The second constituent is akin to the entropic force arising from the deformation of the macromolecular network, which resembles the behavior of rubber elasticity [133–136].

The stress–strain curves of the thermoformed polypropylene cups clearly demonstrate anisotropic mechanical behavior. Figure 4.12, shows that cups cut out strips transversally in the drawing direction exhibit a distinct zero-slope yield point, whereas this is not the case for cups cut along the drawing direction. It is worth noting that the stress-strain behavior in the drawing direction should be regarded as representative of the average behavior of the material's average behavior, as the chain orientation is non-uniform. In the case of polypropylene, the elastic modulus is higher in the drawing direction, and this effect is more pronounced for cups with higher draw ratios.

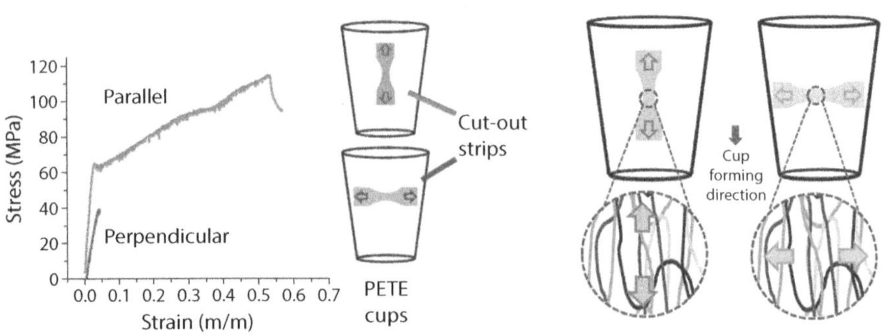

Figure 4.12 Schematic representation of cup cut-out strips drawing and transverse direction [Erk, Reprint with permission from [137]. Copyright © 2015 The American Chemical Society and Division of Chemical Education, Inc.].

Figure 4.13 illustrates the stress–stress curves of the polypropylene cups. The increased stress in the transverse direction (y-direction) is optimum when the depth of the cup is lower. As the depth increases, the stress increases in the transverse direction of the cup, and the polymers can recover their initial shape when exposed to temperatures above the glass transition temperature. This is because the entanglement network of the polymer chains remained intact, even during plastic deformation. When external deformation occurs, the relaxation times of the polymer segments increase, leading to the development of intermolecular strain-hardening stresses instead of the intramolecular stresses associated with entropic strain hardening. Notably, low-density polymers exhibit properties comparable to those of rubber [139].

Thermoformed products exhibit residual stress originating from two primary sources: frozen-in forming-induced and thermally-induced stresses. The former occurs because of the viscoelastic deformation of the sheet during the forming process, whereas the latter occurs during the cooling phase, both within and after the product is removed from the mold. Notably, the thermally induced residual stress is significantly stronger (at least 10 times), than the frozen-in-forming-induced stress [140, 141].

At low extension rates, the extensional viscosity gradually increases until it reaches an equilibrium value, which is three times the shear viscosity at a zero shear rate. Conversely, higher extension rates typically result in rapid stress growth [142]. In the thermoforming process, biaxial deformation of the original material is inevitable. However, the magnitude and direction of the induced strain vary extensively across a deformed surface [143, 144].

To address these challenges, two key concepts can serve as a foundation. The first concept pertains to the structure of matter, whereas the second concept revolves around the nonlinear behavior of sheet materials.

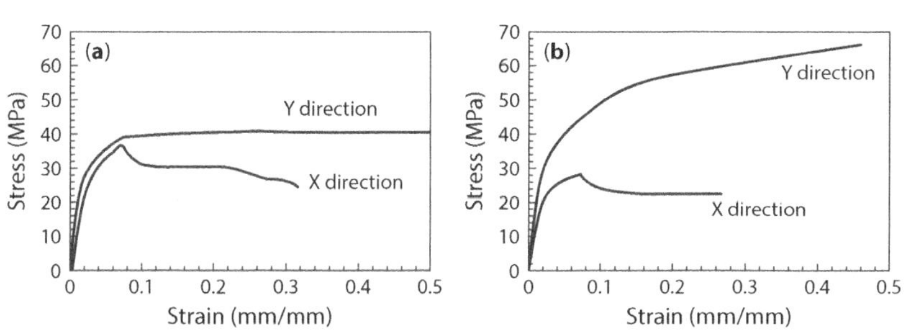

Figure 4.13 Stress–strain curves for PP cups with depths of (a) 58 mm and (b) 100 mm. [Reprint with permission from [138]. Copyright © 1997 Society of Plastics Engineers. All rights reserved].

It is appropriate to categorize nonlinear phenomena into three primary groups: (1) effects resulting from significant deformations without altering the structure of matter, known as geometrical nonlinearity; (2) changes in the structure of matter that give rise to a new order, referred to as physical nonlinearity; and (3) induced instabilities of various types, leading to transformations in the thermodynamic state of matter, known as thermodynamic or phase nonlinearity [145].

The extension process has a significant impact on the microstructure and morphology of both homogeneous polymers, such as PE and PP, and multiphase polymers, such as ABS. This highlights the relationship between the processing, structure, and properties of polymer-based materials in thermoforming, where extension plays a dominant role. Thermoforming operations involving polymeric materials that are primarily subjected to extension result in biaxial stretching deformation. This extension has the potential to modify the morphology of homogeneous polymer systems by inducing orientation phenomena of the macromolecular chains along the flow direction. Consequently, the extension during processing leads to the preferential orientation of the embedded particles along the flow direction, which has beneficial effects on the overall performance of the material.

In thermoforming, biaxial extension involves stretching the material in two different directions, resulting in a decrease in size in the third direction. This phenomenon has been studied through primitive chain network simulations and experiments on the elongational flows of entangled linear chains [146, 147]. Rubber-modified polymers, such as ABS, can also exhibit interesting rheological behaviors, including deviations from linear viscoelasticity at both high and small strains. These deviations can manifest as strain hardening in uniaxial extension, shear stress overshoot, steady shear viscosity, yield stress, and formation of higher-order particle structures [148].

4.26 Modulus and Stiffness

A polymer sheet with a high degree of crystallinity enhances its modulus, resulting in improved strength and self-supporting properties. This is particularly beneficial during the heating stage of thermoforming because it reduces sagging and enhances the strength and rigidity of the final products. Additionally, the presence of small spherulites in the polymer sheet improves its clarity and impact properties, both in its extended form and when thermoformed. However, it is worth noting that semi-crystalline thermoplastics generally have poorer thermoformability than amorphous materials, as they experience a significant decrease in stiffness and viscosity when they melt [149].

One of the challenges in the vacuum thermoforming of semi-crystalline polymers is the prevention of excessive sagging during the heating step. This requires the polymer to maintain a sufficient stiffness over a wide temperature range. Unlike amorphous thermoplastics, semicrystalline polymers exhibit a sharp decrease in modulus at their melting temperatures, resulting in a narrower range of sufficient stiffness. To prevent film damage during pre-stretching and forming, as well as the flow of the polymeric melt in the vacuum holes, the film must possess a certain elongational viscosity over a wide temperature range. This allows for a larger processing window and ensures the integrity of the thermoformed product [150].

4.27 Sag

The sagging of a polymer sheet in thermoforming processes is a significant factor affecting the position and orientation of each sheet element in relation to the heater bank. This, in turn, leads to variations in the heat flux distribution across the sheet, which complicates the subsequent shaping of the sheet. Understanding the behavior of sags in a sheet at a given temperature and time is crucial for selecting appropriate thermoforming processing parameters.

If the polymer has low melt stiffness during the heating phase of the thermoforming process, sagging may cause damage to system components, such as the heater system, or result in an irregular distribution of wall thickness in the thermoformed component. It is important to note that the sagging of a polymer sheet in thermoforming processes is primarily extensional deformation, with no involvement of shear deformation [151].

4.28 Toughness

The toughness of polymer materials limits the design and safety of engineering components and structures. Polymer toughness is influenced by the interplay between its plasticity and the fracture process [152]. The fracture process involves a competition between shear yielding and cavitation, with the resulting fracture of the polymer material being either brittle or ductile depending on whether the shear stress or cavitation stress is lower. Shear yielding occurs when the molecular alignment overcomes intermolecular interactions and serves as the primary source of plastic resistance. However, cavitation arises when plastic resistance becomes dominant, leading to brittle fracture. The phenomenon occurring around the yield point and the chemical structure of the polymer that governs such mechanisms are crucial. The yield point is typically defined as the point at which pure elastic deformation ceases and plasticity or shear motion commences. Numerous known and unknown factors influence shear-yielding behavior and competition between shear yielding and cavitation in polymers [153, 154].

4.29 Effect of Additives

Numerous additives, including nucleating agents, plasticizers, antioxidants, dyes, and cross-linking agents, are commonly used in the production of industrial polymer products. These additives play a crucial role in the development of novel materials with specific desired properties. Nucleating agents have a significant impact on the crystallization kinetics and morphology of semicrystalline polymers.

Variations in crystallization kinetics can influence the nano- and microstructures, as well as macroscopic properties, such as mechanical characteristics. Consequently, understanding the crystallization kinetics is essential for optimizing processing conditions, such as injection temperature, mold temperature, and cooling procedures. Notably, the nucleation ability of a specific filler varies depending on the polymer matrix and temperature range of interest [155].

Additives, such as talc and nucleation additives, are commonly added to semi-crystalline thermoplastics to modify their properties. Talc is used to reduce costs, while nucleation additives improve the processability of thermoplastics. The effectiveness of these additives is influenced by factors such as the particle size, particle geometry, filler-matrix interactions, and particle distribution. The bonding between the filler and matrix is a critical factor in determining the performance of a material. Poor bonding can result in the dissolution of the plastic and lower viscosity, whereas good bonding inhibits expansion and increases viscosity compared to unfilled plastic [156].

The bonding between the filler and matrix is influenced by physical interactions and mechanical bonding. In cases where filler–matrix bonding is poor, various damage mechanisms can occur, including deformations, matrix fractures, detachment of the filler from the matrix, pulling out of the filler from the matrix, or filler fracture. Glass fiber particles are often used as fillers due to their anisotropic properties and ability to increase stiffness. The addition of particulate fillers can increase the stiffness of the resin and reduce the cost; however, it may also decrease the impact resistance. For example, the modulus of polypropylene is 1.4 GPa, whereas 40 wt% talc-filled polypropylene has a modulus of 2.8 GPa [157, 158].

In summary, the search for improved materials for thermoforming technology has involved significant efforts in the development of synthetic polymers that possess strength, stability, and durability. Polymers have emerged as essential materials for the advancement of science and technology, and the attainment of a high standard of living. Reliance on polymers has become indispensable for the functioning and progress of modern societies.

5

Thermoforming Technology

5.1 Introduction

An ideal material for thermoforming should have specific characteristics to ensure optimal performance during heating and forming. It should have minimal sagging when heated and good flow properties during forming. This requires a balance between elastic behavior, which controls sagging, and viscous behavior, which controls flow.

The polymeric sheet used in thermoforming should not easily change in volume or shape when subjected to external forces. The thermoforming process should be conducted under constant temperature conditions to ensure consistent material behavior. Mass transfer within the material should be insignificant during thermoforming to maintain material integrity and stability. The sheet should have negligible toughness, meaning it should not resist deformation or cracking excessively.

To facilitate the forming process, the sheet should not stick to the mold surface. This ensures smooth and efficient production. The sheet thickness should be significantly lower than its other dimensions to allow for easier deformation and shaping during thermoforming. In reality, achieving ideal thermoforming is challenging due to the many variables involved, such as material properties, mold, and process parameters. Technical issues like variation in wall thickness, material sticking to the mold, and temperature gradients must be overcome to obtain satisfactory formed parts.

5.2 Thermoplastic Sheet Materials

Sheet materials require a good balance between elastic behavior (which controls sag) and viscous behavior (which controls flow). Thermoplastic sheet materials exhibit different viscoelastic characteristics, which encompass both elastic and viscous properties. The thermoforming process effectively utilizes the viscoelastic nature of these materials and their response

Muralisrinivasan Natamai Subramanian. Thermoforming: Processing and Technology, (111–178)

to applied pressure, vacuum, or both, to mold plastic sheets into the desired shape [1].

The stability of the plastic sheet material during thermoforming is influenced by several factors, including temperature, moisture content, molecular weight, and thermal history. One critical factor is the glass transition temperature at which the sheet material transitions to a rubbery state.

The molecular weight of thermoplastic sheets also plays a role in the thermoforming process. Sheets with higher molecular weights offer improved heat strength but can present greater challenges in forming. On the other hand, lower molecular weight materials are easier to form. Achieving the desired quality and efficiency in the thermoforming process requires careful consideration of these factors [2]. The thickness of the sheet is a significant factor to consider when maintaining a constant sheet size. To achieve successful thermoforming, it is crucial for the core layer of the sheet to reach an appropriate temperature. High-quality thermoformed parts essentially depend on the standard of extruded sheet feedstock for use.

5.3 Mechanical Characteristics

Sheet material properties are strongly linked to the chain size and polydispersity of the polymer chains. The most influential properties are as follows:

- Resistant to impact and fracture
- Elasticity
- Sag during processing

Thermoformed parts with a significant enhancement in stiffness and tensile strength can decrease the ultimate strain at break.

The mechanical properties of thermoformed parts depend on the chemical and physical structures of the polymer sheet material. An important aspect of chemical structure is the degree of branching or crosslinking. An increase in the degree of branching or crosslinking leads to a higher stiffness and strength. However, an increase in branching or crosslinking decreases the elongation of the material at break [3, 4].

The influence of the processing conditions on the mechanical properties of thermoformed parts is based on the process parameters. Increasing the sheet and mold temperatures reduces the cooling rate and increases the degree of crystallization, leading to higher strength. Stiffness and strength increase with a decrease in elongation at break and cooling rate. Slower cooling can create molecular orientation with relaxation. Relaxation of the molecular orientation reduces the strength. An increased modulus of elasticity during thermoforming leads to a reduction in the elongation at break. Moreover, the mechanical characteristics of the final products are mainly controlled by the final polymer sheet thickness and the orientation of the molecular chains. The drawing and orientation strongly depend on the process temperature and applied pressure profile [5].

5.4 Thermoformability

Formability, within the realm of thermoforming, pertains to the capacity of a material to undergo deformation and adopt a desired shape without defects or excessive strain. This attribute is important for the manufacturing of thermoformed products [6].

In the thermoforming process, the heating stage plays a fundamental role in the preparation of the thermoplastic sheet for shaping. The importance of adjusting the heating process for specific materials, particularly crystalline or semi-crystalline polymers, which require a longer tunnel length than amorphous polymers. This disparity can be attributed to the following three factors.

- Crystalline and semicrystalline polymers generally exhibit lower thermal conductivities than amorphous polymers. Consequently, more time is required for the heat to penetrate and uniformly soften the material throughout its thickness.
- In crystalline and semi-crystalline polymers, excessive sagging must be avoided during heating. Therefore, the polymer must be sufficiently stiff over a temperature range as large as possible to ensure a robust heating step. The polymers showed a drop in modulus at the melt temperature. The region with sufficient stiffness is narrow.
- The nature of crystalline and semi-crystalline polymers necessitates a more gradual and controlled heat input during the heating process. Rapid or uneven heating can lead to uneven softening, which results in defects or inconsistencies in the final product.

To optimize the heating process for materials with lower thermal conductivity by utilizing a longer tunnel, the heating process can be slower and more controlled, reducing the risk of defects and ensuring a more uniform forming process. This approach ensures that the material reaches an ideal temperature for forming without compromising its structural integrity. The objective is to achieve the necessary malleability of the material while avoiding overheating or localized softening, which could lead to defects in the final product [7].

5.5 Thermoforming Cycle

The thermoforming process offers several benefits, including the implementation of faster molding cycles and a quicker part setup. The reduction in cycle times enhances production efficiency by decreasing the overall time required to produce each part. This leads to higher production rates and increased throughput within a given timeframe. Moreover, the expedited setup times contributed to improved dimensional accuracy and integrity of the formed part. However, it is important to strike a balance between swift-part setup and controlled cooling.

Rapid cooling can result in issues such as stress or warping in the final part. Therefore, a cooling system must be meticulously designed to achieve the desired cooling rate without causing any defects in the part. Overall, the ability to extract the part from the mold at a slightly higher temperature due to faster molding cycles proved advantageous in terms of the production efficiency and part quality in the thermoforming process. A schematic representation of the thermoforming cycle is shown in Figure 5.1.

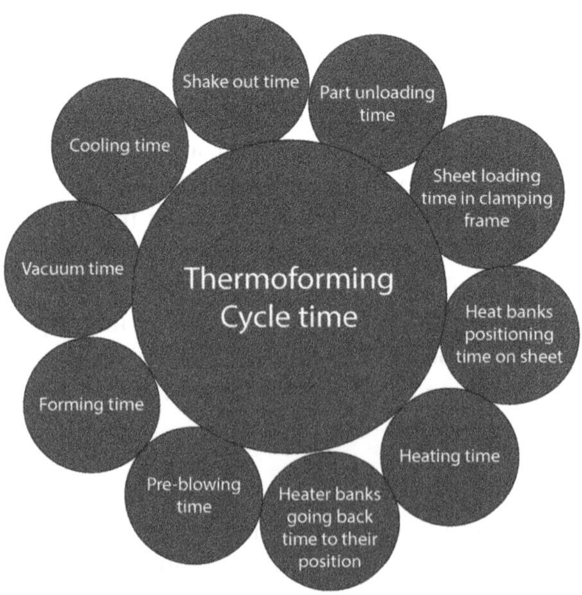

Figure 5.1 A schematic representation of thermoforming cycle time.

The duration of the thermoforming process can vary significantly depending on various factors, including the complexity of the part, material utilized, forming method used, and capabilities of the thermoforming equipment. Typically, the cycle time for thermoforming ranges from a few seconds to several minutes. In the thermoforming framework, the duration of the cycle can vary depending on the complexity of the parts being produced. For simple parts with thin walls and uncomplicated forming requirements, the cycle time can be as short as a few seconds. In such cases, the heating and forming stages can be completed quickly, and the cooling process can be relatively fast.

However, for more intricate parts with thicker walls, deeper draws, and more complex details, the cycle time may be longer. In such situations, the heating and forming stages may require more time to achieve the desired shape and definition of the part. In addition, the cooling process may need to be extended to ensure proper dimensional stability and prevent defects.

Advancements in thermoforming technology have been made to improve the cycle times. These include enhancements in the heating systems, faster cooling methods, and more precise control over the forming process. In additional, efficient material handling and automation can contribute to reducing the downtime between cycles.

It is important to emphasize that the specific cycle time for a particular thermoforming application must be determined through experimentation and optimization. This is because the cycle time can vary significantly depending on the specific requirements of the part and equipment being used [8].

Efficient managing of the cycle time is crucial for maximizing production efficiency in thermoforming. To enhance productivity, the aim is to strike a balance between producing high-quality parts and minimizing cycle times. Shorter cycle times enable higher production rates and reduce costs, making them desirable in most manufacturing settings.

5.6 Draw Ratio

The draw ratio is a metric used to comprehend the extent to which a thermoplastic sheet elongates and thins when pressed against a mold to acquire its ultimate shape. In the context of thermoforming, the draw ratio denotes the proportion of the surface area of the final part to the initial area of the sheet prior to forming. This ratio can be expressed in terms of linear dimensions or cross-sectional areas and is devoid of dimensional units. The draw ratio is significant because it determines the degree to which the material stretches and deforms during the thermoforming. It aids in the anticipation of thinning and enables designers to account for thinning in mold design and material selection. Because different thermoplastics possess varying degrees of stretchability, knowledge of the draw ratio facilitates the selection of appropriate materials. Furthermore, the draw ratio influences the mold design and the ultimate dimensions of the part. Striking a balance in the draw ratio is fundamental for successful thermoforming, as excessively high or low ratios can result in complications. The optimization of the draw ratio necessitates the consideration of material properties, complexity of the shape of the part, mold design, and desired outcomes [9].

The drawing performance is influenced by the drawing ratio and deformation uniformity. Drawability is difficult due to material properties, strain distribution, and instability criteria. Strain uniformity is governed by the material, design, and process variables. The design variables are based on product requirements. The process variables are important in processing. The design and process variables yield the maximum production.

The performance of thermoplastics in deep drawing from a sheet appears to be a viscometric function generated during the rheological aspects of sheet deformation. Sheet deformation includes material parameters, processing temperatures, design parameters, and deformation rate.

During the thermoforming process, the heat-softened plastic sheet is stretched or drawn by a thermoforming mold to achieve the desired shape. This drawing process results in an increase in the surface area and a decrease in sheet thickness [10]. Shape features, such as sharp edges and steep depressions, lead to a high "draw ratio," which is defined as the ratio of the sheet area after forming to that before forming.

The maximum attainable draw ratio decreases with increasing elongational viscosity. The sheet temperature increases with the elongational viscosity. During forming, the mold initially contacts with the rim and then moves to the bottom center of the positive mold. As the process progresses, the material in contact with the cold surface starts to decrease in temperature, resulting in a higher elongation viscosity. Very little additional thinning occurrs in the region surrounding the plug nose. Therefore, the total amount of material required for drawing decreases and the thickness decreases.

5.7 Processing Window

Temperature control is a crucial aspect of thermoforming due to the expansion and contraction of thermoplastic sheet materials. Maintaining proper temperature control ensures the desired forming results. The appropriate temperature range for thermoforming is determined by assessing the performance of the material at different temperatures. The behavior of the materials is significantly affected by the temperature at which they are subjected to processing. Elevated temperatures enhance the material's malleability, whereas lower temperatures make it more rigid and less flexible.

The processing window is the temperature range where a plastic sheet material can be successfully shaped and formed. This is the temperature at which the material becomes sufficiently flexible to be stretched and molded. The specific temperature required for thermoforming varies depending on the type of the material used. Therefore, it is important to determine the appropriate temperature range, or processing window, for successful thermoforming.

A processing window is essential for the thermoforming process of extruded plastic sheet materials. This is because the nature of a material, whether amorphous or crystalline, determines its advantages and challenges. Amorphous or semicrystalline materials have a wide range of melting temperatures, resulting in a broader processing window. This allows for more flexibility in adjusting the processing parameters, such as the heating and cooling rates, cycle time, and forming temperatures. A schematic of the thermal history during thermoforming is shown in Figure 5.2.

In particular, amorphous polymers require lower thermoforming temperatures. This reduces the risk of overheating or material damage during the process. However, these materials may be more prone to creep over time due to their lower stiffness than crystalline materials.

However, crystalline materials maintain their stiffness until they reach the processing window temperature. Once the material reaches this temperature, it becomes pliable. However, there is the risk of warping, which can cause unwanted distortions in the final product.

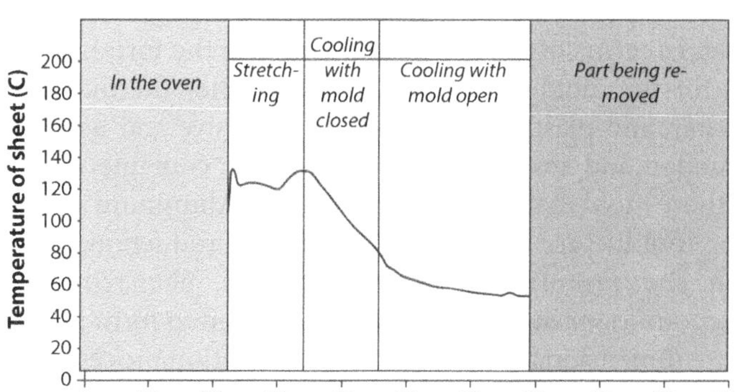

Figure 5.2 Graphical representation of thermal history during thermoforming. Reproduced with permission from X. Haihong, J. Wysocki, D. Kazmer, P. Bristow, B. Landa, J. Riello, C. Messina and R. Marrey in *Proceedings of the ANTEC Conference*, New York, NY, USA, 1999, 872. © 1999, SPE.

Ideally, the material should be formed within the processing window without any failures or defects. Below the processing window temperature, the material remains rigid and difficult to shape. Above the processing window temperature, the material becomes too soft and can undergo undesirable deformation or sagging [11].

5.8 Mold

Thermoforming molds can be made from many materials, ranging from wood or plaster to aluminum or steel. They depend on the number of parts to be formed and the pressure and temperature requirements. The economy of a material is a major factor in mold-making. Costing includes the fabrication, maintenance, storage, and other working aspects. Various materials, such as wood, plaster, and plastics, are commonly used for prototype production and applications, where durability is not a primary concern because of their relatively low cost. For short-run production and prototype purposes, materials such as epoxy, urethanes, fiber-reinforced plastics, and ceramics can be used. However, aluminum and steel molds are preferred for high volume production due to their extended tool life and excellent surface hardness. These materials are relatively easy to fabricate and exhibit excellent surface hardness and low wear. Formed parts that are consistent with dimensions and quality along with efficient production must be achieved with aluminum molds. The effects of the mold material are related to heat transfer, de-molding, surface finish, physical properties, and dimensional accuracy.

Alternatively, plastic molds can be utilized, although they tend to accumulate heat. In terms of the productivity during the forming process, it is important for the mold material to possess better thermal conductivity. Wood, plaster, and plastics are usually inexpensive and useful for prototype production and applications, such as signs, containers, and displays. However, these mold materials are not durable. Aluminum and steel molds with a long tool life are used for high-volume production. For short-run production and prototypes, epoxy, urethanes, fiber-reinforced plastic materials, ceramics or other materials can be used to form parts.

To ensure dimensional consistency, the formation process must be consistent from one production run to another. Any variation in the time that the part spends in the mold directly affects the amount of shrinkage. It is vital to maintain the part in the mold until the sheet material reaches the desired temperature. Materials that are less prone to wear and corrosion may require less frequent maintenance, leading to a reduced downtime.

The choice of the mold material can also significantly affect the quality of the formed product. Opting for a high-quality mold material with a smooth surface finish can result in improved aesthetics. Mold materials with low thermal expansion and good dimensional stability ensure that the desired dimensions of parts are achieved. In situations where the series is limited or when working on a prototype, the use of a wooden mold is deemed acceptable. However, it is important to note that wood tends to accumulate heat within molds. Consequently, this leads to longer cooling times and reduced productivity in the case of continuous production.

The mold material is an important factor in determining the effectiveness of thermoforming processes. Notably, the chosen mold material can have a significant impact on several aspects of the thermoforming process. These include the heat transfer efficiency, ease of demolding, surface finish quality of the formed parts, physical properties of the molds, and dimensional accuracy.

5.9 Mold Design

Thermoforming molds can include regular and freeform surface patches that can be considered shape features because of their characteristic representation of the shape elements of the thermoforming mold. Some properties of the mold material are listed in Table 5.1. These features can be mapped to generic shape patterns and possess functional significance [12]. It has been observed that thermoforming mold makers intuitively utilize this concept of shape features to identify and select regions of interest when manipulating the thermoforming clay model.

Thermoforming features with a low draw ratio are easier to obtain. Therefore, thermoforming mold makers chamfer sharp edges, add fillets to acute corners, and fill deep depressions to prevent the formation of high-draw-ratio features. This practice facilitates the removal of the formed sheet

Table 5.1 Properties of mold material.

		Mold material			
Properties	**Unit**	**Aluminum**	**Wood**	**Mild steel**	**Epoxy**
Density	kg/m^3	2702	350–930	7,863.86	1150
Specific heat	J/kg K	908	1200–2400	511	1100
Conductivity	W/m K	237	0.1–0.2	45–65	0.17–1.34

from the thermoforming mold by allowing the sidewalls of the mold to taper outward. Commonly used taper angles or draft angles range from 31 to 51°. Depending on the surface curvature and orientation, some surfaces do not obstruct demolding; therefore, the incorporation of draft angles on these surfaces may not be necessary.

The mechanical strength of a thermoformed sheet can be enhanced by incorporating various features such as corrugations, webs, and bridges. In some cases, a slot-like feature is utilized as a hinge, enabling the sheet to fold along the slot. Another example involves the use of a protrusion feature on one side of a folded sheet and a depression feature on the other side. This allows the protrusion to fit into the depression, creating a locking effect that securely holds the folded sheet together [13].

The primary purpose of a thermoforming mold is to provide security and protection to packaged items. This is achieved by utilizing an encapsulating surface that closely conforms to the original shape of an item, enclosing and safeguarding it within the surrounding space. Consequently, the encapsulating surfaces are considered the main packaging features, with their shape derived from the original surfaces of the packaged item through a positive offsetting operation. The offset gap between the thermoformed sheet and packaged item serves as a buffer, effectively isolating the item from external forces that may arise from dropping or rough handling.

In thermoforming (Figure 5.3), the importance of efficient thermal management in both heating and cooling molds is widely acknowledged.

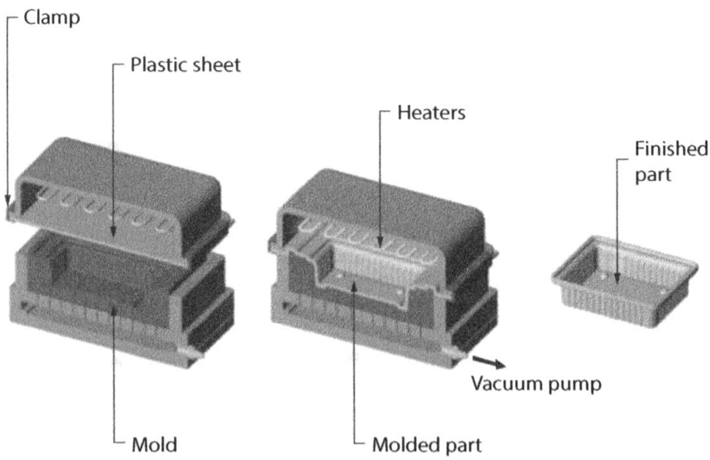

Figure 5.3 Representation of mold for thermoforming process. [Reprint with permission from [13]. © 2020 Copyright. All rights reserved].

The forming cycle constitutes a significant portion of the thermoforming process. Appropriate temperature control of the mold is essential to achieve the desired quality parts. To enhance the forming process, cooling channel systems are commonly utilized to reduce the cooling times and enhance the surface and microstructural quality of the final parts.

The complexity of the mold shape, material, hierarchy, and functionality play a role in the overall process (Figure 5.3). The cooling system of a thermoplastic mold is of utmost importance in the production process because it has a significant impact on both the quality of the final product and the time required for each production cycle. Traditionally, the cooling systems of thermoforming molds have been limited by the constraints of conventional drilling methods. Consequently, a cooling system typically consists of simple parallel straight channels, which are inadequate for effectively cooling complex shapes. As a result, there is currently growing interest in the redesign of molds to achieve improved performance, leading to better products and enhanced overall production efficiency.

The application of cooling systems that are specifically designed to match the shape of a mold offers significant advantages in the thermoforming process. The incorporation of cooling channels has been shown to result in a notable decrease in the mold temperature. To ensure the effective design of cooling channels, guidelines should be followed, such as placing them near the hottest areas, optimizing their section, distributing them appropriately around the mold shape, and ensuring that the structural integrity of the mold is not compromised.

The primary issue encountered in this process arises during the heating stage. Specifically, repeated heating cycles resulted in excessive heat on the mold surfaces in contact with the clamps. Due to the limitations of the conventional cooling system, overheating necessitates halting the thermoforming machine to permit the refrigerant fluid to cool the mold. By reconfiguring the cooling channels, it is feasible to enhance the cooling phase and prevent machine interruptions, expanding productivity.

Numerous factors can affect the lifespan of the molds used in thermoforming processes. One such factor is wear, which is prevalent in thermoplastic composites. In addition, extended production runs can contribute to wear and tear. Deformation of the mold can occur because of various factors, including excessive clamping force, uneven temperature distribution, and inadequate mold design and construction. Surface defects on the mold can arise from polishing problems, and certain plastics can degrade and lead to surface defects. Moreover, corrosive materials, such as PVC, can cause mold surfaces to corrode over time. Aggressive cooling or high humidity in the production environment can also contribute to corrosion.

In the context of thermoforming, various factors contribute to the overall process, including the vacuum bores or orifices, mold surface, mold temperature, and mechanical support temperature. The choice of mold material significantly affects the performance of the mold and various aspects of the tooling performance, including mold life, quality of the plastic part, and productivity. Mold life refers to the duration during which the mold remains functional and can produce high-quality products without significant deterioration.

The design of the mold allows for the adjustment of the drawdown ratio by incorporating a bottom insert with variable thickness. The choice of mold type for manufacturing a specific plastic component is determined through meticulous analysis of various factors and data, with the aim of identifying the most optimal solution to the problem at hand. This is crucial, as the molds must operate in an automated manner to produce a large quantity of pieces.

Thermoforming is a widely used technique for transforming thermoplastics, and is also utilized for thermostable materials. During the design phase, careful consideration must be given to the desired configuration of the plastic pieces because this determines the shape of the mold to be utilized throughout the production process. It is most advantageous to employ a positive mold to produce single curved products with a larger radius. This can be achieved through either sweep-forming using thin steel foil or vacuum bagging.

Vacuum bagging, a cost-effective and straightforward process has certain drawbacks. One such drawback is the time required to apply the vacuum bag, which can be a time-consuming task. In addition, there is a risk of excessive sheet cooling during the process.

In the drape and vacuum forming processes, it is necessary to create and utilize positive and negative molds, respectively. Thermoforming molds can

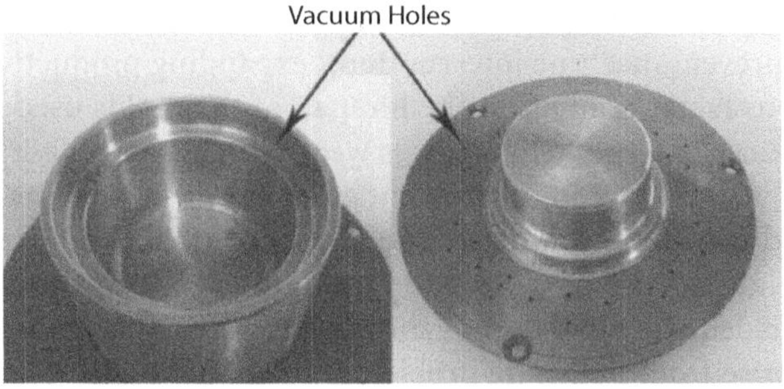

Figure 5.4 (a) negative mold, (b) positive mold. [Reprint with permission from [14]. Copyright © 2014, Springer-Verlag London. All rights reserved].

be constructed using various materials such as wood, plaster, epoxy resins, or reinforced polyesters. However, these molds have a limited lifespan and are best suited to low-volume production batches. Furthermore, the low heat transfer coefficient of these materials results in inadequate control of mold temperature. The recommended mold draft for thermoforming ranges from 3° to 7°, with a 7° draft being preferred for easy ejection in both types of molds.

Notably, the desired optical outcome of the final thermoformed components can be achieved using chromium steel molds. However, aluminum is the preferred material for permanent molds because of its lighter weight than steel and higher strength than basic materials. Therefore, both molds have to be fabricated from aluminum, as shown in Figure 5.4.

To mitigate this issue, one approach is to assemble a vacuum bag on a cold sheet and a mold. Subsequently, the mold should be placed in a hot-air oven, and the temperature inside the sheet should be monitored using a thermal gauge. Once the desired temperature is achieved, vacuum should be applied. Following this, the mold should be removed from the oven and allowed to cool while maintaining the vacuum. It is crucial to carefully consider the factors of time, temperature, and vacuum to prevent any potential creep effect.

When it comes to mold selection, it is recommended to use a negative mold for smaller radii. This can be achieved through either vacuum bagging or matching mold-forming techniques. The latter is particularly suitable for cases where the radius is small, the thickness or density is high, and a substantial load is anticipated. It is essential to use fixed stops to prevent compression of the core. In the realm of thermoforming, numerous methods exist, including, but not limited to, vacuum-assisted, pressure, drape, sweep, match mold, and free forming [14].

5.10 Heating Elements

The heating elements play a vital role in thermoforming. The heaters soften the plastic material, rendering it malleable for shaping. The thermoforming machine is equipped with multiple heating zones to achieve precise temperature control throughout the plastic sheet. Different areas of the sheet may require varying the heating temperatures to achieve consistent shaping outcomes. The heating procedure during thermoforming must be carefully regulated to prevent overheating, which can result in material degradation, inconsistent shaping, or burning. Modern thermoforming machines are often equipped with advanced control systems that accurately monitor and regulate heating parameters.

During the preheating stage of thermoforming, the heaters elevate the temperature of the thermoplastic sheet to a specific softening point. This preheating step is critical for ensuring that the plastic becomes flexible and easily moldable without sustaining any damage. Commonly used heating elements in thermoforming include infrared (IR), calrod, ceramic, tubular, quartz, and nichrome wire heaters.

5.10.1 Infrared Heaters

Infrared heaters are commonly used in thermoforming machines. Infrared radiant heaters are commonly used in industrial and manufacturing processes to heat thin- and medium-thickness sheet materials with efficiency and precision. These heaters emit infrared radiation, which is a form of electromagnetic wave with longer wavelengths than visible light. When radiation interacts with a material, it induces molecular vibrations and generates heat.

The effectiveness of an infrared radiant heater depends on the capacity of the material to absorb infrared radiation. Different materials exhibit distinct absorption spectra, indicating that they absorb energies at different wavelengths. To optimize the heating process, it is crucial to align the peak wavelengths of the emitted infrared radiation with the wavelengths at which the sheet material absorbs the most energy. This ensured that the material efficiently absorbed the radiation, resulting in effective and rapid heating.

Quartz heaters are particularly efficient types of infrared radiant heaters. They employ quartz tubes as heating elements, which emit infrared radiation at specific wavelengths that can be adjusted to maximize the absorption by the target material. Quartz heaters are well known for their prompt response time, uniform heat distribution, and high thermal efficiency. Quartz heaters are characterized by their quick response time and energy efficiency and are commonly used in small-scale thermoforming machines or for specific applications that require rapid heating and cooling cycles.

Radiant heaters are widely recognized for their effectiveness in transferring heat to the surfaces of thin sheet materials, making them suitable for industrial applications that require rapid processing or shaping. However, the efficiency of radiant heating decreases when applied to thicker sheet materials. This is primarily due to the occurrence of "thermal lag," a phenomenon where excess heat supplied to the surface does not propagate quickly enough to the core of the material. Thicker materials possess a higher heat capacity, resulting in a delay in elevating the temperature of the entire sheet, as the heat requires time to penetrate the thickness. Horizontal heating methods, such as infrared radiant heaters, are preferred for continuous forming processes because they offer better control over the heating process and minimize the risk of part defects.

In the field of thermoforming, infrared (IR) heating is frequently used to heat roll-fed plastic sheets in multi-cavity molds. Given the significance of production rates, it is common practice to heat each section of the sheet in rapid sequential steps as it progresses towards the forming stage. Consequently, the length of the heater bank is typically greater than that of the portion of the sheet that undergoes thermoforming in each cycle. The intricate thermal interactions between the radiators and sheet result in the ability to induce variations in the sheet temperature by adjusting the temperature of a single heater.

5.10.2 Flash Heaters

Flash heaters are specialized heating systems used in single-station vacuum-forming machines to rapidly heat thermoplastic sheets before shaping them into desired forms. These heaters offer notable advantages over traditional heating methods such as radiant heaters or convection ovens in terms of cycle times and productivity. Flash heating refers to the swift and intense heating process facilitated by these heaters, which utilize high-intensity infrared radiation to rapidly heat the surface of the thermoplastic material. This enables the efficient heating of both thin and thick materials, making flash heaters versatile for various applications.

Flash heaters offer several advantages including reduced cycle times, increased productivity, and efficient heating across the entire material surface. These heaters expedite the heating of thermoplastic sheets to forming temperatures compared with traditional methods, resulting in shorter cycle times and higher production rates. The uniform heating provided by the flash heaters ensures consistent forming outcomes and minimizes the occurrence of defects.

Nevertheless, certain considerations and limitations are associated with flash heating. Textured materials with a heavily textured surface may present challenges, as the texture can absorb and distribute heat differently, necessitating adjustments in heating settings to achieve uniform heating. Additionally, significant variations in the sheet thickness may require modifications to the heating parameters to ensure uniform heating.

5.10.3 Forced Convection Hot Air Heating

Forced convection hot air heating is a widely used approach in industrial processes, particularly in the thermoforming of sheet materials with higher thicknessness. This method involves the circulation of hot air over the surface of a material using a fan or blower. This is particularly advantageous for applications that demand high optical quality, such as the production of aircraft windshields.

When employing forced-convection hot-air heating, several important considerations should be considered. It requires high optical standards, which aid in achieving uniform heating, minimizing defects or irregularities that could impact optical clarity. Forced convection hot air heating is especially effective for thicker sheet materials, as it enables rapid and even heating. Thicker materials require more heat to attain the forming temperature throughout their entire cross-section, and forced convection efficiently facilitates this process.

When implementing vertical heating with forced convection hot air heating, it is crucial to apply an adequate clamping pressure to prevent sheet slippage. As the material softens due to the heat, there is a risk of sheet slipping or deformation if not properly secured. However, it is important to note that vertical heating is generally not recommended for the continuous manufacturing of forming parts.

5.10.4 Other Heaters

Pattern heating, also referred to as local heating or thermal screen heating, is a widely used method in heating processes aimed at regulating the temperature distribution within a material or component with nonuniform properties or complex geometries. This technique involves the targeted application of heat to specific sections of the part to mitigate undesirable consequences such as excessive thinning. By controlling the temperature distribution through pattern heating, manufacturers can achieve more precise and controlled outcomes, minimize material waste, and prevent defects, such as thinning, warping, or overheating.

Nichrome wire heaters, composed of a nickel-chromium alloy, are occasionally used for localized heating or custom-built equipment. To attain the desired temperature profile, materials possessing excellent heat resistance and conductivity, such as wire meshes, are commonly utilized. The wire mesh is precisely shaped, sized, and strategically positioned on the surface of the part to govern the distribution of heat and avert overheating in particular regions, while enabling other areas to attain elevated temperatures. The implementation of wire mesh or other thermal screens is an effective and practical approach for applying this technique in different industrial processes.

Ceramic heaters are known for their rapid attainment of high temperatures and ability to maintain stable heat output. Tubular heaters are appropriate for applications that require elevated temperatures. The materials used in the thermoforming process possess distinct absorption bands, and ceramic heaters can be regarded as selectively heating only the surface of the sheet. Consequently, there is a notable disparity in the temperature between the surface of the sheet and its center. Calrod heaters are durable and provide consistent heat distribution, making them suitable for use in smaller or medium-sized thermoforming machines.

Regardless of the type of heating element used, precise temperature control and uniform heat distribution are crucial for ensuring consistent and high-quality thermoformed products. Contemporary thermoforming machines often integrate advanced control systems that monitor and regulate the heating process to achieve desired outcomes efficiently.

The arrangement and configuration of heating elements in thermoforming machines can vary depending on the machine design and specific requirements of the forming process. Some machines employ combinations of various heating elements in different zones to achieve optimal temperature control and efficient heating.

5.11 Plug Material

Plugs made from insulating materials such as plastics and wood are considered "passive" as their temperature is dependent on the duration of contact between the sheet and plug. Conversely, plugs made from metals like aluminum are considered "active" as their temperature can be controlled independently and used to regulate variations in component thickness. In recent years, syntactic foams (SF) have been significantly used as plug materials. Syntactic foams consist of hollow spheres made of glass, ceramic, or polymer embedded in a binder or matrix, such as epoxy, urethane, or thermoplastic. The low thermal conductivity of SF foams enables the formation of sheets at low temperatures, resulting in reduced cycle times.

The measurement of the distribution of the interim sheet thickness in the preform at the end of the plug stroke revealed significant variations in wall thickness distribution when different plug materials are used. These variations are attributed to the different frictional properties of the surfaces in contact, which are temperature-dependent. It has been observed that larger, flatter plugs are more effective at drawing material into the base of the product and that changes in the geometry alone can result in substantial alterations in the wall thickness distribution.

The extent of sliding extension during the thermoforming process is greatly influenced by the choice of plug material, which is determined by the frictional properties of the contacting surfaces. Furthermore, these properties are highly sensitive to temperature, leading to considerable changes in the plug temperature.

The plug plays a crucial role in the thermoforming process, as evidenced by its significant impact on the final distribution of wall thickness in the produced part. The interaction between the contact friction and heat transfer is instrumental in determining this distribution. As the plug makes contact, the sheet gradually stretches across the contact surface as the plug descends.

The selection of the plug material, its design, and temperature are particularly important factors because they can result in substantial variations in the wall thickness of the final part. Flat-bottomed plugs are effective in retaining thickness at the base of the product by trapping the sheet material upon contact. On the other hand, more rounded plug shapes promote sliding of the sheet material upon contact, redistributing material from the base towards the sidewalls of the product.

5.12 Plug Design

The inclusion of a plug in the thermoforming process introduces significant complexities in controlling the wall thickness output. This is because of several additional variables that must be considered, such as the shape and temperature of the plug, as well as the thermal and frictional properties associated with it [15].

Thermoforming applications encompass a wide range of products, from simple trays to deep draw pots and tubs. In this process, a heated polymer sheet is rapidly deformed by the application of either air or mechanical pressure, resulting in its final shape. Typically, this deformation is achieved by moving a mechanical plug, followed by the application of air pressure. The equilibrium temperature of the working plug is typically approximately 100 °C when the sheet temperature is 150 °C. Plugs are usually designed to be smaller, rounded versions of the mold shape, and previous designs are often modified when possible [16]. However, in cases involving new products or substantial changes in the materials or operating conditions, entirely new plug designs must be developed to ensure optimal performance. Despite these economic disadvantages, it remains commonplace for companies to employ trial and error methods during production to finalize a new plug design.

During the thermoforming process, the center of the clamped sheet is firmly pressed into the surface of the plug, whereas the material on either side experiences rapid extension as the plug moves downward. This transition between the two regions is commonly referred to as the "plug mark," which creates a visible discontinuity in the wall thickness distribution [17].

To ensure effective thermoforming, it is important that the material does not simply adhere to the plug, as this would result in an excessively thick base in the final product. Therefore, low-friction plug materials are intentionally chosen to allow the sheet to gradually slide back up and redistribute the material to the sidewall. However, this sliding mechanism is not completely free because deformation can only occur through the extension of the sheet across the contacting surface. Consequently, friction reduces the extent of extension of the sheet in contact.

The size of the plug relative to the mold plays a role in the distribution of the material towards the base of the product. This effect is enhanced by having a flat area on the base of the plug; however, it can be mitigated by increasing the base radius or sidewall taper of the plug. Minor deviations from the initial plug shape tend to result in more balanced wall thickness distributions, whereas significant changes tend to cause the base and corners of the product to become excessively thin.

In the industrial sector, one of the most complex decisions is the selection of an appropriate plug design that aligns with a specific mold shape. Unlike other variables, this aspect is particularly challenging, costly, and time-consuming to modify once the production is underway. Consequently, there are significant economic advantages associated with achieving the correct plug design in the first attempt.

In packaging applications, plugs are typically made of polymer materials and are not heated. Instead, they gradually absorbed the heat from the hot sheet during the initial start-up period. The working plug reached an equilibrium temperature of approximately 100 °C when the sheet temperature is 150 °C [18]. Syntactic foam, acetal (or POM), and lacquered felt are commonly used plugs. These materials are chosen primarily due to their low heat transfer and low surface friction properties. The selection of materials varied depending on the type of sheet material used.

5.13 Product Design

Product design is a deliberate and systematic process that encompasses multiple stages, starting from the initial concept and ending with the final manufacturing. Therefore, a comprehensive approach that considers all aspects of the design process is required. The process begins with the development of a concept and progresses through iterations and refinements to create tangible and functional objects. When working with engineering plastics, it is crucial to use a diverse range of materials. Determination of part geometry is a critical step in optimizing its functional integration.

In addition, material selection plays a pivotal role, considering factors such as functionality, cost-effectiveness, and environmental resistance. Once the material and geometry are established, the subsequent steps involve designing the necessary tools and processes for manufacturing the part. It is essential to ensure that the final product is not only functional and aesthetically pleasing but also cost-effective and environmentally sustainable.

While a smooth mold surface is generally preferred, excessive polishing can lead to complications, such as air entrapment during the forming process, resulting in imperfections on the surface of the part. This is because a highly polished surface lacks micro-roughness features that aid the release of trapped air. To address this issue and introduce the desired texture on the formed part, it is necessary to employ various techniques, including controlled roughness, texture application, chemical etching, and bead blasting. These techniques regulate the surface finish, facilitate air escape, and enhance the overall surface quality of a part. In addition, the selection of appropriate materials and modifications to the part design can enhance the feasibility of the forming process. It is important to carefully consider the tolerances specified for the parts, because excessively tight tolerances can significantly increase the cost of the forming process. Furthermore, it should be noted that thermoforming primarily allows for detailed features on one side of the part; therefore, complex detailing should be concentrated on that particular side.

Draft angles are an important aspect in the process of removing a formed part from a mold, as shown in Figure 5.5. Additionally, the inclusion of radii in the design aids in achieving a more consistent wall thickness and reduces the occurrence of molded-in-stress. The depth of the formed part becomes crucial when there are limitations in the availability of forming equipment. Moreover, the wall thickness of a part is influenced by various

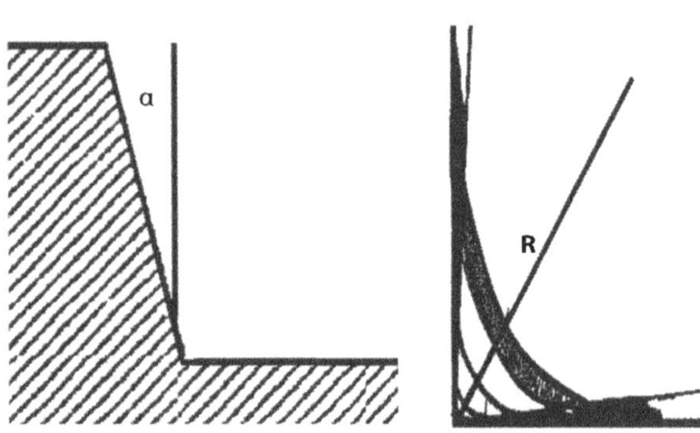

Figure 5.5 Schematic representation of draft angle and draw ratio vs. radius.

factors that must be considered during the design phase. By conducting a comprehensive analysis of these factors, one can improve the efficiency of the thermoforming process and produce superior components that meet specified criteria and quality standards.

The assessment of the complexity of the thermoforming process and the viability of forming specific components depend heavily on the depth of the draw. Components with smaller depth draws can be easily formed using simpler thermoforming techniques such as vacuum forming or pressure forming. Conversely, components with larger depth draws require more intricate thermoforming processes that often involve multiple steps. As the depth of the draw increased, the material underwent greater stretching, resulting in thinner sidewalls in the formed component. Therefore, when designing components with deep draws, careful consideration must be given to the ability of the material to elongate and the potential thinning of the side walls. Furthermore, the selection of the thermoforming method and tooling is also influenced by the depth of the drawing.

By considering these design factors and implementing appropriate techniques, product design can improve the thermoforming process, produce high-quality parts, and achieve the desired surface finish and dimensional accuracy. Efficient thermoforming relies on the presence of well-designed vacuum holes, which play a crucial role in removing air between the thermoplastic sheet and the mold surface. Determining the appropriate size and quantity of these vacuum holes is of utmost importance to avoid complications during the forming process. It is essential to select an optimal hole diameter to ensure swift and efficient air evacuation while minimizing surface defects on the formed part. It is generally recommended that the diameter of the holes should not exceed the thickness of the thermoplastic material. Careful planning of the location and distribution of vacuum holes is necessary to achieve uniform air evacuation and to reduce the likelihood of inconsistencies in the forming process or surface imperfections. When determining the placement of vacuum holes, areas where visible marks or blemishes on the surface are undesirable should be avoided. Optimization of the vacuum hole design for specific part designs and materials often requires testing and iterative refinement.

5.14 Clamping

Ensuring adequate clamping is crucial in the thermoforming process to achieve accurate and successful formation of sheet materials. Inadequate clamping can lead to incomplete or incomplete forming or complete failure of the clamping process.

Pneumatically operated clamps are commonly utilized in thermoforming due to their reliability and consistent clamping force. The inclusion of teeth or ridges on the edge of the clamp enhances its grip on the sheet material, preventing slipping or shifting during forming, particularly when dealing with large or complex shapes.

The level of clamping engagement required may vary depending on the specific thermoforming setup and material being formed. Certain materials may require a higher clamping force to prevent undesired deformations or distortions during forming.

Heating the clamping system is critical for thermoforming. Heated clamps help to maintain the temperature of the sheet material at an optimal level throughout the forming process. Conversely, cold clamps can act as heat sinks, drawing heat away from the edges of the sheet resulting in poor detail or forming failures.

To achieve high-quality thermoformed parts with precise details and consistent outcomes, careful consideration and optimization of the clamping system are necessary. Factors such as material properties, forming setup, and use of heated clamps should be considered to ensure proper control over the forming process. Moreover, proper clamping or securing of sheet material is vital during the thermoforming process to ensure successful outcomes.

5.15 Process Control

The process variables are the most important in thermoforming. The thickness and temperature of the plastic sheet material are directly related to the processing conditions used during the extrusion of the sheet. These effects are critical during the thermoforming process. Plug material, shape, temperature, and speed of movement are important in plug-assisted thermoforming. The thermal properties of the plug material and its surface and friction properties are also important considerations. In addition, the air pressure, plug timing, air stages, mold shape, surface of the mold, and thermal properties of the mold are also examined during thermoforming [19].

In simple vacuum or pressure forming, the presence of the plug dominates output control. In the field of industrial thermoforming, it is necessary to establish a machine in a standard manner until a stable output is attained. Throughout the process, control of the operation consistently yields a continuous and dependable output. The effectiveness and reliability of process control are evident and should be capable of being replicated in relation to the intricate functioning of the thermoforming process. To ensure accuracy, it is imperative to calibrate the sensors using recommended procedures. In the case of temperature sensors, the thermocouples should be calibrated, whereas for the non-contact IR sensor, an appropriate thermal emissivity should be selected.

5.16 Process Variables

In the field of thermoforming, the quality of the final products is significantly affected by various factors, such as raw materials, ingredients, equipment, and processing conditions. The factors influencing the process variables are shown in Figure 5.6. These factors can lead to common defects and problems in the external appearance and interior performance of products. To address these issues, identifying the root causes and implementing appropriate countermeasures are recommended. It is advisable to change one processing parameter or production condition at a time to eliminate defects effectively. Simultaneously changing multiple parameters

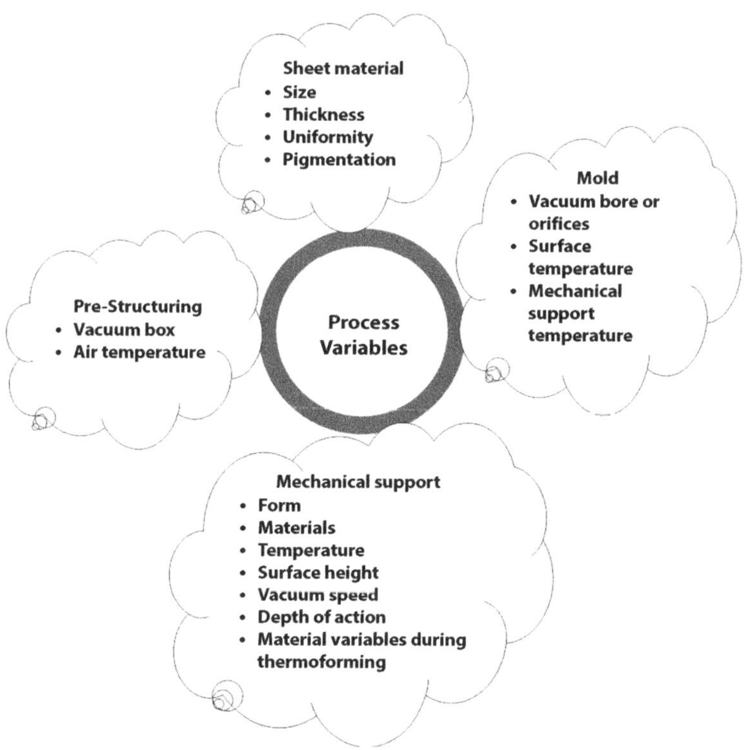

Figure 5.6 Factors influencing product quality.

or conditions should be avoided to ensure accurate evaluation of the impact of each change [16, 20].

Thermoforming encompasses various machines, processes, and quality variables. The diagram visually depicts the factors that affect product quality and the corresponding variables associated with them. In thermoforming control, process variables refer to dependent variables that are influenced by various machine settings, such as the mold, plug, melt viscosity, air pressure, and vacuum. These process variables include the following:

- The melt temperature of the sheet material,
- The adjustment of melt to prevent sagging,
- The velocity of the plug, and
- Rate of heat dissipation and cooling.

Mold variables, such as positive or negative and their combinations. Single-sided molds are used because the local wall thickness of the part cannot be controlled directly by the mold geometry. To effectively control the thermoforming process, it is common to utilize machine settings such as the mold, air pressure, vacuum, melt temperature, and mold separation. These process variables, including the melt temperature, sag, plug advancement, strain hardening, and rate of heat dissipation and cooling, depend on the collective effects of these machine settings [16, 21].

Figure 5.7 shows the deformation acting on the network assistance. Sheet thickness is the main parameter and temperature dependence during thermoforming, as shown in Figure 5.8(a). This also depends on the radially applied pressure profile. Upon applying pressure, the sheet thickness profile shows a minimum at the center of the mold and maximum between the wall and clamp due to the radial distribution of the applied pressure. The sheet touches with mold surface, and the surface contact force increases because of the temperature difference and temperature reduction of the

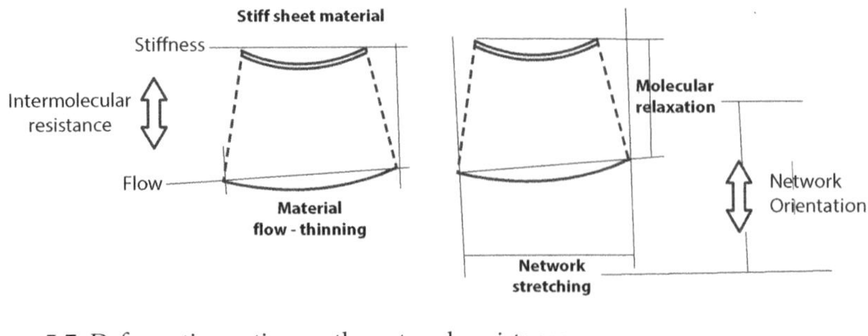

Figure 5.7 Deformation acting on the network resistance

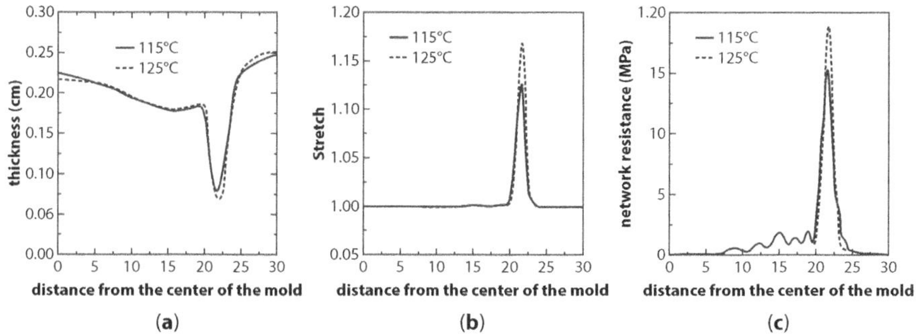

Figure 5.8 (a) Predicted profile of polymeric sheet thickness at different temperatures from the center of the mold to the clamp. (b) Predicted profile of the polymer molecular chain equivalent stretch at different temperatures from the center of the mold to the clamp. (c) Predicted profile of the polymer molecular chain equivalent stretch at different temperatures from the center of the mold to the clamp. [Reprint with permission from [23]. Copyright © 2007 Wiley Periodicals, Inc. All rights reserved].

sheet material, which prevents the sheet from stretching freely. The sheet contact and mold surface changed the temperature at the mold center and spread to the mold wall. The effects of the contact force, temperature deviation, and applied pressure resulted in the nonlinear decrease of the sheet thickness towards the mold wall. At the edge between the mold base and mold wall, the applied pressure pushed the soft polymer to the edge. Due to the applied force, the pressure pushed the hot-sheet material. The system results in stabilization and a slight increase in the sheet thickness in that region.

On the mold wall, the thickness of the sheet shows a dramatic decrease due to the last part of the sheet being in contact with the mold surface. On the mold wall surface, the sheet stretched freely under the applied pressure without any contact with the mold surface. This large variation in the sheet thickness on the mold wall surface induced an important orientation of the polymer chains. Figure 5.8(b) shows equivalent molecular chain stretching.

An increase in temperature results in a slight reduction in the sheet thickness on the mold wall. This leads to a significant increase in the orientation of the polymer chains that region. The molecular orientation is dependent on temperature. Therefore, the orientation may not be as important as seen in Figure 5.8(c). In fact, Figures 5.8(a–c) do not account consider the temperature dependence of the chain equivalent stretch rate.

The high molecular orientation induces an important contribution of network resistance to the mechanical properties of the final product [22]. Figure 5.8(c) shows that the network resistance is maximum in the mold wall region due to the high orientation of the polymer chains in that region. The variation in the network resistance between the mold center and mold sidewall is due to the contact of the polymer sheet with the mold base surface. Figure 5.7 shows the deformation acting on the network resistance represents the molecular network orientation and the molecular relaxation processes.

Several variables that can affect the properties and performance of sheet materials exist. These variables included sheet thickness, sheet pigmentation, sheet size, and temperature uniformity. In the solid state, crystalline materials exhibit enhanced malleability and reduced deformation resistance. This property diminishes the probability of sagging or distortion during thermoforming procedures [23, 24].

Figure 5.9(a) shows that as the bubble height increases, the thickness decreases at the center and the bottom and top of the side increase. A larger bubble height means more deformation and less material in the center. The bubble height increases to a certain value and exhibits a more uniform thickness distribution along the symmetry line. Figure 5.9(b) shows that the thickness distribution with the perfect slip condition is significantly different from that with frictional attractiveness, particularly in the side top and corners. The higher thickness in the top and center areas and the smaller thickness in the bottom and corners may occur because of the faster movement of more material to the top areas resulting from a lower coefficient of friction. The initial contact area with the sheet is the corner between the top and back areas, and the final contact occurs at the corner of the bottom area. The use of a larger coefficient of friction resulted in an increased average wall thickness.

A more uniform thickness distribution for the coefficient of friction (COF) must be in the range of 0.05–0.2. Overall, the frictional properties between the sheet and the mold exert a significant influence on the average wall thickness and its distribution. The effects of other process variables, such as sheet, mold temperature, and thickness, may influence the COF. The transfer of material from the sheet to the mold surface and the oxidation of the mold surface may change the COF over time.

Figure 5.9(c) shows that the thickness distribution of the top and back regions shows a similar trend, where the part thickness decreases as the temperature increases. The influence of the biaxial elongation viscosity and bubble shape on the thickness distribution may be consistent. The primary influences of increasing the sheet temperature are a decrease in the uniaxial

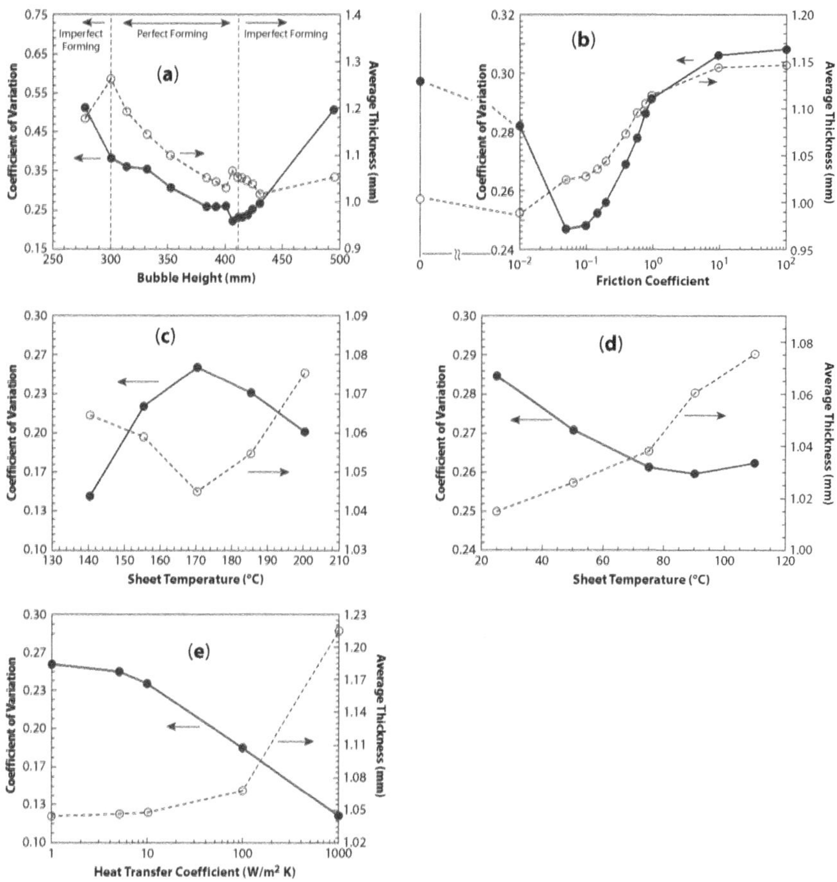

Figure 5.9 (a) Coefficient of variation and average thickness vs. bubble height; (b) coefficient of variation and average thickness vs. coefficient of friction; (c) coefficient of variation average thickness vs. sheet temperature; (d) coefficient of variation and average thickness vs. mold temperature; (e) coefficient of variation and average thickness vs. heat transfer coefficient (HTC). Reproduced with permission from [25].

viscosity and strain-hardening character, and an increase in the COF. The effect of temperature depends on the elongational viscosity.

The effect of mold temperature on the thickness distribution is shown in Figure 5.9(d). The effect of mold temperature on the thickness distribution is similar to that of the COF. The optimum mold temperature provided a uniform distribution consistent with the COF. As the mold temperature increases, the thickness also increases because of the high COF. More material is retained in the center of the part at the expense of other regions.

The change in the thermal conductivity of the polymer has little effect on the thickness distribution. The sheet and mold temperatures plays a less important role in determining the uniform wall thickness distribution. However, higher sheet and mold temperatures result in fewer special defects, such as webbing. Figure 5.9.(e) shows the effect of the heat transfer between the sheet and air on the thickness distribution. As the heat transfer coefficient (HTC) increases, the sheet temperature decreases. The rheological properties and COF change. As the HTC increases, this results in a more uniform thickness distribution and a larger average part thickness. This is due to the higher uniaxial elongation viscosity and strain hardening. The thermoforming process with forced cooling provided a more uniform wall-thickness distribution.

Three primary factors contribute to the local deformation of the sheet during thermoforming. These factors include the stress–strain response of the sheet polymer when exposed to specific levels of biaxial tensile strain, strain rate, and temperature. Additionally, the frictional properties of the plug and sheet materials when in contact play a role in reducing the local deformation in the areas of contact, while introducing additional material through slip in other areas. Furthermore, the thermal properties of the materials influence heat transfer and result in reductions in the local temperature, causing the sheet to stiffen because of changes in its stress/strain response and lowering its coefficients of friction. These effects are interconnected, conflicting, and challenging to isolate in practical applications, and lower the coefficients of friction [25].

5.17 Thermal History

Before thermoforming, the extruded sheet material already imparts a particular thermal history. Therefore, the thermal history dictates the thermoformability of the sheet material and the properties of the final formed part. The crystallinity and morphology of the sheet are strongly influenced by the cooling rate used during extrusion. Therefore, the quenching of the sheet material during extrusion results in a mostly crystalline structure, based on the cooling temperature. After the die exits, the extruded material undergoes rapid quenching and the sheet material passes through a series of highly polished chill-rollers to maintain the mechanical properties of the material.

A highly crystalline sheet material enhances the modulus; therefore, the strength and self-supporting properties help to minimize sagging of the sheet material during thermoforming. Hence, the strength and rigidity of the final product are improved. In glassy polymers, the presence of small spherulites improves both the clarity and impact properties of the extruded sheet and thermoformed part.

The choice of chill-roll temperature is important during sheet extrusion. The high chill temperature promotes slow cooling rates with concurrent increases in sheet crystallinity. Therefore, a substantial relaxation of the residual sheet stress arises from the extrusion process. However, low cooling rates also produce large spherulitic structures that exhibit inherently low impact strength and poor clarity. Low chill-roll temperatures (i.e., high cooling rates) are accompanied by the development of a delicate spherulitic morphology, which ultimately results in lower crystallinity.

This biaxially orientation results in the stretching of the morphological structure in both directions, resulting in a large spherulitic morphology that is aligned and becomes smaller. Stretching on both sides improves the clarity because of the small spherulitic structures and the improvement in the high tensile strength in the traverse direction. The biaxially oriented sheet or film on thermoforming under stress and strain during formation improves the product quality with high mechanical properties.

Figures 5.10(a,b) show the ultimate tensile strength and modulus of elasticity at a strain rate of $0.01s^{-1}$ at temperatures below 115 °C.

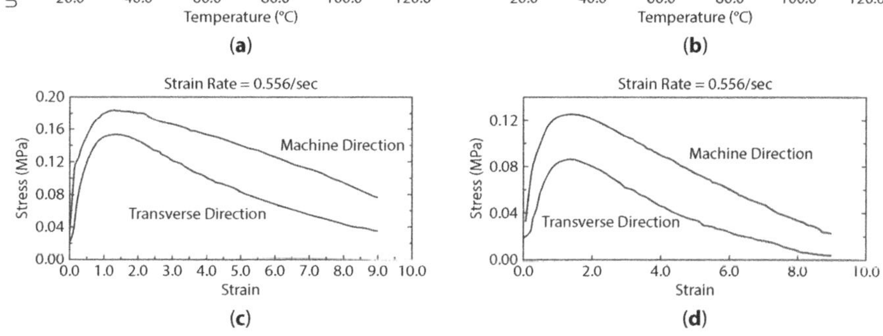

Figure 5.10 (a) Variation in ultimate tensile strength with temperature. (b) Variation in the modulus of elasticity with temperature (c) Engineering stress/engineering strain curve at 130 °C. (d) Engineering stress/engineering strain curve at 140 °C. [Reprint with permission from [26]. Copyright © 1997 Society of Plastics Engineers. All rights reserved].

In thermoforming, as the sheet material approaches the softening temperature, the strength and modulus decrease as the difference between the machine and transverse direction diminishes due to the relaxation within the material. Typical engineering stress–strain curves at 130 °C and 140 °C are shown in Figures 5.10(c,d). Thermoforming at high strain rates and differences in the machine and transverse directions are observed at temperatures close to the softening point. However, these differences are negligible in the thermoforming window [26, 27].

5.18 Pre-Drying

Pre-drying is a necessary step for certain materials such as nylon and polycarbonate, which have a propensity to absorb moisture. The presence of moisture in these materials can lead to various complications during the forming process, including blistering, reduced mechanical properties, and an inferior surface finish.

The main objective of pre-drying is to eliminate absorbed moisture from the sheet materials. After pre-drying, it is essential to safeguard the materials for the subsequent thermoforming process. Drying methods, such as forced circulation drying or vacuum ovens, are commonly used for this purpose. The duration of the drying process depended on the specific material and thickness. To enhance the efficiency, it is advisable to minimize the overall cycle time by pre-drying the sheet separately and transferring it directly to the forming oven [28].

5.19 Plug Movement

As the plug is inserted into the sheet, the exerted force increases steadily in a linear manner until it reached a certain point, referred to as discontinuity. At this point, the rate of force increase diminishes, but the force continues to rise until it reaches a final peak towards the end of the plug's travel [29].

The rate at which the plug or mold moves during the sheet forming process, known as "plugging," is a critical factor in thermoforming processes. The speed at which the plug moves significantly affects the physical properties of the formed products. Alterations in plug speed can affect the distribution of material thickness at the interface between the plug and sheet, as well as near the entrance of the cavity, particularly at the lip of the container or receptacle being formed.

Insufficient formation in thermoforming processes can be attributed to the combination of low plug speed and low plug temperature. However, the utilization of plug assist, particularly when increasing the plug speed, has been shown to enhance material distribution. This contrasts with vacuum-only formation. The increased plug velocity leads to an improved distribution of wall thickness distribution, which can be attributed to the elastic deformation of the plastic sheet during the thermoforming process [30].

5.20 Plug Speed

When the speed at which the plug is inserted into the sheet starts to thin, the forces experienced during the entire cycle also increased. Moreover, higher plug speeds led to steeper initial and secondary gradients of the force response curve, and the peak force is reached earlier. Similar to the impact of the sheet temperature, even minor adjustments in plug speed can have significant effects on the force response. The plug speed determines the rate at which the sheet is deformed during plugging. Therefore, increasing the plug speed is equivalent to increasing the strain rate. Consequently, it is anticipated that higher rates of deformation resulting from an increased plug speed will lead to higher plug forces. Considerable changes in plug speed can result in the following effects:

- A substantial decrease in plug speed can cause the sheet to retain more material in the base or central area of the container. This occurs because a slower plug speed restricts the free flow of the material, leading to its accumulation near the base.
- Conversely, slower plug speeds can result in increased thinning of material near the lip or rim of the receptacle. This is because of the stretching and thinning of the material as it flows into the mold cavity at a slower pace.
- Changes in plug speed can lead to a non-uniform distribution of material thickness throughout the formed part. Regions with greater material accumulation may have thicker walls, whereas those areas near the lip may have thinner walls.

5.21 Sheet Temperature

Temperature plays a significant role in the properties of polymer materials, particularly in thermoforming processes. In the sheet material, the surface may attain high temperatures, while the core remains relatively cooler, potentially leading to blistering or other undesirable effects.

The temperature of the sheet material is a vital factor in the thermoforming process because it significantly affects the flow and deformation characteristics of the material during forming. Each thermoplastic material has its own specific temperature range that is most suitable for thermoforming. To achieve uniform forming and consistent part dimensions, it is necessary to evenly heat the sheet material across its entire surface.

Figure 5.11 shows the temperature distribution at the center of the sheet and a constant temperature gradient across the thickness using numerical modeling during heating. The temperature decreases as soon as heating is switched off. The best method to ensure uniform heating is to heat the sheet from both sides. Ideally, this process is performed in a closed air-circulating oven.

The choice of heating method, such as infrared heaters, radiant heaters, hot air ovens, or contact heating, depends on factors such as material type, sheet thickness, and production setup. The optimal temperature range for thermoforming varies depending on the thermoplastic material used. It should be sufficiently high to render the material pliable, but not high enough as to cause degradation or overheating.

The temperature of the sheet material is the primary factor that determines its internal energy. The temperature has a significant impact on the internal energy and subsequent mechanical characteristics of the material. Specifically, as the temperature increases, the internal energy of the

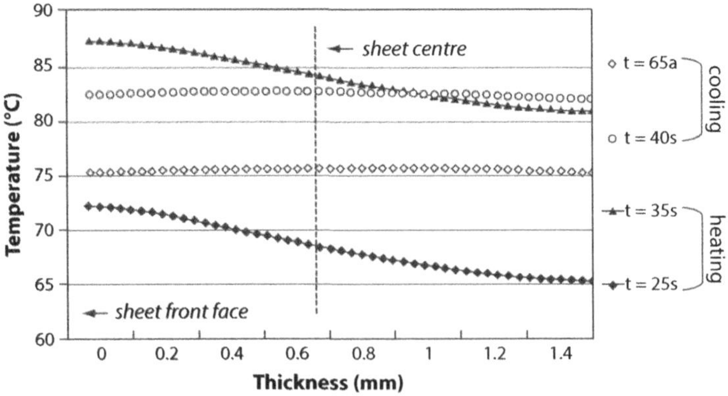

Figure 5.11 Numerical temperature distribution through the thickness. Reproduced with permission [31], © 2001, Elsevier Science. All rights reserved.

thermoplastic sheet increases, resulting in a transition from the glassy state to the rubbery state.

Precise control of the sheet temperature is essential for achieving the desired part quality, including consistent wall thickness, accurate dimensions, and a smooth surface finish. It is important to note that the material is heated to a state of pliability but not to the point of melting.

5.22 Mold Temperature

The temperature of the mold is a critical factor in the thermoforming process because it directly influences the behavior of the heated thermoplastic material during shaping. It is imperative to regulate the mold temperature effectively to achieve consistent and high-quality shaping outcomes. The mold temperature affects the rate of heat transfer to the sheet, which subsequently affects the flow, deformation, and crystallization behavior of the material. Furthermore, the mold temperature affects the cooling rate of the shaped part, which can have implications for the cycle times and production efficiency.

Ensuring a consistent mold temperature is essential for attaining uniform shaping results and a consistent part quality. Variations in the mold temperature can result in discrepancies in the wall thickness, surface finish, and dimensional accuracy of the shaped parts. The choice of mold material also plays a role in determining the appropriate mold temperature because different materials possess specific thermal properties that influence heat transfer.

Controlling the mold temperature can also affect the surface finish of the shaped part, preventing surface defects and undesired marks. Proper optimization of the mold temperature can lead to reduced cycle times, enhancing production efficiency. It is necessary to employ temperature sensors and controllers to monitor and adjust the mold temperature, enabling process control and optimization. However, maintaining a uniform mold temperature can be challenging, particularly for large or intricate molds.

To maintain a consistent mold temperature, it is necessary to implement temperature control measures such as internal cooling channels. Aluminum is often chosen as the material for molds because of its high coefficient of thermal conductivity, which allows for consistent cooling cycle times throughout the production of the components. However, it is important to account for the shrinkage that occurs when casting molds with aluminum. Therefore a typical tolerance must be added.

5.23 Forming Temperature

The term "forming temperature" in thermoforming using polymeric sheets refers to the precise temperature at which a thermoplastic sheet reaches a malleable state, allowing it to be shaped into the desired form. It is of utmost importance to carefully manage this temperature to achieve optimal material behavior during the forming process. Insufficient malleability resulting from low temperatures can lead to incomplete shaping or excessive stress, whereas excessively high temperatures can cause the material to become excessively soft or even melt, resulting in a substandard surface finish or loss of intricate details. The required forming temperature varies depending on factors such as the type and thickness of the thermoplastic material and the complexity of the part being formed. Thermoforming machines are equipped with heating elements and temperature control systems to regulate the forming temperature. It is necessary to conduct tests and experiments to determine the ideal forming temperature for each material and product configuration to ensure consistent and high-quality outcomes.

The polymer used in the thermoforming process should exhibit a suitable combination of viscous and elastic properties at the forming temperature. It should exhibit enough viscosity to flow and accurately replicate the intricate details of the mold, while also possessing enough elasticity to withstand the stresses encountered during biaxial deformation and retain its structural integrity.

5.24 Wall Thickness Distribution

The measurement of the product output quality is determined by the distribution of its wall thickness. Due to the nature of the thermoforming process, achieving a uniform wall thickness in a specific part is nearly impossible despite being desirable in most cases. The variation in the thickness at the sidewall of the formed films increases as the process parameters, such as sheet temperature, plug temperature, mold temperature, plug displacement, and forming air pressure, increase. However, the opposite effect is observed with plug speed [32].

In addition to process parameters, design parameters also influence the quality of the product output. These include plug design, mold design, depth of draw, plug material, mold material, sheet material, and sheet thickness [33, 34].

The plot presented in Figure 5.12 illustrates the distribution of the wall thickness for the production container. As is common with thermoformed

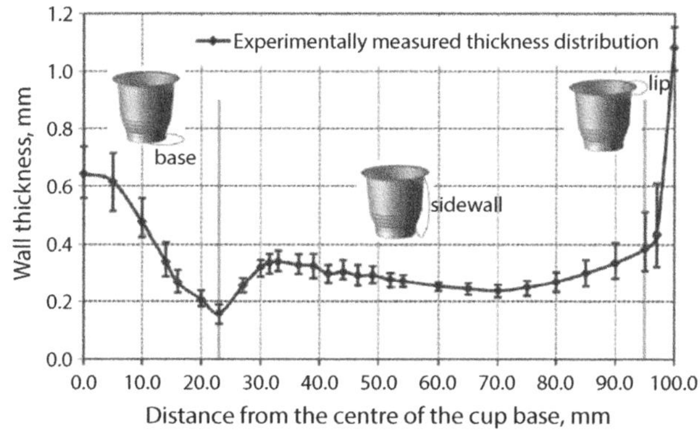

Figure 5.12 Wall thickness measurement for the product. [Reprint with permission from [35]. Copyright © 2013 Elsevier B.V. All rights reserved].

products, there are significant variations in the thickness across different locations. Notably, the thinnest part of the container is found in the bottom corner, measuring less than 0.2 mm. This is in contrast with the measurements taken from the center of the base of the container along a vertical profile terminating at the flat lip of the receptacle, which indicates a much greater thickness (>0.6 mm) at the base center. Throughout much of the sidewall, the material distribution is relatively balanced, with thickness values slightly above and below 0.3 mm. The relative thicknesses are influenced by the plug, as previous research has emphasized the critical role of friction during contact between the plug and sheet [17, 35, 36]. Lower friction and increased slip promote the movement of the material from the base towards the corner and sidewall of the container [37]. However, this change must not be excessive because it can lead to quality issues in the final product due to an inappropriate thickness distribution. Understanding the behavior of a material under different conditions is crucial for accurately predicting the thickness distribution of thermoformed products [38]. The thickness variation reflects the inhomogeneity of strains on the sheet. The inhomogeneity may be due to non-isothermal conditions across the thickness and area [35].

5.25 Sheet Deformation

Three primary factors interact to determine the localized deformation of the sheet during thermoforming.

The following factors are of interest for sheet deformation:

1. Response of the sheet polymer to biaxial tensile strain, strain rate, and temperature, specifically in terms of stress and strain.
2. Frictional properties of plug and sheet materials upon contact. This contact reduces local deformation and introduces additional material through slip.
3. The thermal properties of the materials affect heat transfer and can lead to reductions in the local temperature.

These factors collectively contribute to the stiffening of the sheet due to changes in its stress–strain response, as well as a decrease in its coefficients of friction. It is important to note that these effects are interconnected, conflicting, and challenging to isolate in practical applications [17, 21, 39].

5.26 Heat Transfer

In the deformed sheet, there are a significant variation in the local strains and strain rates. Typically, the strain is predominantly unidirectional on the sidewall, whereas it is equally biaxial at the base. However, the situation becomes more complex due to the influence of local conductive and

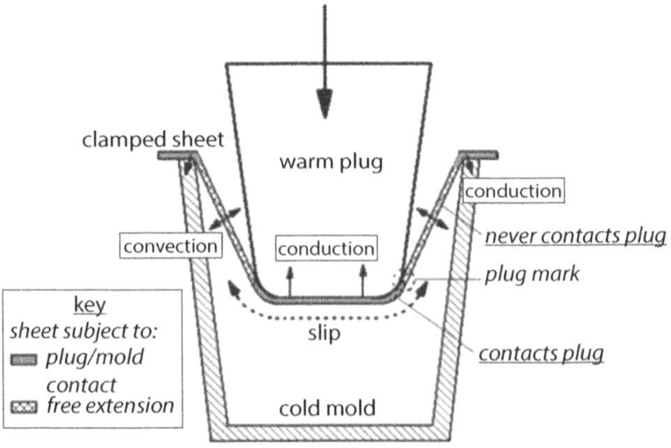

Figure 5.13 Heat transfer and slip during plug assist. [Reprint with permission from [17] Copyright © 2007 Society of Plastics Engineers. All rights reserved].

convective cooling effects, which results in a reduction in the initial sheet temperature. These phenomena are illustrated in Figure 5.13.

It is anticipated that losses resulting from conduction will be predominant in regions where the sheet is secured to the cooler mold and where it comes in contact with the warmer, although the lower temperature plug. Conversely, in areas that do not make contact with the plug, the combination of a rapid increase in the surface area of the sheet and a rapid decrease in its thickness is expected to result in substantial losses due to convection [40, 41].

5.27 Effects of Temperature Distribution

The temperature distribution affects the thickness distribution of the final products. It is difficult to decrease the temperature distribution of the surfaces and center of the sheet material due to its low thermal conductivity. The differences between the surface and center sheet temperatures are time dependent and very large after pre-heating. To decrease the temperature difference, infrared preheaters are used in the preheating process [42, 43]. The thinning or deformation of the sheet material is purely based on the temperature distribution. The hotter areas of the sheet deform first and yield thinner parts, whereas the cooler areas deform slower and yield thicker parts [44].

5.28 Effect of Drawing

During the drawing process, thermoforming causes stretching in different directions, resulting in uniform thickness distribution in the part. This uniformity allows for homogeneous deformation during thermoforming. Large-part thermoforming helps minimize the impact of local temperature variations on the sheet. Thermoforming molds are inexpensive because they only require a single-sided mold, with the other side being a sheet that does not come into contact with the shaping surface.

The drawing performance of a material in thermoforming is influenced by factors that control the draw limit and deformation uniformity. Optimum drawing performance can be achieved with a material that has a high rupture stress and is processed at a low extensional strain rate, resulting in a uniform profile. Viscoelastic functions are not utilized in industrial applications [45, 46].

5.29 Effect of Frictional Force

The friction increases drastically near the glass transition temperature. In the molding process, when the sheet is overheated or when the mold is too hot, frictional force is extremely high in the area where contact first occurs. As a result, the sheet loses its overall mobility and results in localized elongation at the side wall or when the friction force is as its minimum. The role of sliding friction between the sheet and mold in the molding operation remains constant. The reduced friction between the sheet and the positive mold produces a good thermoformed part. The friction between a sheet and mold is a function of temperature [47, 48].

5.30 Effect of Plug-Assist/Vacuum

During the process of thermoforming a polymer, mechanical properties, such as strength and stiffness are improved in the direction of orientation. This orientation is achieved by aligning the molecules parallel to the expected stress direction, often using plug-assist or vacuum drawing techniques. Consequently, the resulting material displays a notable degree of anisotropy in its physical characteristics, which can be attributed to the alignment of molecular chains [49, 50].

Figure 5.14 illustrates the thermal profile of a sheet subjected to heating. After a duration of 40 s, the temperature distribution in (a) reveals a significant variation of 20 °C between the central region and periphery

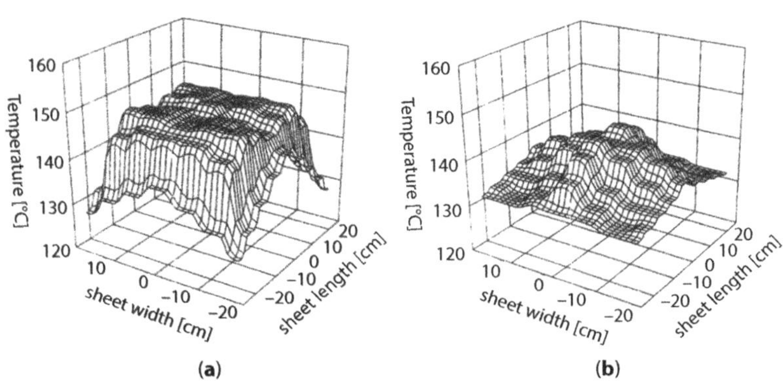

Figure 5.14 Measured temperature distribution of the sheet (by infrared thermometer) (a) 40 s of heating without plug assist (b) with plug assist, [Reprint with permission from [50]. Copyright © 2000 Society of Plastics Engineers. All rights reserved].

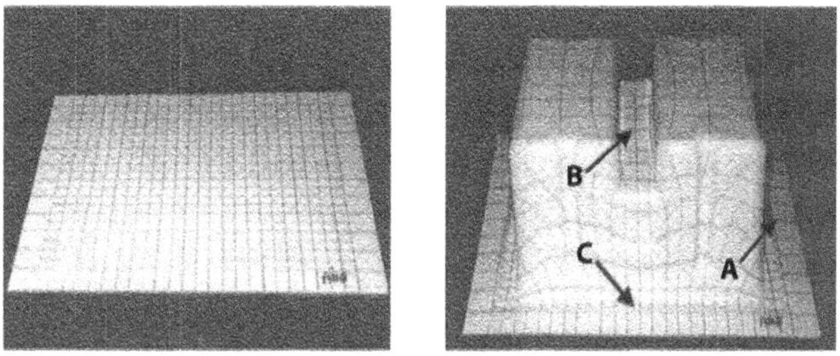

Figure 5.15 (a) ABS sheet and (b) thermoformed product. [Reprint with permission from [50].].

of the sheet. Notably, the edge portion experienced rapid cooling due to clamping, resulting in lower temperatures. In contrast, in (b), where plug assistance is used, the temperature distribution became relatively more uniform [50–52].

Figure 5.15(a) presents a grid annotated to represent the initial condition of the material before undergoing the thermoforming procedure. Additionally, the grid in Figure 5.15(b) highlights the fluctuations the thickness along the central path line (A, B, and C). It is evident that the speed at which the plug is inserted displays an inverse relationship with both dimensional and thickness variations, indicating that an increase in plug speed leads to a reduction in these variations. Furthermore, the temperature of the mold influences the stretching ratio of the end product and the rate at which the thickness of the sheet fluctuates [50].

5.31 Effect of Applied Pressure

The heat-softened sheet is then subjected to positive or negative pressure. Pressure is applied at different rates. The thermoforming speed determines the strain rate to which the sheet is subjected. Typical thermoforming strain rates are in the range of 0.1 s^{-1} to 10 s^{-1} [53–56].

Pressure is applied either to the upper face of the plastic sheet or by evacuation of the air between the plastic sheet and the mold. The behavior of polymeric materials is strongly affected by stress level, loading rate, and temperature. Appropriate control of the process strain rate and temperature leads to improved control of the material.

At the onset of the cycle, there is a temporary decrease in the pressure in the upper cavity as the plug descends. This decrease is a result of the volume expansion when the plug enters the lower cavity. However, this minor change quickly reversed as the blowing air enters the cavity. Although the air pressure increased steadily throughout the cycle, the rate of increase is relatively modest and occurrs within a short period. During the thermoforming process, the rate of pressure increase is significantly lower.

The rate at which pressure increases in a particular machine is influenced by a complex combination of machine-specific variables such as the characteristics of the air supply and the dimensions of the cavities and channels involved. It is reasonable to expect significant variations in these variables between different machines. The desired shape of a material can be achieved by applying pressure or loads to its surface, or by creating pressure differentials. In certain cases, the surface configuration can be maintained using a die or former to control the shape. Subsequently, the material is cooled and returned to its solid state while retaining the newly acquired form.

5.32 Effect of Heating

The sheet temperature is proportional to the heater temperature and heating time. Failure occurrs at lower temperatures due to crack initiation and growth at the corners of the parts. The normal yield stress is subjected to a uniaxial load, which shows that the craze region is occupied by drawn fibrils and voids. In glassy polymers, crazes initiate cracks that are responsible for overall brittle fracture. This indicates that a higher sheet temperature prevents cracking. However, the draw ratio increases with increasing temperature up to a certain point, beyond which the sheet does not have enough melt strength to hold its place.

The control of sheet reheat is complicated by the uncertainties associated with radiative and convective heating. This is further complicated by the high level of coupling between the individual sheets and heater zones. Sheet sag results in changes in the radiation view factors and sheet thinning, both of which directly affect the heating of the sheet material. The sheet temperature is proportional to the heating temperature and time. At lower temperatures, crack initiation and growth occur at the corners.

The failure mode occurring at lower temperatures was due to crack initiation and growth at the corners of the products. Normal stress yielding in sheets or films is subjected to uniaxial loading, and it has been shown that the craze region is occupied by drawn fibrils and voids. In a glassy polymer, crazes initiate cracks that are responsible for the overall brittle fracture at a 5% elongation level. Consequently, a higher sheet temperature is required to eliminate these cracks. The draw ratio increases with increasing sheet temperature up to the point where the sheet will not have sufficient melt strength to hold its place [57, 58].

5.33 Effect of Cooling

The viscoelasticity of polymers manifests as creep under constant stress, and creep recovery during stress removal. The desired shape is formed by imposing stress (creep) using air and/or a mechanically driven plug. A sufficient cooling time and removal of the part from the clamps release the imposed stress to zero, which is known as creep recovery. Low-creep recovery, i.e., high-plastic-strain polymers, can be unclamped from the thermoforming machine faster, reducing the number of cycles. A polymer with creep recovery would cause poor mold replication if unclamped sooner, necessitating longer cycling times.

Cooling plays a vital role in the thermoforming process, specifically in solidification and maintenance of the shape of the formed part. The primary objective of cooling is to solidify the thermoplastic material to its desired shape to ensure dimensional stability. Effective cooling methods, such as natural cooling, forced air cooling, water cooling, and cooling with chilled molds, are used to prevent the deformation or warping of the formed part. The rate at which cooling occurs is a critical parameter in thermoforming because both rapid and slow cooling can have drawbacks. Rapid cooling may not allow sufficient relaxation of internal stresses, whereas slow cooling can prolong the cycle times and reduce productivity. Thermoforming machines are equipped with cooling systems and temperature controls to regulate the cooling processes. After the initial cooling phase, secondary cooling or trimming can be performed to remove excess material and further refine the final shape. The cooling phase is essential for achieving high-quality final products, and it is necessary to optimize this process to ensure consistent quality and productivity, while minimizing defects and part rejection rates. The use of cooling lines, coolants, and compatible mold release agents is crucial for ensuring a smooth and efficient molding process [59].

Figure 5.16 displays light-microscopic images depicting the fine and smooth surfaces of 1.25 mm thick sheets at different cooling-roll temperatures, namely, 35, 40, 60, and 90 °C. The images reveal that different cooling roll temperatures have a significant effect on the extruded sheets.

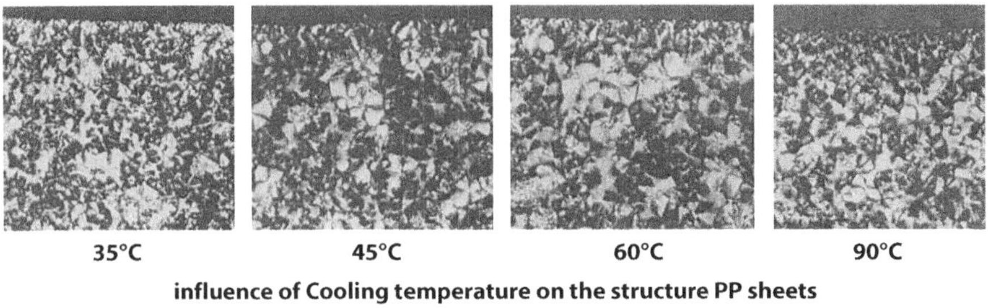

Figure 5.16 Effect of cooling temperature on the structure of PP sheets. [Reprint with permission from [60]. Copyright © 1982 John Wiley & Sons, Inc. All rights reserved].

The structural characteristics of the polypropylene (PP) sheets became apparent during the cooling process, as indicated by the stretching and drawing observed in the thermoformed products [60].

Rapid cooling and solidification are critical in thermoforming because they facilitate the easy removal of parts from the mold. This can be achieved through several factors. First, rapid cooling occurs when the heated sheet comes into contact with the cold mold surface, causing the material to cool and solidify rapidly. This rapid cooling helps to achieve a well-defined shape and dimensional stability of the part. Additionally, higher forming temperatures contribute to faster setup times. At elevated temperatures, the material becomes more pliable and easily conforms to the shape of the mold. Once the desired form is achieved, cooling the part quickly helps to retain its shape and prevent any deformation [61].

In the context of mold cooling, a decrease in the mold temperature results in a shorter cooling cycle. However, it is important to acknowledge that reducing the mold temperature also leads to an increase in internal stress within the formed part. It is observed that thinner regions of the part demonstrate a higher resistance to further deformation, suggesting that convection cooling plays a significant role in areas that have undergone significant deformation and do not directly come into contact with the plug. The initial thickness of the sheet is significantly reduced during contact with the plug, indicating that the sheet could deform through sliding across the contact surface. Consequently, the frictional properties of the materials involved are crucial for the successful execution of the process [17, 62].

The cost effectiveness of the molding process is primarily determined by the duration of the molding cycle, with the cooling phase being the most crucial step. The time spent in the cooling cycle directly affects the production rate of the parts. In modern industries, time and costs are closely intertwined because longer production times result in higher costs. Therefore, reducing the time required for cooling the parts would significantly increase the production rate and decrease the costs. Consequently, it is imperative to understand and optimize the heat-transfer process within a typical molding process [63–65].

The efficiency of the thermoforming mold is heavily influenced by the rate of heat exchange between the thermoformed plastic and the mold. Heat must be extracted from the plastic material until it reaches a stable state, allowing demolding. The time required to achieve stability is known as the cooling time. To facilitate optimal heat transfer between the melted plastic material and the mold, a well-designed cooling system is essential. Traditionally, this has been accomplished by incorporating multiple straight holes within the mold core and cavity, through which a cooling fluid (typically water) is circulated to dissipate excess heat from the molten plastic [66, 67].

5.34 Rate of Deforming

For thermoforming, the forming time is typically shorter than the heating and cooling times. The deformation rate is a critical parameter in thermoforming that significantly affects the quality and outcome of the process. This pertains to the velocity at which the heated thermoplastic sheet is elongated or pulled over the mold to achieve the desired shape. Several factors must be considered when determining the rate of deformation [68, 69].

- The behavior of different thermoplastic materials during the forming process varies at different rates due to their viscoelastic properties. Consequently, the rate of deformation should be meticulously selected based on the specific material used to ensure optimal formation without any defects.
- The forming temperature is closely related to the deformation rate. Higher temperatures make the material more pliable and easier to stretch, enabling a faster deformation rate.
- The complexity of the part being formed also influences the deformation rate. Parts with intricate shapes or deep draws may require a slower deformation rate to prevent material thinning or excessive stretching.

- Maintaining a consistent rate of deformation is of utmost importance to ensure uniform material distribution and thickness throughout the part. Variations in the deformation rate can result in uneven material distribution and defects in the final product.
- Thermoforming machines are equipped with mechanisms to control and adjust the rate of deformation, such as variable-speed drives, adjustable clamps, or programmable control systems.

The rate at which a material is formed during the thermoforming process has a significant impact on the overall cycle time. By carefully balancing the rate of deformation with other process parameters, it is possible to enhance productivity while maintaining the desired quality of the formed parts. Following the forming stage, it is necessary to allow sufficient time for the formed part to cool and solidify to prevent any potential warping or distortion. Therefore, it is crucial to control the cooling process. It is necessary for thermoforming processes to adjust the rate of deformation according to factors such as the material being used, the design of the part, and the desired quality of the final product. By conducting trials and experiments, they can determine the optimal rate of deformation that ensures efficient production and the production of high-quality parts. The rate of deformation is just one of several process parameters that are carefully adjusted to achieve the most favorable outcomes during the thermoforming process [69–71].

5.35 Rate of Sagging

The term "rate of sagging" in thermoforming refers to the degree of deformation or drooping that occurs in a heated plastic sheet prior to or during the forming process. This deformation is primarily caused by the weight of the sheet or the force of gravity. The rate of sagging is significant in thermoforming, because excessive sagging can result in various complications. These complications include inconsistent thickness of the formed part, incorrect dimensions of the part, and the development of wrinkles or folds on the surface of the plastic sheet [72, 73].

It is necessary to employ several strategies to mitigate the rate of sagging and achieve optimal outcomes in thermoforming. These strategies involve pre-stretching the plastic sheet before forming to diminish the tendency for sagging and ensure a more uniform material distribution. It is crucial to control the heating temperature and duration to bring the plastic sheet to the appropriate softening point without overheating, which can exacerbate the sagging. Support structures such as frames or grids, can be utilized to provide additional support to the heated plastic sheet during the heating and forming stages. Minimizing the time interval between heating and forming can also aid in reducing the sagging rate. Furthermore, the selection of a plastic material with suitable thermoforming characteristics can influence the sagging rate and overall formability [74, 75].

By effectively managing the sagging rate, it is necessary to enhance the efficiency and quality of the thermoforming process, resulting in consistent and accurate plastic parts. It is imperative to carefully consider the material properties, heating processes, and forming conditions to control sag and ensure successful thermoforming outcomes [76].

5.36 Effect of Air Temperature

During the forming process, the air temperature within the upper cavity experiences a significant increase compared to the surrounding ambient conditions. This can be attributed to convective heat transfer from the hot surfaces of both the plug and the polymer sheet. As the plug descends, the air temperature within the cavity initially decreases and then becomes variable as the blowing air enters the cavity. Subsequently, the temperature gradually increases and then gradually decreases as the overall cavity and formed part lose heat. At the end of the cycle, the air temperature within the cavity remains elevated compared to its initial state [77, 78].

5.37 Effect of Air Pressure

The role of air pressure in thermoforming, particularly in shaping heated plastic sheets, is of utmost importance. This pressure is applied to the sheet to induce conformity to the mold or tool, ultimately achieving the desired shape of the final product. The precise control of the forming air pressure is critical as it significantly impacts the quality and precision of the formed part [79, 80].

The influence of the forming air pressure on the thermoforming process can be understood from several key aspects.

- The amount and distribution of the forming air pressure directly affect the stretching and conformity of the plastic material to the mold. By properly controlling the air pressure, uniform material distribution across the mold can be ensured, resulting in a consistent wall thickness in the final product.

- Second, in certain thermoforming processes, plug assistance is utilized in conjunction with the forming air pressure. Plug assist is an additional tool that aids in the initial stretching of the heated sheet before applying air pressure. The combination of plug assist and forming air pressure allows the production of more complex shapes and ensures better material distribution.
- Furthermore, the level of forming air pressure affects the surface quality of the formed part. Excessive pressure may cause the material to stretch too thinly, resulting in a rough or textured surface. Conversely, insufficient pressure can lead to surface imperfections or wrinkles.
- The duration required for the plastic sheet to fully conform to the mold is influenced by the forming air pressure, which affects the cycle time. Higher pressures can expedite the forming process, reduce cycle times, and enhance production efficiency.
- Finally, the forming air pressure can affect the thickness of the formed part. A higher pressure may result in thinner walls, whereas a lower pressure may lead to thicker walls.

Determining the optimal air pressure for forming in thermoforming processes is a complex task that depends on several factors such as the type of plastic material, mold complexity, desired wall thickness, and specific thermoforming equipment. It is often necessary to conduct trials and experiments to determine the ideal combination of forming air pressure and other process parameters to consistently produce high-quality parts [81–83].

Maintaining precise control over the air pressure during the forming process is crucial for achieving the desired results and preventing defects or inconsistencies in the products. The rate at which the air pressure is applied during the blowing stage of thermoforming is particularly important. This rate determines the strain rate experienced by the plastic sheet, which refers to how quickly the material is stretched or deformed. The strain rate has a significant impact on the final properties and quality of the formed part [84, 85].

During the blowing stage, the heated plastic sheet is forced against or into a mold using various methods, including vacuum, pressure, and mechanical forming. Vacuum forming involves applying vacuum to the backside of the heated sheet, causing it to tightly conform to the mold due to the difference in air pressure. On the other hand, pressure forming utilizes compressed air to apply pressure to the front side of the heated sheet, forcing it to stretch and take the shape of the mold. Mechanical forming involves the use of devices such as plug assists or pressure boxes to mechanically push the heated sheet into the mold, aiding in the stretching and forming processes [86].

In thermoforming, metal molds are often preferred for long production runs or when higher pressures are required because of their durability and longevity, which makes them suitable for high-volume production. The control of the rate at which the forming air pressure is applied is crucial in vacuum forming, pressure forming, and mechanical forming to achieve consistent part quality and prevent defects. Proper control of the straining rate is necessary to avoid material thinning, nonuniform stretching, and other issues that could result in variations in the formed parts. It is necessary to carefully adjust the rate of applying the forming air pressure, as well as other process parameters such as heating temperature, cooling rate, and mold design, to ensure optimal results based on the specific requirements of the thermoforming application [87].

5.38 Effect of Crystallinity and Morphology

During the extrusion process, the crystallinity and morphology of the sheet are strongly influenced by the cooling rate used. Therefore, the quenching of the sheet material during extrusion results in a mostly crystalline structure, depending on the cooling temperature. After the die exits, the extruded material undergoes rapid quenching and the sheet material passes through a series of highly polished chill-rollers to maintain the mechanical properties of the material [88].

The choice of chill-roll temperature is important during sheet extrusion. A high chill temperature promoted slow cooling rates, leading to concurrent increases in sheet crystallinity. Therefore, substantial relaxation of the residual sheet stress arises from the extrusion process. However, low cooling rates also produce large spherulitic structures that exhibit inherently low impact strength and poor clarity. Low chill-roll temperatures (i.e., high cooling rates) are accompanied by the development of a delicate spherulitic morphology, which ultimately results in lower crystallinity (Figure 5.17).

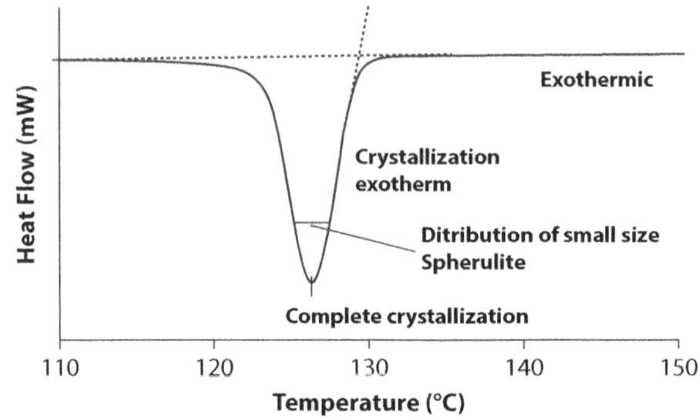

Figure 5.17 Illustration of a typical crystallization exotherm from the controlled cooling of polypropylene. [Reprint with permission from [89]. Copyright © 1998 Society of Plastics Engineers. All rights reserved].

The biaxial orientation results in the stretching of the morphological structure in both directions, resulting in a large spherulitic morphology that is aligned and becomes smaller. Stretching on both sides improves clarity due to the small spherulitic structures and improvement in high tensile strength in the traverse direction. The biaxially oriented sheet or film when subjected to stress and strain during thermoforming improves the product quality with high mechanical properties.

A highly crystalline sheet material enhances the modulus; therefore, the strength and self-supporting properties help to minimize sagging of the sheet material during thermoforming. Hence, the strength and rigidity of the final product are improved. In glassy polymers, the presence of small spherulites improves both the clarity and impact properties of the extruded sheet and thermoformed part.

During thermoforming, these materials experience heating and cooling effects once again. As the polymer is stretched after heating, the molecular chains can be aligned in the direction of stretching where a regular structure of the melt can be formed and the elasticity of the melt can be enhanced.

5.39 Processing Technology

Thermoforming is a multistage process that involves the following steps [90]: heating, forming, and cooling (Figure 5.18).

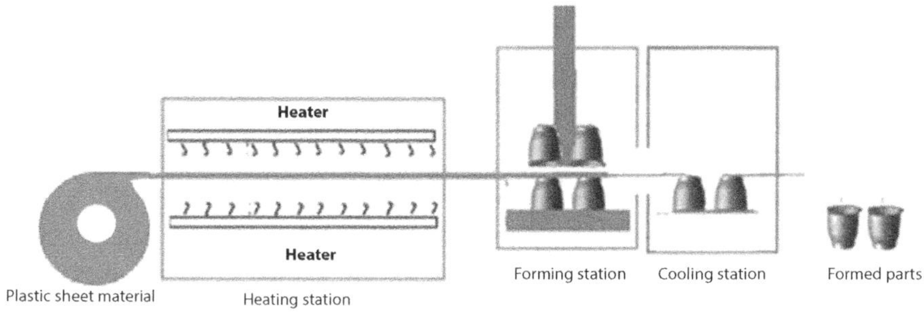

Figure 5.18 Schematic diagram of thermoforming technique.

In certain cases, additional punching or trimming may be required to remove the excess material and finalize the shape of the formed product.

The thermoforming process is highly advantageous for efficient and cost-effective production of high-quality products using thermoplastic sheet materials. This process involves four stages: heating, softening, forming, and cooling of the materials, allowing for the creation of intricate shapes and custom products to meet specific requirements [91].

5.39.1 Heating Stage

The first stage involves heating the thermoplastic sheet to its processing temperature. This temperature is within the thermoforming window where the material undergoes a transition from a glassy state to a rubbery state. In this state, the material becomes more pliable, can undergo significant deformation without fracturing and is easily deformable [92].

The quality of thermoformed parts essentially depends on the standard of extruded sheet feedstock used. The extruded sheets impart a particular thermal history. Secondary heating during thermoforming determines the thermoformability of the material and the properties of the final formed part. Thicker sheets require a greater amount of heat energy to soften and a longer duration to attain the desired formation temperature for the core layer [93].

5.39.2 Sheet Heating

Sheet heating involves the selection of heaters, the machine, and parameter settings, taking into consideration the material to be heated. Heating is performed using two heater banks that heat the sheet from both sides.

Therefore, the temperature gradient across the sheet thickness should be minimized. During thermoforming, the sheet material is heated for a specified time. Once it exits the oven, it moves to a forming station. The forming process depends on the temperature of the heated sheet and, consequently, on the sheet reheating phase.

The glass transition temperature is fundamental in thermoforming, as it determines the temperature range within which the material can be effectively shaped without causing permanent damage or deformation. Once the sheet reaches the appropriate temperature within the thermoforming window, it becomes soft and flexible [94].

5.39.3 Forming Stage

The sheet is placed over or within a mold cavity, and with the application of pressure, it stretches and conforms to the shape of the mold, taking the desired form. In the forming stage, a hot-sheet material with a converging flow pattern on a mold and higher stress can be developed.

5.39.4 Draw Ratio

By increasing the draw ratio of a hot-sheet material with a converging flow pattern on a mold, a higher stress can be developed in the hot-sheet material. The high stress developed on the hot polymer sheet in the forming stage results in a higher melt strength. For hot-sheet materials with low-melt regions, the melt strength increases more rapidly at higher draw ratios [95].

5.39.5 Cooling Stage

Cooling is the subsequent stage of the process. After the forming process, the shaped material is cooled. This cooling process solidifies the plastic, effectively locking it into the desired shape and providing the necessary mechanical strength and stability. Before removing the applied pressure, the freezing step is performed to cool the polymer to a temperature below its Tg in order to preserve the desired part geometry and molecular structure. The behavior of polymeric materials during the cooling stage is strongly affected by the stress level. With the cooling mechanism, the material solidifies and assumes its final shape. Blowing fans on the top of the machine, either water spray or chilled air accelerates cooling [96].

5.40 Thermoforming—Processing of Thermoplastic Sheet Material

Thermoforming techniques involve subjecting a material to substantial deformation to achieve the desired stretching and shaping outcomes. Materials that possess high elongation properties are well suited for thermoforming processes because of their ability to undergo substantial stretching without encountering excessive strain or tearing. This characteristic allows these materials to deform and adapt to intricate shapes [97].

Thermoforming is a manufacturing process that involves the manipulation of a thermoplastic sheet through the application of heat and force to achieve the desired shape. The process consists of three main steps: heating, forming, and cooling [98].

- During the heating stage, the thermoplastic sheet is subjected to elevated temperatures in an oven or tunnel equipped with heaters. This heating process renders the material soft and malleable, facilitating subsequent shaping.
- The forming stage involves the application of forces to the heated and pliable sheet to mold it according to the contours of a specific mold. This can be accomplished through various methods, including vacuum, pressure, or drape forming.
- Following the forming stage, the thermoplastic sheet is allowed to cool while in contact with the mold. The cooling solidifies the material and ensures that it retains its desired shape.

Figures 5.19(a,b) show schematic diagrams of thermoforming with positive and negative molds.

Various techniques can be used in the thermoforming process, including positive and negative mold vacuum forming, bubble and plug assist, and snap-back. These techniques provide additional support, promote the uniform distribution of materials, and ensure precise dimensions of the final product.

Thermoforming encompasses a range of processes such as drape, vacuum, and pressure forming. The selection of a particular process is based on specific requirements of the manufactured product. The choice of a specific thermoforming technique depends on factors such as the shape and complexity of the product, material properties, production output, and the required tolerances. By utilizing different thermoforming methods, it is

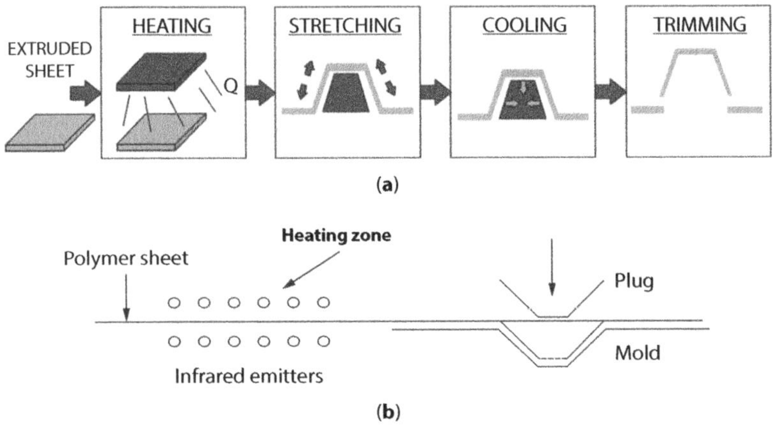

Figure 5.19 (a) Basic operations in thermoforming process - male mould. Reprint with permission from [99]. © 2000, Elsevier. All rights reserved. (b) Plug assist thermoforming - female mould. Reproduced with permission from [100]. © 2003, Elsevier Ltd.

possible to produce a diverse array of products with various shapes and properties [101].

Sheet forming can be accomplished using two distinct techniques that are selected based on the thickness of the sheet. One-step forming is used for thin-gauge sheets and is considered straightforward due to the malleability of the material, allowing for shaping with basic tools. This technique is suitable for parts with low complexity, and the uniformity of wall thickness is not a significant concern [102].

Multiple-steps forming is utilized for thicker gauge sheets. This technique involves stretching the sheet using plug assist or a billow to evenly distribute the material and minimize thinning. Subsequently, the sheet is shaped against a mold using a punch-and-die system to achieve the desired shape with precision and uniformity. Multiple-step forming facilitates the attainment of greater uniformity in wall thickness and enables the production of parts with intricate shapes [103].

The selection between one-step and multiple-step forming is contingent upon the requirements of the part, the thickness of the sheet, and the desired uniformity of wall thickness. Thicker gauge sheets typically require multiple steps to ensure the production of high-quality and uniform parts. In multiple-step sheet metal forming processes, a combination of mechanical pre-stretching and vacuum or pressure forming is commonly used to achieve complex shapes and enhance the uniformity of wall thickness. This combination optimizes material properties and improves the mechanical characteristics of the formed products. By integrating these methods, it is necessary to achieve precision and quality in formed parts, particularly for thicker gauge sheets. This process has applications in various industries, including automotive, aerospace, and consumer goods manufacturing [104, 105].

5.41 Methods of Thermoforming

Depending on the method used, forming can be categorized into two types: low-pressure air forming (vacuum forming) and high-pressure air forming (bubble and drape assisted vacuum forming) [106]. The stretching process during forming can occur in two different parts of the mold: negative mold forming, where stretching occurs in the mold cavity, and positive mold stretching, where forming is done in the positive mold [107].

5.42 Low-Pressure Forming Technology

Low-pressure forming technology utilizes negative atmospheric pressure to shape a preheated sheet of thermoplastic material against a cold mold surface. In the vacuum-forming process, the gaseous medium within the enclosed working cavity is removed, resulting in a pressure differential between the external and internal surfaces of the flat blank. Consequently, the blank undergoes deformation due to external pressure and makes contact with the surfaces of the shaping tool used to shape the final product.

5.42.1 Basic Vacuum Forming

Basic vacuum forming (Figure 5.20) is a well-established thermoforming technique. Once the plastic sheet reaches the desired temperature, it is positioned over the negative mold. A negative mold is typically made of aluminum or other materials with effective heat-conduction properties. Subsequently, a vacuum is applied between the negative mold and the plastic sheet. This vacuum pressure pulls the heated sheet tightly against the contours of the mold, shaping it into the desired form. After the sheet

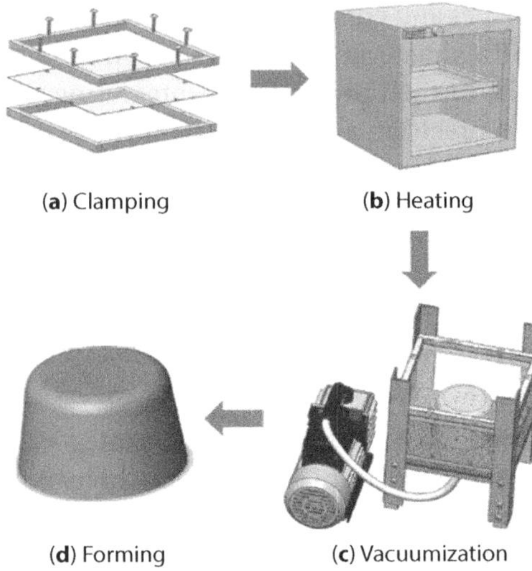

Figure 5.20 Schematic representation of vacuum forming. [Reprint with permission from [109]. Copyright © 2015 American Chemical Society].

has taken on the shape of the mold, it is cooled to solidify the plastic into a formed shape. Finally, the vacuum is released, and the formed part is removed from the mold. Any excess material or flash may be trimmed during the demolding process [108].

Although basic vacuum forming is a well-established technique, advancements in thermoforming technology have resulted in the development of other variations, such as pressure forming and twin-sheet forming. These variations offer enhanced capabilities to achieve finer details and more intricate shapes. The choice of thermoforming technique depends on the specific requirements of the product and the desired level of complexity and precision.

This process can be summarized as follows:

1. Plastic Sheet Heating: The plastic sheet is subjected to a controlled heating process, typically within an oven or heating station, until it reaches a malleable or pliable state.
2. Placement of the Plastic Sheet: Once the desired temperature is attained, the softened plastic sheet was carefully positioned over the mold, which is commonly constructed from durable materials such as aluminum.

3. Application of Vacuum: A vacuum is created between the mold and heated plastic sheet, generating a pressure differential that causes the sheet to be drawn tightly against the contours of the mold, assuming its shape.
4. Cooling and Solidification: Following the successful formation of the desired shape, the plastic sheet is cooled to facilitate the solidification of the plastic material, ensuring the retention of the desired form.
5. Demolding: Once the cooling process is completed, the vacuum is released, allowing for the removal of the formed part from the mold. Any excess material is trimmed as necessary.

Vacuum forming is a widely used and cost-effective method for manufacturing plastic components. It offers a moderate level of precision and surface-finish capabilities. One notable advantage is the ability to produce parts with varying thicknesses. However, it is important to note that certain materials may not be suitable for vacuum-forming. In large-scale production, tools made of aluminum are commonly used due to their thermal conductivity and durability.

Once positioned, a vacuum is applied to the backside of the mold, causing the sheet to adhere tightly to the mold's surface. This vacuum effectively removed any trapped air between the sheet and the mold, ensuring accurate shaping.

Additionally, the removal of air allows the plastic sheet to be evenly drawn across the mold surface, maintaining a uniform wall thickness throughout the formed part. This is crucial for achieving complex shapes and preventing uneven thicknesses in specific areas. The absence or improper application of vacuum can cause significant issues during the thermoforming process. Insufficient or no vacuum can result in incomplete shaping and inadequate definition of the part.

5.43 Plug-Assist Thermoforming

Plug-assist thermoforming is a technique (Figure 5.21) that involves using bubbles and plug assists. It entails using a bubble to push the heated sheet into the mold cavity or employing a plug to aid in the forming process. This method ensures a more uniform material distribution and reduces thinning in specific areas.

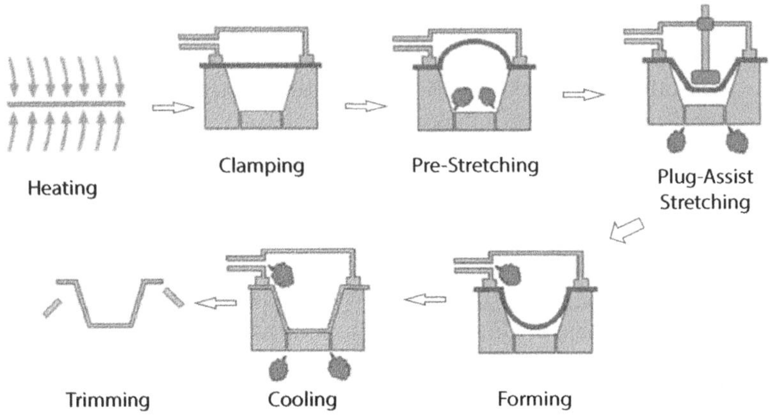

Figure 5.21 Plug-assist thermoforming. [Reprint with permission from [110] Copyright © 2014, Springer-Verlag London. All rights reserved].

5.43.1 Plug-Assist Vacuum Forming

Thermoformability explores different thermoforming techniques such as positive and negative mold vacuum forming. In positive vacuum forming, a sheet of material is heated and subsequently subjected to vacuum pressure, causing it to conform to the shape of a positive mold. Conversely, negative vacuum forming involves heating a sheet and pulling it into a negative mold using vacuum pressure to achieve the desired shape [111].

Plug-assisted thermoforming (Figure 5.22) is a mechanothermal shaping method used to produce hollow components from flat (sheet or film) polymer components. When using this method, the thermoplastic sheets are heated to a temperature corresponding to the highly elastic state of atmospheric thermoplastics or to the temperature corresponding to the start of melting of the crystalline phase of the thermoplastics with different degrees of crystallinity. The shaping of the completed components occurrs under the effect of a gradually moving shaping tool (plug). Therefore, the strains in the blank during its shaping into the component are the strains of pure shear and, usually, their values are high, whereas the shaping time is relatively short.

Plug-assisted vacuum thermoforming is a widely utilized technique in the manufacture of food packaging items such as pots, tubs, and trays. This method involves the use of a mechanical plug to stretch a heated thermoplastic sheet before completing the forming process with the application of air pressure. Its popularity stems from its ability to produce parts with

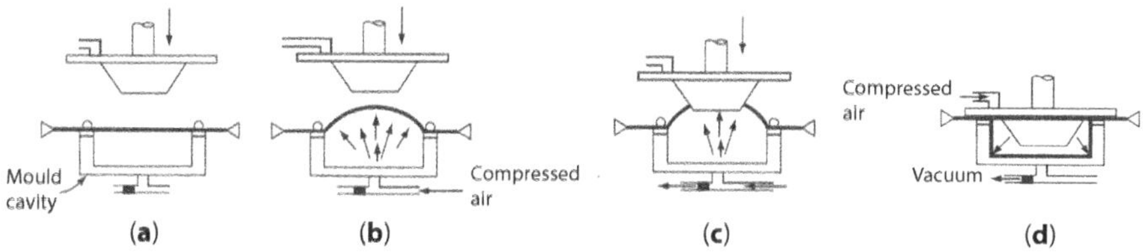

Figure 5.22 Schematic representation of plug-assist vacuum forming. (a) clamping (b) pre-forming (c) pressing with mold and (d) vacuum forming. [Reprint with permission from [112] © 2002, Elsevier Science].

consistent wall thicknesses, making it suitable for applications that require precise dimensions.

Plug-assisted vacuum thermoforming is a technique commonly used for manufacturing hollow polymeric materials. This method involves heating a polymeric sheet to a temperature in the range Tg < T < Tm. Evaluating this property allows for the estimation of the physical characteristics of polymeric products, such as their strength and toughness [113, 114].

One of the processes involves initially creating free blow bubbles, which lead to a relatively uniform deformation of the entire plastic sheet and an increase in its dimensions (although the surface area is always smaller than that of the mold cavity). This is followed by vacuum-assisted drape forming, where the plastic sheet is forced against the positive mold at room temperature or higher. The bottom of the finished part corresponds to the part of the positive mold that directly impacts the plastic. Subsequently, the plastic undergoes stretching due to the impact of low-pressure air (vacuum forming), resulting in the formation of the side walls of the finished part, which undergo free stretching [115].

5.44 Pressure Forming

Pressure forming, also referred to as pressure thermoforming, is a manufacturing technique that employs positive air pressure to shape a plastic sheet against a mold. In Figure 5.23, the diagram shows the applied pressure on the sheet material during forming. In contrast to vacuum forming, which uses vacuum to draw the plastic sheet onto the mold, pressure forming applies air pressure to forcefully press the sheet against the mold, resulting in the desired shape. The process involves several stages, including heating the plastic sheet until it becomes pliable, positioning the sheet

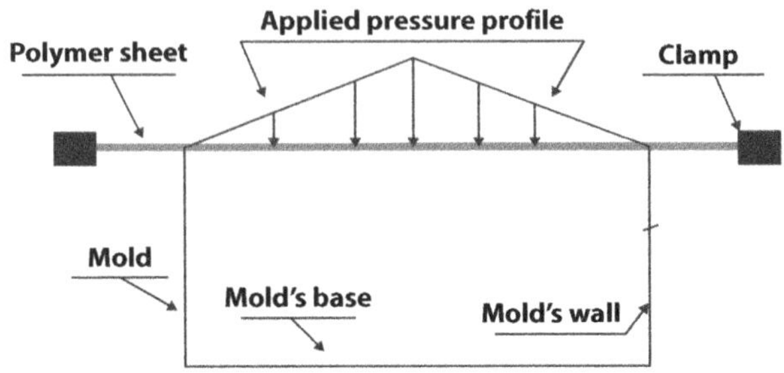

Figure 5.23 Schematic diagram of pressure forming. [Received permission from [23] Copyright © 2007 Wiley Periodicals, Inc. All rights reserved].

over a female mold, applying positive air pressure through a pressure box to push the sheet against the mold, cooling the sheet to solidify it, and finally, removing the formed plastic part from the mold.

Pressure forming offers several advantages over vacuum forming. This ensured a uniform distribution of the material, leading to a more consistent wall thickness and part dimensions. Additionally, pressure forming can reproduce the intricate details and features of the mold more accurately than vacuum forming. Less time is required to achieve the desired shape, making it more suitable for high-volume production. Furthermore, pressure forming can be effective for thinner plastic sheets, resulting in reduced material costs.

Although pressure forming typically utilizes moderate air pressure, higher pressures may be used to form complex or larger parts, as well as when tighter tolerances are required. Figure 5.24 shows the pressure

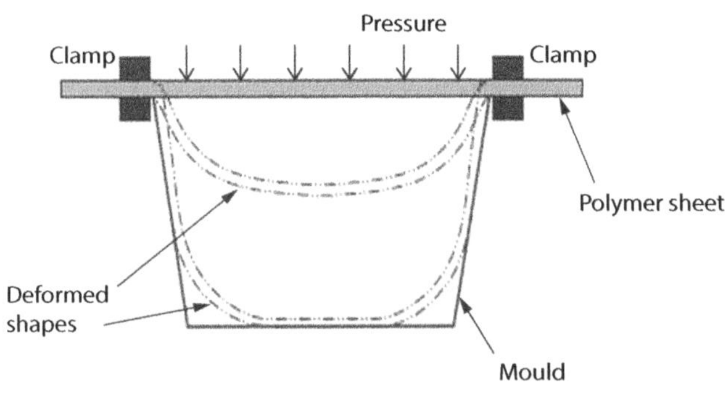

Figure 5.24 Pressure forming. Reproduced with permission from [116]. Copyright © 2003, Elsevier Science BV. All rights reserved.

forming of the thermoplastic sheet material. However, it is important to note that pressure forming does not achieve intricate details or high surface quality.

Pressure forming is a manufacturing process that combines the cost-effectiveness and versatility of thermoforming with the precision and quality of injection molding. This is particularly advantageous for medium- to high-volume production runs that require intricate shapes and fine details beyond the capabilities of vacuum forming alone. Moreover, pressure forming can offer faster production cycles and lower tooling costs than injection molding in certain applications.

5.44.1 Plug Assist–Air Pressure Thermoforming

The plug-assisted thermoforming process is depicted in Figure 5.25, which presents the typical configuration used to produce thin-gauge products. Initially, the plastic sheet is heated and subsequently conveyed to the mold where it is securely fastened. A preheated mechanical plug is then lowered into the sheet, which is referred to as the Plugging Phase, during which the sheet is pre-stretched to approximately 90% of its ultimate deformation in the mold. Finally, in the Pressure Phase, air pressure, typically ranging from 5 to 9 bar, is applied from above, causing the plastic to separate from the plug and conform to the wall of the mold. The part is then rapidly cooled, resulting in the creation of the final product.

Plug-assisted pressure forming is a technique commonly used in the field of thermoforming, wherein a mechanically driven plug is utilized to push a heated plastic sheet into a mold. This process aids in the stretching of the material and ensures even distribution. Once the material has been

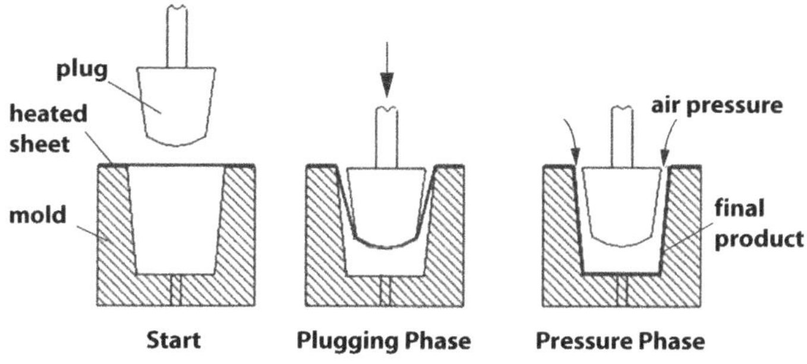

Figure 5.25 Plug assist–air pressure forming. [Reprint with permission from [17]. Copyright © 2007 Society of Plastics Engineers. All rights reserved].

adequately pre-stretched, air pressure is applied to the backside of the plastic sheet, while the plug continues to exert force. The combined action of the plug and air pressure effectively shapes the material tightly against the mold walls, resulting in the desired part shape.

5.45 Snapback Thermoforming

Snapback refers to the elastic recovery of the material after the forming process. This characterizes the tendency of certain materials to slightly revert to their original shape after undergoing thermoforming. A comprehensive understanding of snap-back is essential to obtain accurate dimensions of the final product.

The vacuum snap-back forming is shown in Figure 5.26. Following the heating cycle, the sheet is subjected to pre-stretching by lowering the top vacuum box onto the sheet and creating a vacuum. The height of the bubble and the speed at which vacuum is applied could be adjusted.

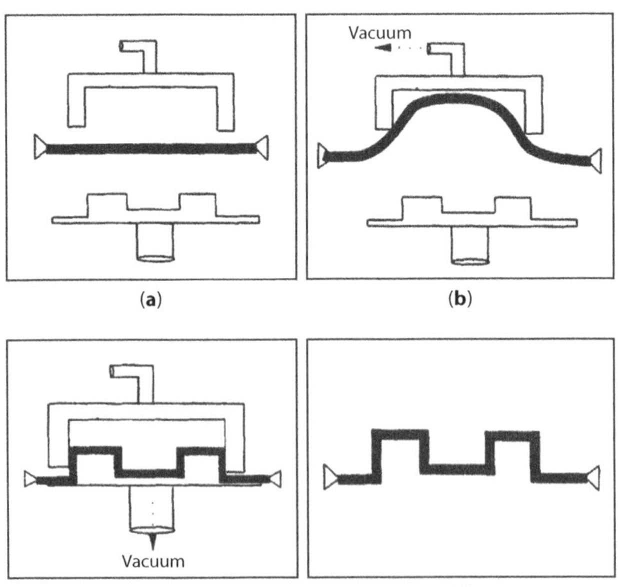

Figure 5.26 Schematic illustration of the vacuum snapback thermoforming operation. [Reprint with permission from [117]. Copyright © 2001 Society of Plastics Engineers. All rights reserved].

The vacuum snapback forming technique offers the advantage of producing components without altering the external surface. This is possible because the process does not involve direct contact with the mold and can be managed using alternative technologies.

5.46 Drape Forming

Drape forming is a thermoforming technique that involves heating a plastic sheet and then placing it over a mold. The material is then shaped onto the contours of the mold with the help of gravity and/or mechanical means. The plastic sheet is heated until it becomes soft and flexible, allowing it to conform to the shape of the mold. This process is suitable for parts with complex shapes; however, it needs be analyzed and optimized to meet the stringent specifications of the aeronautical industry. Figure 5.27 shows the equipment used for hot-drape forming equipment.

Various aspects of drape forming using a thermoforming machine equipped with heating elements on both sides are explored. Additionally, the role of the draft angle and the significance of the corner radii in achieving strong corners are explored. Drape forming is a thermoforming process (Figure 5.28) that relies on draping a heated plastic sheet over a mold to achieve the desired shape. The utilization of a thermoforming machine with heating elements on both sides ensures uniform heating and minimizes the risk of tearing.

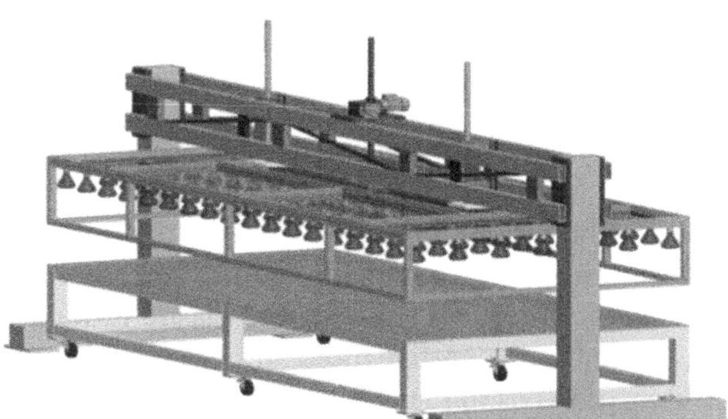

Figure 5.27 Hot drape forming equipment. [Reprint with permission from [118]. Copyright © 2015, Springer-Verlag London. All rights are reserved].

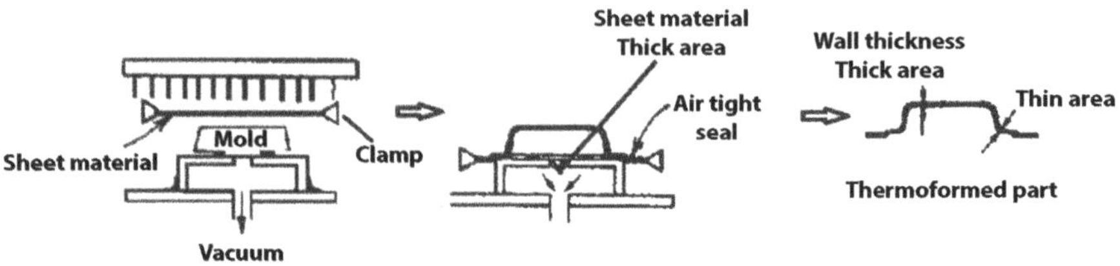

Figure 5.28 Schematic representation of drape forming.

Draft angle and corner radii are crucial factors that significantly affect the quality of a product. Drape forming is a commonly used technique in industries where there is a demand for large quantities of plastic parts with simple shapes. This process involves the application of heat, stretching, and the use of vacuum or air pressure to create various shapes and products. Drape forming allows for precise control over the internal dimensions of the formed part, often resulting in thicker walls at the center and thinner walls at the edges. It is a more intricate process than basic vacuum forming as it involves mechanical stretching.

Positive molds are generally more cost-effective and easier to build for the construction of molds for drape forming. However, they are also prone to damage. In contrast, negative molds enable the formation of multiple cavities and require less space. In addition, drape forming can be achieved solely using gravitational force.

Molds for drape forming can be crafted from wood or aluminum, with a felt covering. The application of only slight pressure is necessary, and it is important to remove any masking tape before heating to prevent unwanted marks. The heated sheet is promptly placed on the mold after heating and forced cooling with air is used.

5.47 Matched Mold Forming

The process of matched mold formation involves the preparation of positive and negative molds, typically made of metal or other suitable materials, based on the desired final shape of the plastic product. The plastic sheet or film is placed on a negative mold and heated using methods such as infrared heaters or ovens to make it more pliable. Once the plastic sheet reaches the appropriate temperature, the positive mold is brought down to the negative mold, enclosing the plastic sheet between them. The molds are designed to match the desired shape of the final product. This process, a

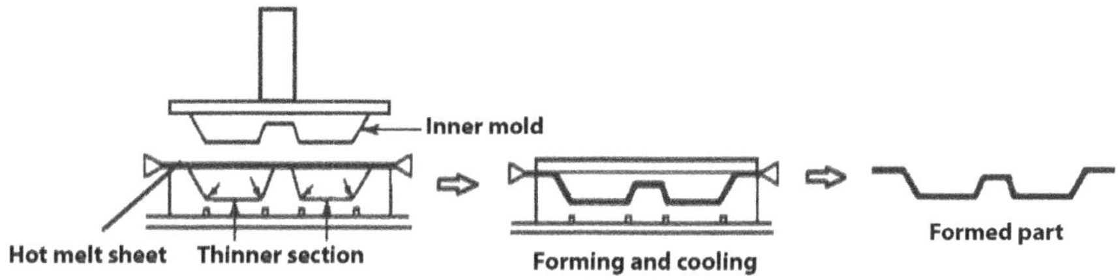

Figure 5.29 Matched mold forming.

variation of thermoforming, is known to produce accurate parts with close tolerance. However, it can be more expensive than other thermoforming methods because of the complexity and precision involved in creating and cooling molds. Despite the higher initial investment, matched mold forming (Figure 5.29) is suitable for producing parts with intricate details, thick walls, and high accuracy requirements, making it commonly used in industries where precise specifications and tight tolerances are essential.

Another similar process is matched die forming, which involves mechanically pressing the material into a shape using matched molds without the application of vacuum or air pressure. Matched die forming is particularly suitable for producing products with complex shapes, thick walls, and precise details, offering advantages such as reduced material waste, faster cycle times, and the ability to produce high-quality parts with tight tolerance. However, it may require more expensive molds than other forming methods because of the need for higher mechanical pressure and robust tooling.

5.48 Foam Sheet Forming

The process of shaping foam sheets involves the utilization of vacuum pressure and quenching techniques. Initially, a foam sheet composed of amorphous or semicrystalline materials is used. To facilitate the shaping process or create void spaces, an “assist-plug” is positioned on or near the surface of the foam sheet. Subsequently, vacuum pressure is applied to the opposite side of the foam sheet, causing it to be drawn towards the mold and assume the desired shape. While still under vacuum pressure, the foam sheet is brought into contact with a low-temperature mold, which rapidly cools the foam sheet and solidifies its desired shape. The cooling process is commonly referred to as quenching. The overall procedure involves heating the foam sheet to the appropriate temperature range, applying vacuum pressure, and quenching it on the mold to achieve the desired shape.

The specific temperature ranges and materials used depend on the intended application and desired properties of the final product. However, the thermoforming of polymeric foam sheets can be challenging due to the intricate and nonlinear effects of processing parameters involved. The behavior of polymeric foam sheets during thermoforming is complex and does not adhere to a straightforward relationship with the processing parameters. The interactions between different parameters are interconnected, making it difficult to accurately predict the final outcome.

5.49 In-Line Thermoforming

In-line thermoforming is a highly advantageous technique compared to traditional methods due to its energy efficiency, material savings, and process control. One notable advantage is its energy-saving capability, as it utilizes the residual heat in a freshly extruded web, reducing the need for additional heating sources and improving overall process efficiency. Moreover, the cooling process in in-line thermoforming can be precisely regulated, resulting in a favorable crystallite structure in the extrudate. This controlled cooling facilitates the attainment of desired material properties and enhances product performance. Additionally, in-line thermoforming eliminates the requirement for foil wind-up and off-wind stations, which are necessary in traditional thermoforming processes. This simplifies the production process, reduces complexity, and costs. Material and contamination savings are also achieved through in-line thermoforming because the web scrap generated during the process can be directly returned to the extruder for recycling. This not only leads to material savings but also mitigates the risk of contamination from external sources.

In-line thermoforming is a manufacturing process that offers several advantages in terms of material properties, temperature control, and material distribution. By preventing post-crystallization issues that may occur during storage, this process ensures consistent and predictable material properties. The use of double-sided heaters with zoned temperature control allows precise temperature control, resulting in a uniform thermoforming temperature and consistent product quality. Additionally, the use of a stenter chain with a spreader compensates for the sag in the sheet or foil, ensuring even and controlled material distribution during the thermoforming process.

Furthermore, in-line thermoforming offers versatility in materials, allowing to produce thin-walled moldings using both mono- and multilayer sheets. This versatility enables the manufacturing of various products, including disposable packaging and containers for ready meals.

Overall, in-line thermoforming is a technologically advanced and efficient process that offers numerous advantages. It not only saves energy and optimizes material utilization, but also ensures high-quality products. Consequently, it has gained popularity in industries where these advantages are highly valued, such as in the production of packaging materials and food containers.

Thermoforming is a versatile set of techniques used for shaping thermoplastic sheets into various products and components. The process involves several essential steps, as shown in Figure 5.30. Initially, a roll of the thermoplastic sheet material is unwound and introduced into the thermoforming machine. Subsequently, the sheet undergoes preheating, where it is heated to a specific temperature to soften the material. Following preheating, the sheet enters the main heating area, commonly referred to as the "tunnel," where it is exposed to high temperatures to attain the thermoforming temperature necessary for shaping.

Once the sheet reaches the appropriate thermoforming temperature, it is fed into a forming station, where it is shaped using a mold or a former specifically designed to achieve the desired shape of the thermoplastic sheet. As part of the thermoforming process, the excess material surrounding the

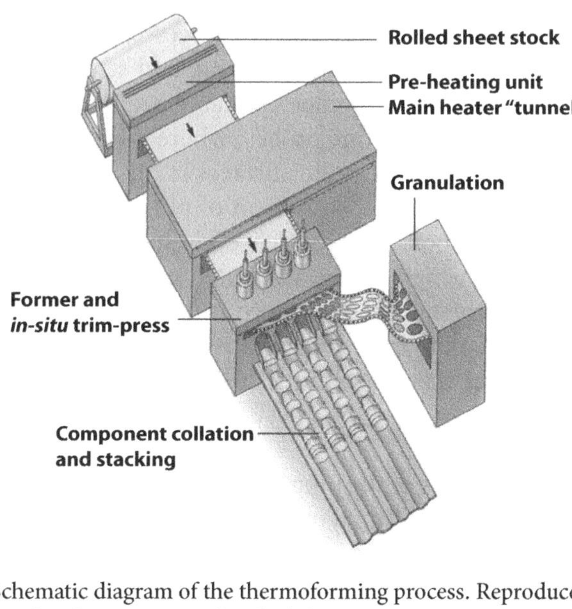

Figure 5.30 Schematic diagram of the thermoforming process. Reproduced with permission from [119]. © 2005, Basell Polyolefins.

formed product, known as "flash" or "trim", is trimmed away while still in the forming station. This *in situ* trimming ensures that the final product possessed clean edges and the desired dimensions.

After the forming and trimming stages, the thermoformed products are gathered and stacked for further processing or packaging. Thermoforming machines often incorporate mechanisms to automate collection processes. The trimmed material, also referred to as scrap or trim, is directed to a granulator, which breaks it down into small particles that are suitable for recycling. This recycling process enables the reuse of excess materials, reducing waste and promoting sustainability.

Thermoforming is extensively used across diverse industries to produce a broad spectrum of items, such as packaging, automotive components, consumer goods, and medical devices. Its widespread adoption can be attributed to its versatility, efficiency, and capacity to accommodate various thermoplastic materials, making it a preferred method for manufacturing plastic parts and products.

5.50 Industrial Versus Laboratory Thermoforming

Unlike industrial thermoforming, laboratory thermoforming operates over a given time interval and does not account for the cycle time, leading to different curves. This consequently alters thermoforming, such as sheet temperature and mold temperature, and processes such as heat release, cooling time, volume change due to drawing, and strength development. Industrial thermoforming of complex and large-scale products sometimes requires a stretch-assisted forming process where the polymer sheet is prestretched before contacting the mold surfaces [120].

Figure 5.31(a) illustrates the utilization of plug force sensors and plug thermocouples in thermoforming. Each cycle generated a peak force of approximately 145 N, which can be attributed to the characteristics of the polymer sheet at the forming temperature and its thermoforming capability. It is important to note that this force is considerably lower than that observed in industrial thermoforming during laboratory tests. However, it should be emphasized that the sheet used is thinner, hotter, and not stretched.

The temperature results of the plug indicated that the plug temperatures attained are relatively high. The temperature of the plug bulk, denoted by T_{under}, is approximately 80 °C. Although this value remains constant, it gradually increases at a slow rate, suggesting that the system has not yet reached thermal equilibrium conditions despite achieving stable production in the

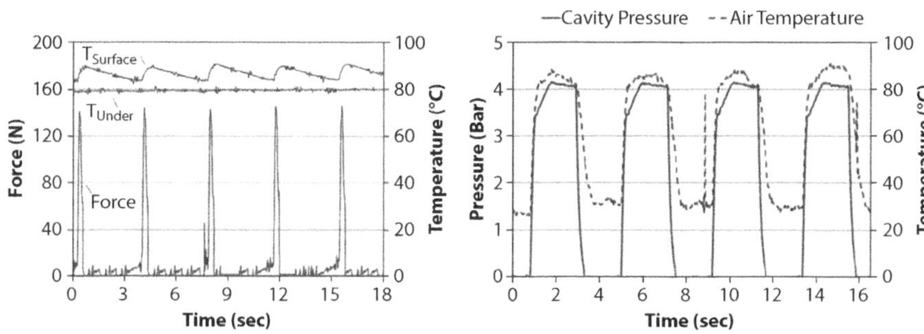

Figure 5.31 (a) Variations in plug force and plug temperature with time during industrial thermoforming. (b) Variations in upper cavity air pressure and temperature over time during industrial thermoforming. [Reprint with permission from [121]. © 2014 Society of Plastics Engineers. All rights reserved].

context of industrial thermoforming. On the surface of the plug, the temperature ($T_{surface}$) exhibited a regular cyclic pattern, rising to approximately 90 °C upon contact with the polymer sheet and then decreasing to approximately 83 °C for the remainder of the cycle. This predictable behavior is valuable for validating thermal properties and temperature-dependent relationships. Temperature plays a crucial role in understanding friction during plug contact because it is known to be highly dependent on temperature. Furthermore, the results confirm that the plug temperature of 85 °C selected for the laboratory tests closely aligns with the temperature observed in industrial processes [122, 123].

Figure 5.31(b) shows the air pressure and temperature measurements. A comparison between the laboratory tests and industrial thermoforming process reveals that the pressure system used in the latter is significantly more efficient. Initially, the pressure in the upper cavity rises rapidly but gradually slows down as the cavity fills, eventually reaching a peak of approximately 4.15 bar. Subsequently, it gradually decreases to approximately 4 bar for the remainder of the cycle before returning to zero as the mold opens for part ejection. This pattern demonstrates a high level of consistency across multiple cycles. The number and dimensions of the vent holes in the mold, as well as the magnitude and timing of the vacuum assist, in the case of utilization will determine the pressure in the lower cavity. Closely related air temperature is more susceptible to cyclic variability. Upon entry, the air rapidly reached a temperature of approximately 83 °C and further increased to approximately 88 °C as it filled the cavity. Subsequently, it gradually decreased as the formed part cooled within the mold and dropped rapidly to approximately 30 °C when the mold opened for part ejection. This final temperature reduction is more pronounced compared with the laboratory tests, primarily because of the larger and more open shared upper pressure cavity in the industrial process, which limits the potential for air entrapment and stagnation. The significance of these findings lies in confirming that the air within the cavity remains relatively hot during the forming process, minimizing the impact of convective heat losses on the polymer sheet. The thickness variation that exists is sacrificed to some extent [124–126].

In conclusion, the thermoforming method offers advantages such as flexibility, cost-effectiveness, faster iterations, reduced lead times, and the ability to quickly incorporate design changes. However, they may have limitations in terms of durability and suitability for high-volume production. As thermoforming technology continues to advance, it is likely that more innovative solutions have emerged to address these limitations and enhance the capabilities of production and parts.

6
Troubleshooting Thermoforming

6.1 Introduction

Instrumentation and control tools significantly improve production and resource processes in the field of thermoforming [1]. The monitoring of processes is crucial for identifying instabilities during thermoforming and taking immediate action to prevent serious issues. Restoring production after such problems can be a complex task, as it may involve addressing complications, such as blisters, sheet tearing, and webbing. Various factors, including changes in the sheet material and composition, temperature variations, and mold conditions, can contribute to process instabilities.

In industrial processes, monitoring parameters can be categorized into two groups: parameters that characterize the processes and early indicators that represent them. Product quality can be affected by a number of external factors such as the physical environment, process parameters, and material properties. Control devices are widely used in thermoforming to monitor the various aspects of these processes. Monitoring the diverse parameters involved in industrial processes can lead to process failures. It is important to recognize that these parameters may fluctuate and require troubleshooting measures. It is crucial to understand that problems can arise not only from changes in sheet material and process failures but also from equipment malfunction, inadequate maintenance, or design deficiencies. Therefore, technicians and operators must have a comprehensive understanding of the primary equipment structure during thermoforming to ensure a stable process [2].

Troubleshooting is a critical aspect of the optimization process, as it involves identifying and understanding the fundamental components of a machine's conceptual framework. These components are organized in a prescribed format that includes an objective function. The objective function represents economic criteria, such as profit, cost, energy, and yield, and it is evaluated in relation to the key variables of the process. The process and constraints outline the interdependencies among key variables.

Muralisrinivasan Natamai Subramanian. Thermoforming: Processing and Technology, (179–208)

It is important to note that different troubleshooting approaches may arise from the analysis depending on the underlying assumptions. The selection of an appropriate troubleshooting method is crucial for resolving an optimization problem, which requires a certain level of expertise, judgment, and experience [3].

Two key factors must be considered to address issues in thermoforming. A comprehensive understanding of the behavior of the material and the occurrence of defects during the thermoforming of thermoplastics is essential. Efforts should be made to optimize the design of thermoplastic products produced through thermoforming with the goal of reducing production costs [4].

6.2 Product Quality Analysis

Examination of quality management and assessment of the production of plastic parts through thermoforming offers valuable insights into the production characteristics and data collection of these products. This study explores the technology and characteristics of thermoforming, as well as the common quality defects that occur during the production of thermoplastic parts. A series of actions can be implemented to address and resolve product and quality issues by collecting data.

The evaluation of production includes a crucial assessment of product quality. Failure to meet quality standards can have significant consequences for consumer safety. Therefore, ensuring product quality is a major concern throughout the production process, starting from the product design stage as part of production preparation. Quality is defined as a combination of product features that aims to meet customer expectations to a certain extent. These features, along with other aspects of the product, contribute to competitiveness. However, achieving the desired quality level according to design assumptions is not common. Therefore, permissible deviations are established to determine whether the product complied with accepted quality standards [5].

The quality features of a product can be categorized based on their measurability within the quality dimensions [6]. Measurable features can be quantitatively expressed by using specific units of measurement. These measurable characteristics can be further classified as either continuous or discrete. Continuous measurable characteristics have measurements limited by the resolution of the measurement method, whereas discrete measurable characteristics allow for a finite number of states.

However, nonmeasurable features cannot be numerically measured or expressed. They can be observed, studied, and qualitatively described [7].

To analyze the quality of a product, it is necessary to consider various variables that can be categorized into four distinct groups. The first group pertains to the manufacturing process, which includes factors such as feeding, preheating, force applied during formation, pressure exerted, and the rate of cooling. The second group is related to the design variables used in the input and output stages. The third group involves material and mold design, which encompasses aspects such as plug size, design pattern, drawing ratio, and the type of material used. The fourth group focuses on the control process of the system, which includes heating, punching, and ejection [8].

Advancements in thermoforming machinery have led to improvements in the overall process. These advancements have been achieved in various areas, such as the heating stage, flexible production units, utilization of motors, and trimming. It has been demonstrated that optimizing the heating process, along with consistent implementation of controllers, can significantly enhance part optimization through improved machine engineering. This is because these measures allow for precise coordination and automatic optimization of all parameters that influence the thermoforming process. Consequently, shorter cycle times can be achieved, and operators can receive specific guidance throughout the process. In addition, there are no issues [9].

6.3 Product Quality

Defining quality in a concise and structured manner can pose difficulties given the dynamic nature of this concept. The process of ensuring the quality of a product requires consideration of various factors. The manufacturer must ensure the product's competitiveness, adhere to quality standards established during the design phase, comply with safety regulations, ensure reliability to enhance user satisfaction, reduce maintenance costs, and ultimately meet customer expectations [10, 11].

6.4 Product Defects

Plastic processing replaces multiple components in day-to-day activities for consumers, as well as engineering products made of metals, ceramics, etc., with plastic alternatives. Consequently, there is a requirement for systematic regulation of mass production within the domain of plastic products. This is particularly critical because of the manufacturing procedure of thermoforming, which involves the utilization of plastic materials. However, it is noteworthy that the thermoforming process can result in the formation of flawed plastic products, which can be attributed to both internal and external environmental influences. Visual inspections are performed to detect these defective products.

The occurrence of defects depicted in Figure 6.1 in plastic parts can be attributed to the instability of the thermoforming process. Hence, the integration of a defect detection system is of great significance in improving productivity, particularly during the inspection stage. It is important to highlight that the defect rate tends to escalate gradually if a single defective product is identified during the thermoforming process. Therefore, a thorough analysis of all the thermoforming processes is imperative to ascertain the underlying causes of these defects [12]. Figures 6.2–6.4 serve as visual representations of the challenges pertaining to materials, processing, and product quality, respectively.

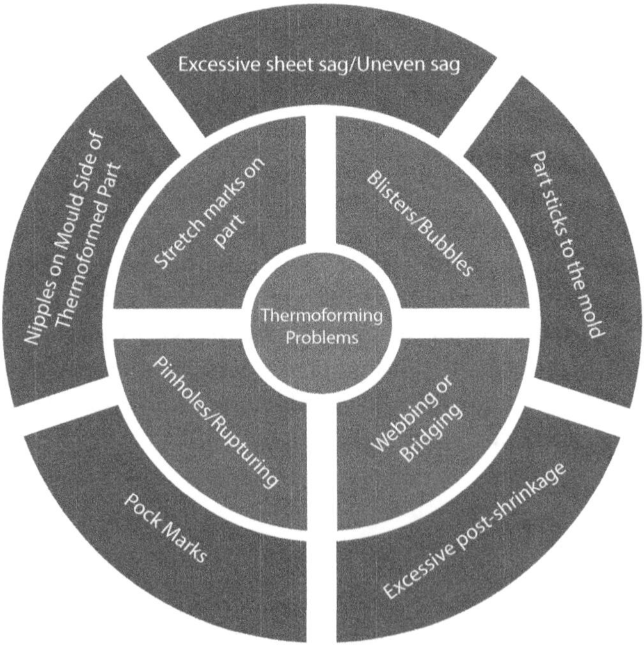

Figure 6.1 Thermoforming problems.

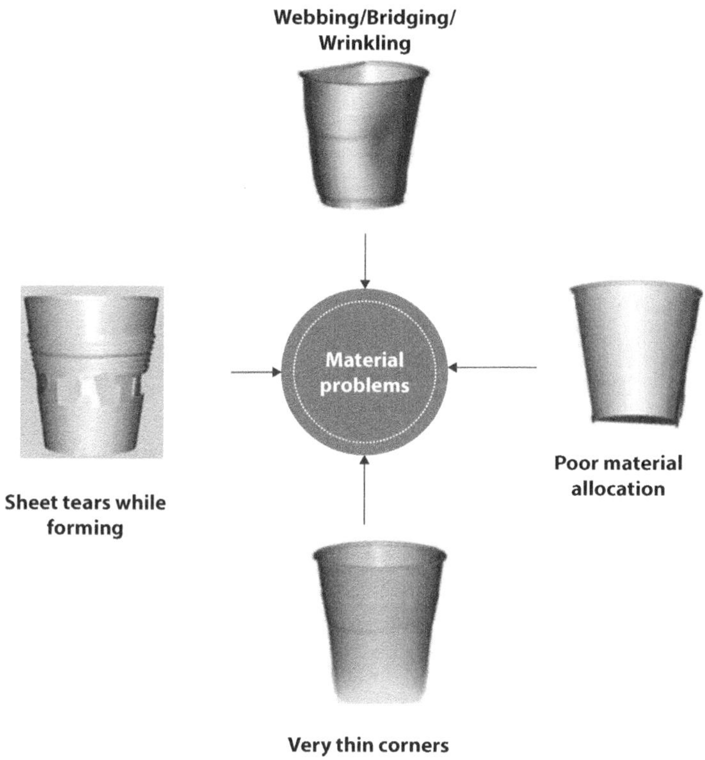

Figure 6.2 Thermoforming defects related to material.

6.5 Fundamental and Inherent Defects

An inherent limitation of thermoforming technology is the occurrence of sheet rupture, shrinkage, and product warpage during the thermoforming process. This issue becomes more pronounced at high temperatures. Product warpage refers to the non-uniform and heterogeneous alteration of the geometric dimensions of products over time, leading to deviations from their original form [13]. In the production of plug-assisted vacuum thermoformed products, two types of defects can arise: fracturing of the polymeric sheet during the deformation process and warpage (Figures 6.5(a,b)).

Fracture of thermoplastic sheets is a significant issue encountered in plug-assisted vacuum thermoforming. This defect can manifest during both plug-assist and vacuum-forming stages [16]. These imperfections are primarily attributed to the inherent characteristics of the

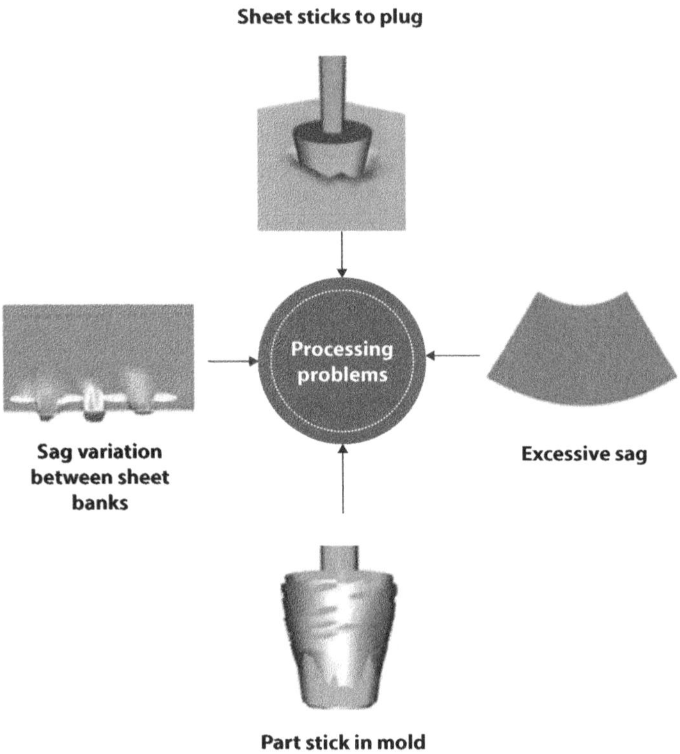

Figure 6.3 Thermoforming defects related to processing.

thermoforming process. Warpage is defined as a nonuniform change in the geometrical dimensions of the products over time, resulting in a deviation from their original form. Another significant defect associated with thermoforming technology is the warpage of the products during their application, particularly at high temperatures (Figure 6.5b). Warpage refers to a non-uniform change in the geometrical dimensions of products over time, leading to a deviation from their original form [17].

The occurrence of defects in the products during their application can be attributed to the thermoforming technology, as illustrated in Figures 6.5(a,b), which highlights the challenges encountered in the thermoforming process. These defects are particularly prominent in thermoforming processes. Although it may seem that controlling the state parameters of the thermoforming process can rectify these flaws, it is a complex undertaking. To effectively address these flaws, it is essential to first identify their origin. Once the origins have been established, the process parameters can be adjusted to prevent the occurrence of these defects. However, a

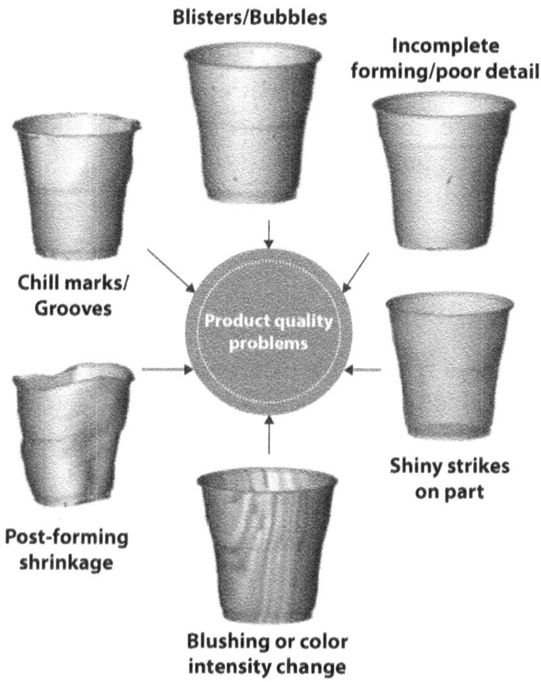

Figure 6.4 Thermoforming defects related to product quality.

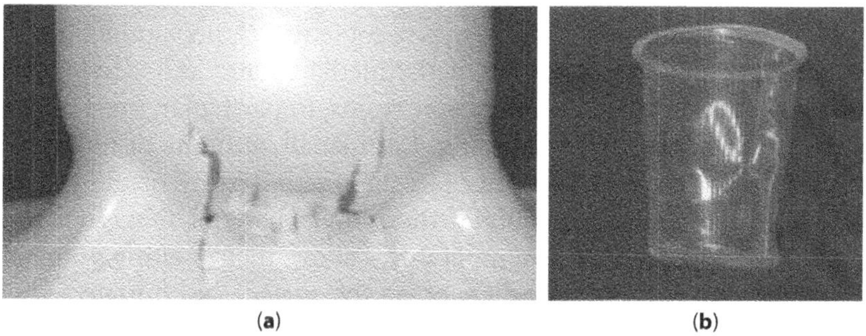

Figure 6.5 (a) Fracture of polymeric sheet during plug-assisted vacuum thermoforming, (b) warpage of a polymeric product during its application. [Reprint with permission from 6.5(a) [14]. Copyright © 2010 Wiley Periodicals, Inc. All rights reserved. 6.5(b) [15]. Copyright © 2006 Elsevier Ltd. All rights reserved.].

comprehensive understanding of all the potential causes of flaws is necessary to successfully carry out these steps. By employing systematic reasoning, appropriate process parameters can be determined to eliminate defects effectively.

6.6 Troubleshooting

To improve the overall quality of thermoformed products, it is possible to employ a range of techniques to enhance their properties and address the limitations and defects in these products. These techniques involve modifying of the mechanical, thermal, and processing properties of thermoplastic materials.

6.6.1 Blisters or Bubbles

Inadequate moisture levels can have detrimental effects on the surface finish of a product, including roughness, scaliness, and formation of bubbles and voids. The introduction of air during the compression process in automatic machines can lead to the formation of blisters or bubbles as well as the emergence of moisture as bubbles within the sheet material. The occurrence of blisters or bubbles can be attributed to excessively rapid heating or lower temperatures as well as uneven heating.

Elevated temperatures also contribute to the formation of blisters and excessive smoke. To address this issue, it is recommended to decrease the amount of heat and increase the dwell time. The absence of rippling and the presence of uniform sagging can serve as indicators of the need to consider thermoforming.

Maintaining low moisture content is crucial for the proper handling of certain types of resins, such as hygroscopic resins, polyamide, polycarbonate (PC), polyurethane, polymethylmethacrylate, and acrylonitrile–butadiene–styrene (ABS). Hygroscopic resins, in particular, must undergo adequate drying prior to processing. Even small amounts of moisture in most materials can lead to changes in the melt viscosity of the polymer, affecting its processing behavior. Furthermore, when polymers such as PC and polyester are exposed to temperatures above their melting points, the presence of moisture in the pellets or on the surface can initiate a chemical reaction. This reaction has the potential to degrade polymers, resulting in alterations in their molecular weight, melt viscosity, and mechanical strength.

Proper storage of sheets intended for thermoforming is crucial to prevent moisture absorption. Various factors come into play, including the method of heating (on one or both sides), the rate of heating, the thickness of the sheet, the design of the mold, the material used for construction and the heating pattern, the flow of evacuation, control mechanisms, and

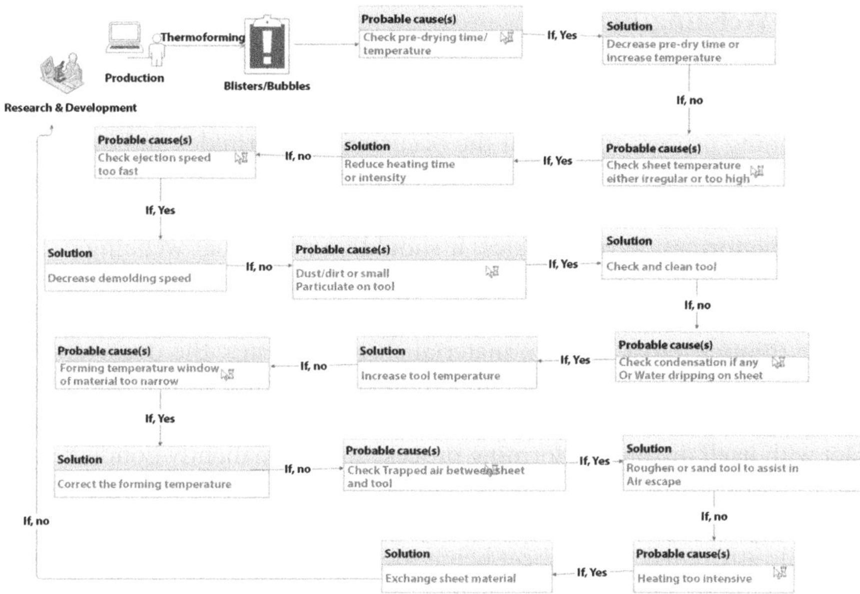

Flowchart 6.1 Troubleshooting process diagram for blisters and bubbles in the product.

the level of vacuum applied. The presence of moisture in the sheet is typically the primary cause of bubbling, although this issue can sometimes be resolved by increasing the heating rate rather than pre-drying the sheet. In addition, the occurrence of bubbling can be minimized by employing a lower forming temperature. This issue is particularly prevalent for blisters and bubbles, as illustrated in Flowchart 6.1.

Elevated temperatures can result in the blistering of the sheet and/or excessive smoke generation. In such instances, the sheet is heated too rapidly, necessitating a reduction in heat and an increase in the dwell (cycle) time. The absence of rippling and the presence of a uniform sag are indicative of successful thermoforming.

When the temperature exceeds the specified process conditions, blistering may occur due to the vaporization of absorbed moisture, leading to degradation of the polymeric sheet. Insufficient heating of the sheet or excessive cooling prior to or during the forming process can also cause issues caused by the induced stress. Blisters or bubbles can be attributed to excessively rapid heating or lower temperatures; uneven heating can also contribute to this problem.

6.6.2 Webbing/Bridging

Webbing, bridging, and folding are critical defects in thermoforming processes. It is of utmost importance to establish appropriate conditions for the forming process to prevent the occurrence of wrinkles and folds. The term "webbing" refers to a phenomenon that occurs when a heated material comes into contact with itself and forms a permanent bond, resulting in the formation of wrinkles. It should be noted that webbing occurs exclusively when a plug mold is used. The decision-making process for this fundamental design concept includes considerations such as the location of the thinnest cross-section, material shrinkage rates, the occurrence of a defect known as "bridging," and the determination of the draw ratio. Bridging, another defect, occurs when the heated material comes into contact with itself during the forming process and permanently bonds, resulting in a bridging effect. It is important to emphasize that webbing occurs only when a plug mold is used. The occurrence of webbing or bridging is typically attributed to the sheet being too hot at the center, whereas wrinkling can be attributed to the low melt strength of the sheet. The guidance provided in Flowchart 6.2 can assist in addressing and eliminating these problems. Webbing is a commonly encountered problem in thermoforming caused by excess material.

In the thermoforming procedure, increasing the temperature of the sheet can lead to a reduction in the internal stress within the thermoforming packages. However, excessive heating can cause the sheet to sag, resulting in the formation of bridges or webs. To enhance the forming process and minimize or eliminate bridging in blank-fed male-mold thermoforming machines, the utilization of wire-helper grids can be beneficial.

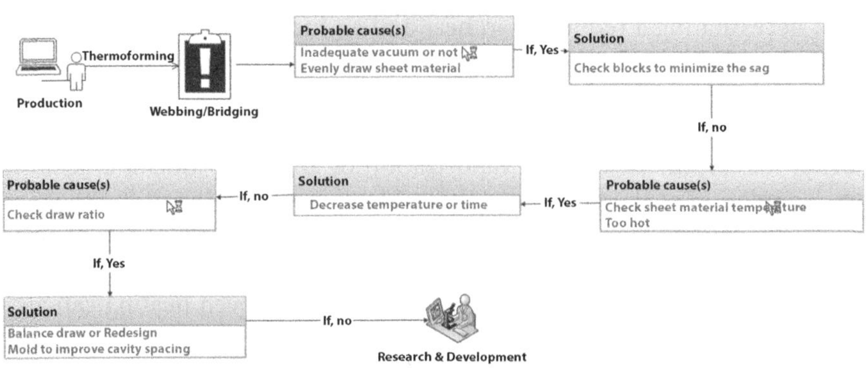

Flowchart 6.2 Troubleshooting process diagram for webbing/bridging in the product.

6.6.3 Excessive Sheet Sag

The overheating of sheet materials can lead to various issues, such as sagging, crystallization, whitening, embrittlement, excessive sag, and webbing. To address these problems and minimize waste during thermoforming, a flow chart is provided with measures to eliminate the issues and observations to be made.

One method to prevent sagging during heating is to use an air pillow, which reduces pre-stretching. However, this technique has limitations as it does not allow for controlling the sag temperature and restricts the heating source to one-sided heating, resulting in a longer processing time. Alternatively, take-up blocks and pushers are more effective in removing the excess material around the web. The extent of sagging is directly influenced by the oven temperature and dwell time of the sheet in the oven.

During the softening process, the sheet may excessively extend under its own weight, causing a concave shape known as excessive heat sag. This can affect the interaction between the sheet surface and heater bank, leading to variations in the heat flux distribution across the sheet. Additionally, it can cause differences in the heat flux over and under each element of the sheet, resulting in temperature gradients across the sheet or through its thickness, which can complicate the thermoforming process. To mitigate sagging, practitioners stretch the sheet laterally using cambered transfer rails to maintain tension in the sheet [18].

Decreasing the duration of oven exposure and increasing the oven temperature can result in a higher temperature of the sheet without excessive sagging. Reducing the dwell time offers the advantage of reducing the overall cycle time, which allows for the use of a smaller oven in the thermoforming machine and reduces energy consumption.

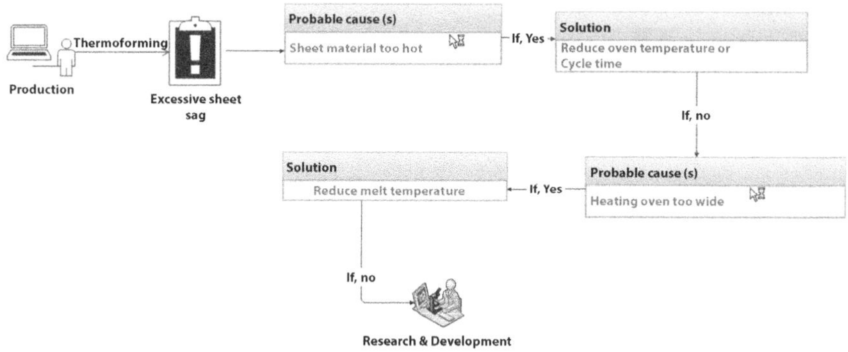

Flowchart 6.3 Troubleshooting process diagram for excessive sheet sag in the product.

To address this issue, according to Flowchart 6.3, the heating time or intensity can be reduced, the air pressure can be increased, the tool temperature can be lowered, and the dimensions of the plug and/or ring, tool geometry, or cycle time can be changed. If the molding is inadequate, changing the material or implementing technical aids can also be effective.

Excessive sagging of the sheet is primarily caused by overheating of the polymeric material. The defective appearance, characterized by material stretching or sagging, can be controlled by reducing the heating time or intensity. If the molding is insufficient, changing the material can also be considered. A defective appearance can also result from differences in shrinkage across the width or excessive growth in the transverse direction, which can be addressed by changing the defective material.

Bridges or webbings can occur because of various factors such as excessively hot materials, insufficient compressed air, overheated tools, material sagging, improper spacing between the plug and mold, and inadequate tool construction. Adjusting the cycle time, tool temperature, or plug-mold spacing can help prevent the formation of bridges or webs.

6.6.4 Pinhole or Rupturing

During thermoforming, the sheet material undergoes deformation, and during subsequent reformation, uneven heating can lead to rupture. The occurrence of ruptures in polymeric sheets is a significant issue encountered during plug-assisted thermoforming. This defect can manifest during the mechanical stretching stage with a plug as well as in vacuum or pressure thermoforming. This problem is particularly problematic for pinholes or rupturing and solutions, as depicted in Flowchart 6.4.

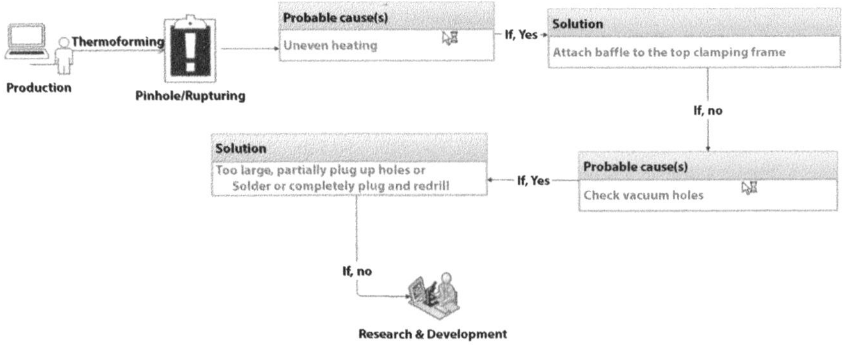

Flowchart 6.4 Troubleshooting process diagram for pinholes or rupturing in the product.

The presence of dirt on the surface of the sheet may be attributed to the development of an electrostatic charge. To prevent the accumulation of dirt and contamination, it is advisable to store the sheet at an appropriate location. Electrostatic charges can be mitigated by incorporating antistatic additives during the sheet manufacturing process. The pinholes on the sheet are caused by the presence of either contamination or dirt particles. These pinholes can result in an increase in the flow trend in the material. When this flow trend comes into contact with the mold, it can cause plastic deformation in the lower corner of the mold's sidewall, ultimately leading to rupture in the area where the pinhole is located.

6.6.5 Uneven Sag

Sag formation is a crucial factor that requires careful management during thermoforming. However, accurately predicting a sag is challenging. The uneven sag of the sheet material can significantly impact the positioning and orientation of each element in relation to the heater bank. This, in turn, leads to variations in the heat flux distribution across the sheet, complicating the subsequent shaping of the material. Developing an understanding of the sag behavior of sheet materials under specific temperature and time conditions is essential for selecting appropriate thermoforming processing parameters [19, 20].

Sagging, thermal buckling, and inhomogeneous temperature distribution are common challenges encountered in sheet materials after heating but prior to the application of pressure. To address these issues, the utilization of Flowchart 6.5 can be advantageous. By carefully assessing the heaters and implementing appropriate corrective measures, a significant reduction in waste can be achieved. It is important to acknowledge that the presence of regrind in the material may not be consistent throughout the heating process, resulting in uneven sag. This uneven sag can be attributed to either non-uniform heating or malfunctioning heaters.

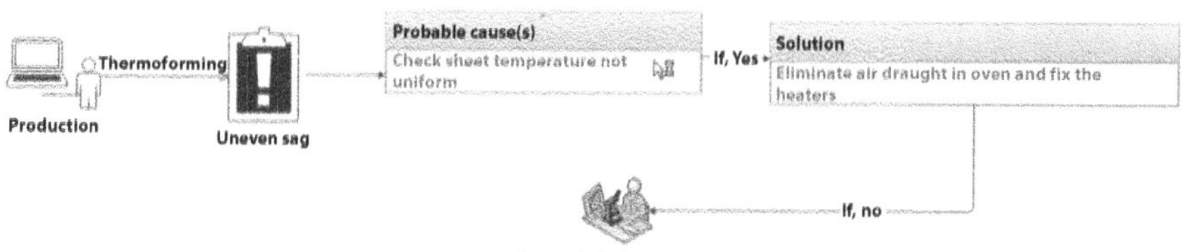

Flowchart 6.5 Troubleshooting process diagram for uneven sag in the product.

During the heating process, noticeable indications, such as pronounced vibrations and softening of the material, are observed. Furthermore, the material may begin to exhibit small undulations, which is commonly referred to as oil canning. The presence of smoothed sags and ripples signifies that a substantial portion of the stresses is relieved, and the material gradually regains its smoothness. The recommended tools include thermo labels or an infrared pyrometer to ascertain the surface temperature of the sheet.

6.6.6 Part Sticks to the Mold

During the molding process, it is expected that the hot polymeric material will instantly solidify upon contact with the cold surface of the mold, causing the sheet material to adhere to the mold. This adherence can result in a decrease in production speed. To address this issue, Flowchart 6.6 provided to offer guidance on how to eliminate this problem.

When the distance between the thermoplastic sheet and the sheet frame is small, it is necessary to significantly reduce the thickness of the sheet. This is because, in such cases, the high temperature of the molten sheet material tends to cause it to stick to the mold.

Several factors can contribute to the formation of a part sticking to the tool. These include insufficient or absent ejection air, sharp edges or corners, undercuts in the mold, a rough surface on the tool that hinders material sliding, inadequate demolding of the film, poor mold release, excessive heat in the tool, or excessively high material temperature.

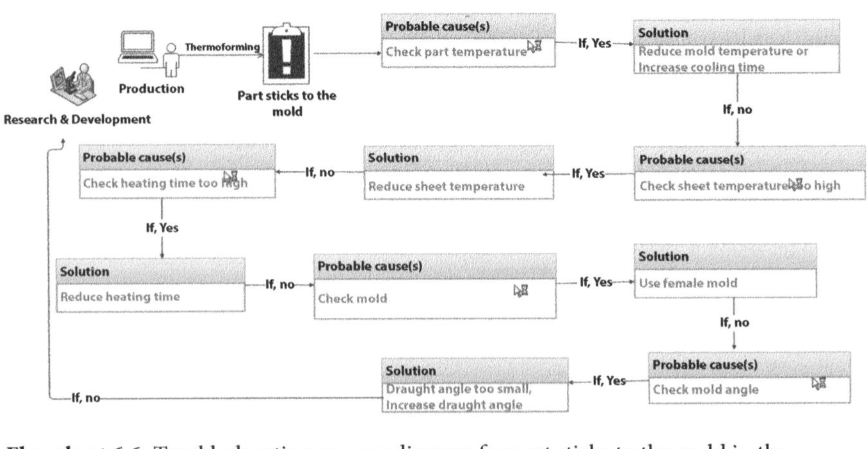

Flowchart 6.6 Troubleshooting process diagram for part sticks to the mold in the product.

6.6.7 Stretch Marks

The thermoforming process involves significant strains and variations in thickness, resulting in the stretching of the plastic sheet. To address this issue, the implementation of Flowchart 6.7 is beneficial. Stretch marks, which cause physical discomfort in the thermoformed part, manifest as linear scratches. Troubleshooting techniques can be used to resolve this problem. Monitoring internal stresses using a polarized light table is recommended to effectively eliminate stretch marks. Additionally, it is important to maintain a consistent temperature of either 40 °C or 60 °C on the mold surface to prevent the formation of chill lines or sticking.

6.6.8 Nipple on the Mold Side of the Thermoformed Part

The formation of nipples (elevated portions on the mold side of the thermoformed sheet) can occur due to the presence of vacuum holes and the application of vacuum pressure to the sheet. Flowchart 6.8 provides guidance on preventing the formation of nipples on the mold side of the thermoformed product. One approach is to reduce the sheet temperature, which may also help in reducing the occurrence of nipples on the mold side.

Another solution is to increase the distance between the heaters and sheet, as this can eliminate the problem. It is important to ensure that all heaters function properly and to use screening to balance the heat

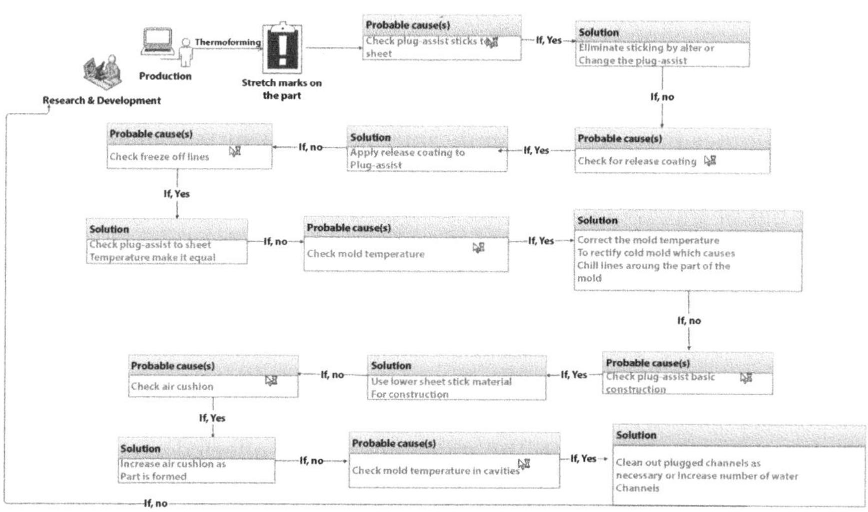

Flowchart 6.7 Troubleshooting process diagram for stretch marks on the part.

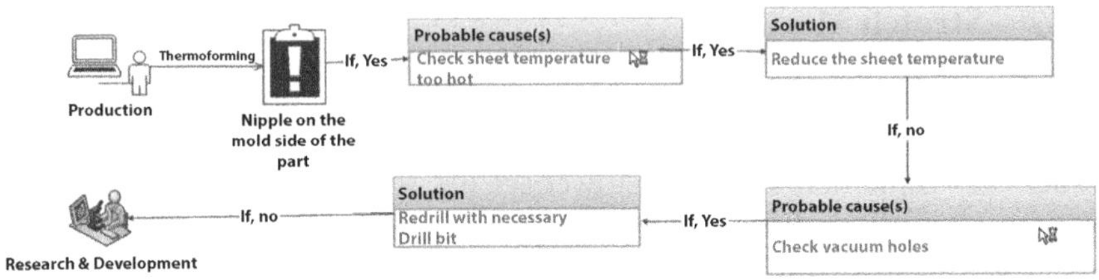

Flowchart 6.8 Troubleshooting process diagram for nipple on the mold side of the thermoformed part.

distribution. If there is excess moisture, the sheet material should be predried or preheated.

During the heating process, the material softens and bulges slightly because of the heat from below. As it approaches the thermoforming temperature, ripples may begin to form. These ripples are a result of inherent stresses from the extrusion process and are common in extruded plastics. As the ripples smooth out, the material formed into a part may have lower quality or higher internal stresses, potentially leading to poor part definition and thinning of the side walls.

To achieve better-quality parts with fewer rejects and cost savings, it is important to ensure that the surface temperature is at an appropriate thermoforming temperature. If the surface temperature is heating too quickly, the oven settings should be lowered, and the dwell time should be adjusted. This results in a smooth material that is free from ripples along the clamping frame with a slight sag. Proper surface and core temperatures are key factors for achieving better parts and reducing rejection.

Cleanliness of the thermoforming area is crucial for obtaining high-quality parts. It is essential to always maintain a clean thermoforming area, as dust and other particles can significantly affect the surface quality of the finished product. Additionally, excessive vacuum pressure can lead to the formation of nipples on the mold side of the thermoformed part, which may also be caused by a vacuum hole.

6.6.9 Pock Marks

Pock marks are the result of air becoming trapped in a material during the manufacturing process. To address this issue, a solution is proposed in the form of a Flowchart 6.9. The appearance of the affected part is characterized by variations that have limited significance, and descriptions of its appearance tend to be dark, such as hazel, brown, or black. The presence of

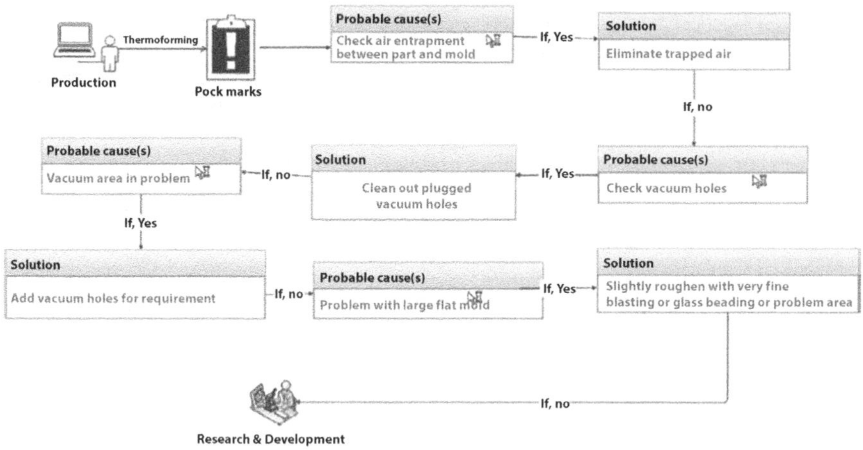

Flowchart 6.9 Troubleshooting process diagram for pock marks of the thermoformed part.

trapped air between the part and mold generates heat, which leads to burns or pock marks on the surface.

To mitigate this problem, it is recommended to either apply a soft rubber coating to the mold surface or fabricate the part using a material with a matte finish. In the latter case, it is important to ensure that the matte side of the part is positioned next to the mold during the manufacturing process.

6.6.10 Poor Wall Thickness

Sheet materials that exhibit significant nonlinearity, resulting in large stretches and strains, can lead to uneven thickness. Additionally, the thickness variation is influenced by the duration of the forming stage. To address this issue, a troubleshooting Flowchart 6.10 can be used to identify and eliminate the problem. The poor wall distribution can be attributed to malfunctioning pressure or vacuum systems. Rectifying the issues with vacuum or pressure effectively resolves this problem.

During thermoforming, the plastic material is subjected to various stresses, resulting in deformation, particularly in the bottom and sidewalls of the final product. The thickness distribution also plays a role in determining the rigidity of packaging. Although achieving a completely uniform thickness distribution is not feasible in thermoforming, a more uniform distribution can contribute to the production of rigid and stable packages [21].

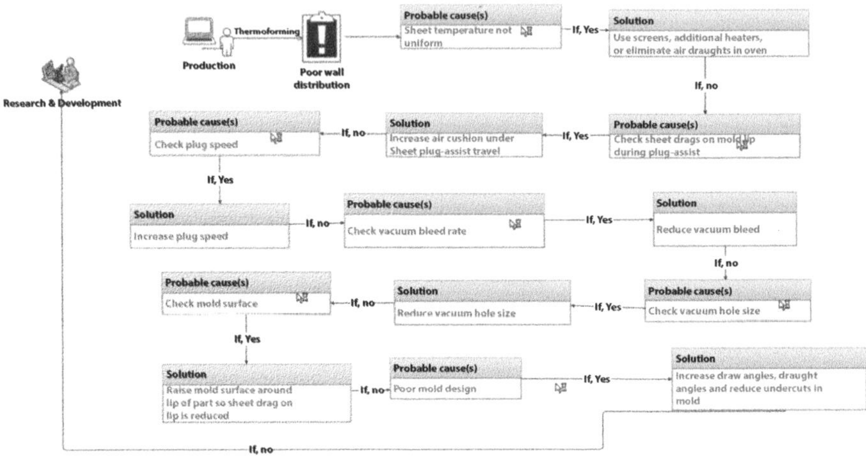

Flowchart 6.10 Troubleshooting process diagram for poor wall thickness.

6.6.11 Uneven Edges

During thermoforming, a plastic sheet is heated and stretched into a mold cavity by applying pressure, often with the assistance of direct mechanical loading. However, if a vacuum is applied too early during this process, uneven edges are formed in the final product.

To address this issue, Flowchart 6.11 is provided as a troubleshooting guide to help eliminate the problem. Additionally, uneven edges that may arise during the forming stage can be easily identified and corrected by adjusting the temperature of the sheet material. This is because the temperature difference in the polymeric sheet can lead to the formation of uneven edges in the thermoformed part.

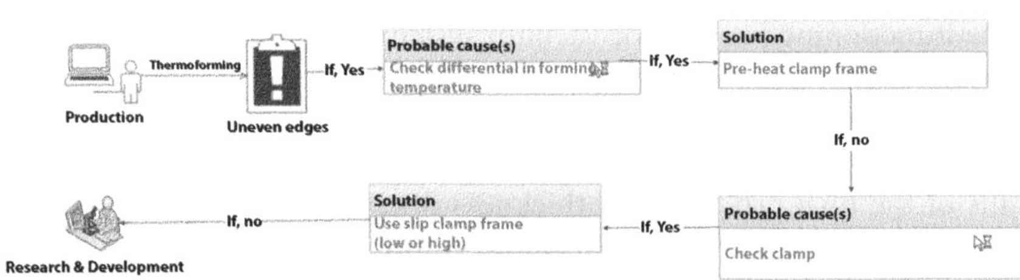

Flowchart 6.11 Troubleshooting process diagram for uneven edges.

6.6.12 Tearing Sheet When Forming

The thermoforming processing temperature must be carefully controlled to prevent tearing of the sheet material. This temperature should be between the glass transition temperature and the crystallization temperature, ensuring that the material is in a semi-solid state and that the molecular chains do not have enough mobility to cause tearing.

One potential cause of tearing at the mold edge is inadequate clearance between the frame and the mold. To address this issue, a troubleshooting process diagram, such as Flowchart 6.12, can be used to identify and minimize the problem.

The production of consistent thermoformed parts relies on the understanding and control of various materials and process variables. Among these variables, the quality of the sheet feedstock is of the utmost importance. Sheets of the same material can vary in terms of polymeric contamination, thickness, thermal stresses, amount of regrind, volatiles, color, gloss, and grain. Additionally, the mold material and mold temperature can vary, further affecting the quality of the thermoformed parts.

Several measures can be taken to eliminate tearing. These include reducing the heating time or intensity, decreasing the level of pre-stretching, optimizing the timing or speed, choosing a thicker material, checking the thickness profile over the width, verifying the material specification, and adjusting the vacuum timing or pressure. It is important to note that as the sheet or film stretches around the returns on form, it becomes thinner. The degree of thinning is influenced by the steepness of the draught. Therefore, it is advisable to minimize the depth of the draw to avoid excessive thinning of the film.

Tearing or blowing holes in a material can occur when it comes into contact with the tool or during the forming process. This can occur if the material is either too hot or cold. In such cases, the heating time or intensity should be adjusted accordingly. Additionally, tearing can also be

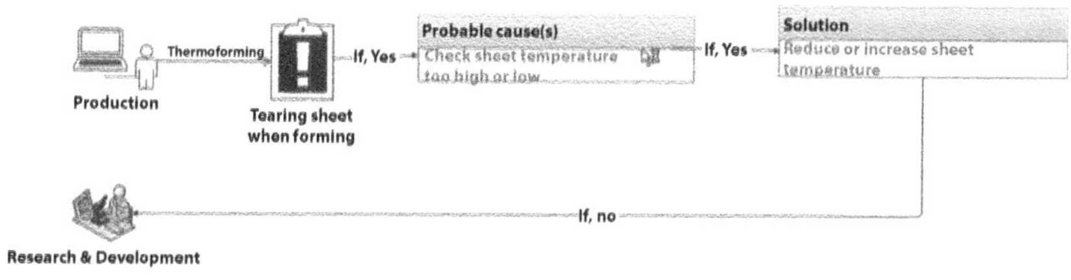

Flowchart 6.12 Troubleshooting process for tearing sheet when forming.

caused by applying compressed air or a vacuum too early. By adapting the delay in applying these forces, this problem can be eliminated.

The temperature of the tool should be increased in cases where tooling is excessively cold. If the edges and corners of the tool are extremely sharp, it is advisable to modify the geometry of the formed part. In situations where the tool surface is rough, polishing is recommended, as a rough tool surface can impede material sliding. To address issues, such as excessive material sag or tool dragging on the index, it is necessary to reduce the heating time or increase the line speed. Insufficient clearance between tooling components can be rectified by adjusting the tool to create greater distances between male and female components. If the speed in the forming press or plug is too fast, adjustments can be made accordingly. In instances where the material gauge or review tooling draw ratios are inadequate for the intended draw, increasing the material gauge is a viable solution.

Polymeric contamination, thickness, thermal stress, amount of regrind volatiles, color, gloss, and grain are factors that can manifest during thermoforming, resulting in tearing, wall thinning, shape distortion, fading, pinholes, and grain distortion [22].

Material tears at the bottom of the formed part or excessively thin formed cavities can occur because the materials become too hot, excessive pre-stretching, late or slow application of the upper plug, excessively thin material, uneven thickness profile over the width, material inclusions, premature application of vacuum, or excessive pressure.

6.6.13 Bad Definition at the Edge

The term "bad definition at the edges" refers to the uneven thickness of the thermoformed sheet material. This issue arises when the hot-sheet material is subjected to vacuum during the forming process, resulting in poor definition at the edges of the forming area. To address this problem, Flowchart 6.13 can be used.

This may be caused by the material being too cold or the part being ejected too quickly, resulting in insufficient dwell time. This can be resolved by increasing the cooling process and adjusting the heating parameters. By increasing the processing temperature, the stress response at the equivalent strain level is weakened, and the onset of strain hardening is significantly delayed. Consequently, this leads to of poor definition at the edge of the product [23].

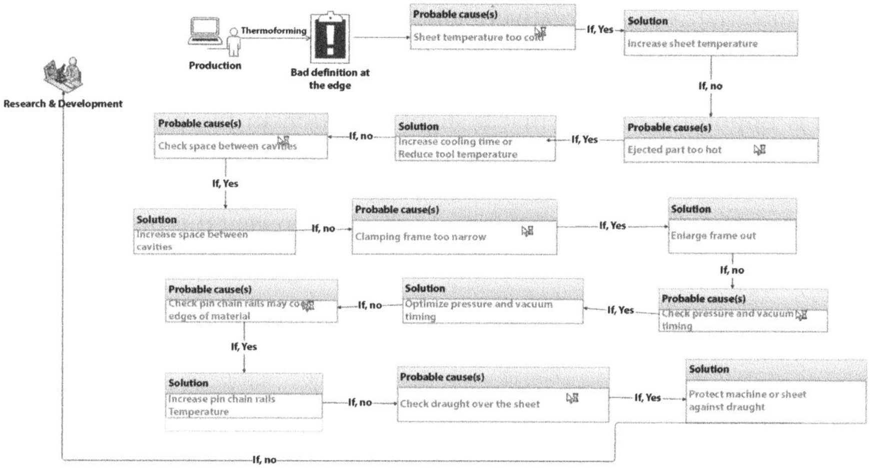

Flowchart 6.13 Troubleshooting process for bad definition at the edge.

6.6.14 Glossy Spots

During thermoforming, deformation of the material causes a reduction in the wall section thickness. During the forming stage, the melt sheet material may not be subjected to sufficient pressure, resulting in a glossy appearance in a particular area. Additionally, even areas with excess wall thickness can develop glossy spots due to shrinkage. These glossy spots are caused by the presence of overheated melt particles in the thermoformed

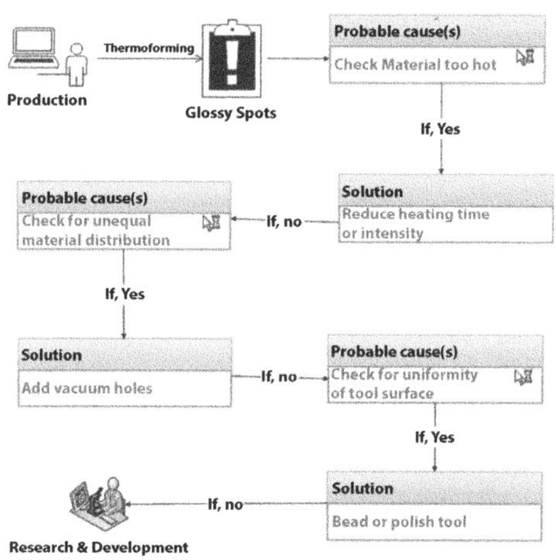

Flowchart 6.14 Troubleshooting process diagram for glossy spots.

sheet material, which negatively affected the overall appearance of the product. The use of Flowchart 6.14 can help address and eliminate this issue [24].

It is important to note that a slight deviation in the heating process can lead to degradation of the polymeric material and the formation of glossy spots. Therefore, it is crucial to maintain the temperature within the appropriate range during thermoforming to prevent this problem from occurring.

6.6.15 Warpage

The occurrence of warpage in plastic products is influenced by various factors, including the physical and mechanical properties of the polymer, contraction and cooling conditions, and the angle effect in the mold and mold wall. However, warpage is highly sensitive to changes in the process parameters, making it necessary to systematically optimize these parameters for plastic products. The primary issue in thermoforming technology is the warpage of products when they are exposed to high temperatures. This problem arises because of the release of internal stresses during the thermoforming process, leading to nonuniform changes in the geometric

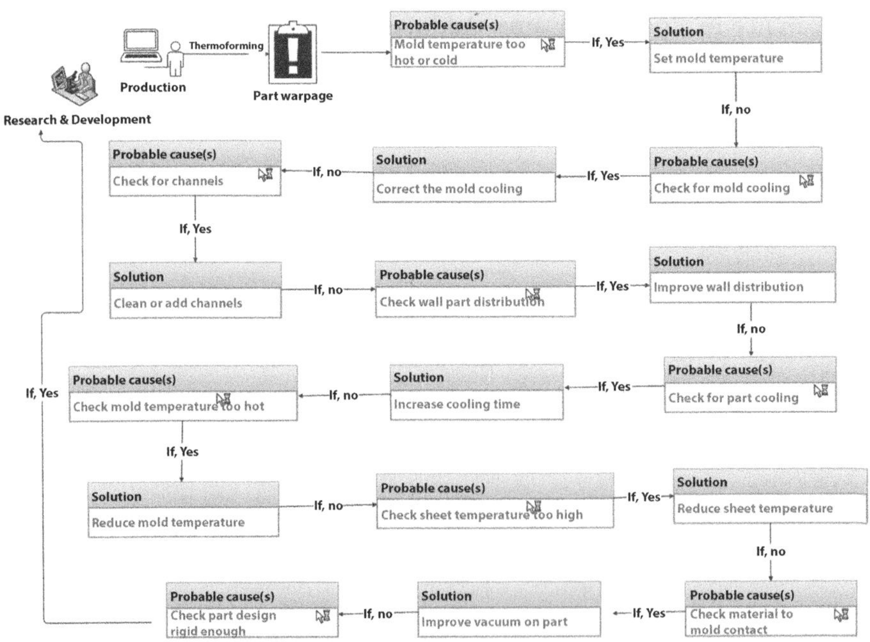

Flowchart 6.15 Troubleshooting process diagram for part warpage.

dimensions of the products. Implementing Flowchart 6.15 can help mitigate this issue.

Warpage refers to the non-uniform alteration of the geometric dimensions of products over time, leading to a distortion of their original shape. Consequently, it is feasible to develop effective strategies to mitigate or eliminate these defects [25, 26]. It is preferable to achieve uniform cooling to preserve the integrity of the design, minimize warpage, and ensure that the residual stress is both minimal and evenly distributed. Warpage is a fundamental flaw inherent to thermoforming technology, and its effects are particularly pronounced when the products are subjected to high temperatures.

6.6.16 Cracking in Corners

The occurrence of cracking in the corners during heating and thermoforming operations can be attributed to the high levels of stress generated. In the thermoforming process, it is necessary to heat the sheet to the appropriate temperature to achieve uniform drawing into the mold. This must be performed at a rate and in a manner that balances the accurate reproduction of the mold with the attainment of adequate wall thickness and minimal stress in the thermoformed parts [27].

One approach for mitigating the risk of cracking in corners is to specify a stress–strain relationship based on the hardness of the plug material. It is important to note that stress–strain can occur during the plug-assist stage, which may contribute to the development of cracks at the corners.

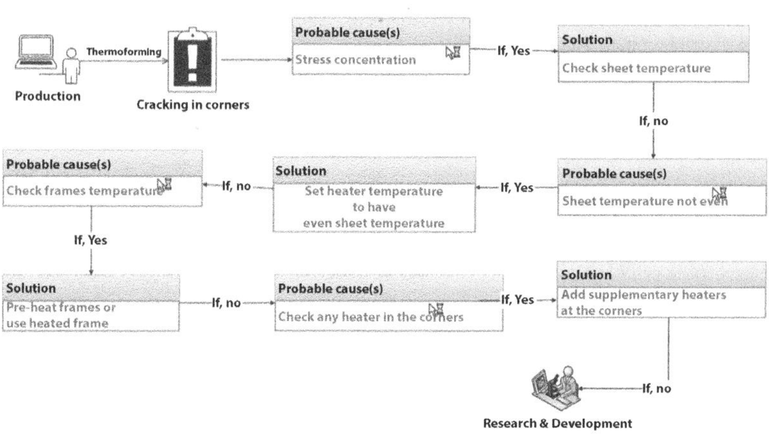

Flowchart 6.16 Troubleshooting process for cracking in corners.

A Flowchart 6.16 is provided to illustrate a troubleshooting method for addressing this issue.

It is crucial to ensure that the sheet is not heated too rapidly, as excessive heat accumulation can damage the sheet and render it brittle when formed. Similarly, the cooling process should not be conducted too quickly, as it can generate stress that leads to cracking in the formed part. Inadequate heating or excessive heat loss during the forming process can result in uneven and high residual stress in the final product. If not properly managed, this residual stress can cause cracking and warping during subsequent reinforcing processes or over an extended period [28].

6.6.17 Raised Corners

The presence of raised corners in the plastic sheet materials can be attributed to the temperature of the frame and the resulting stress on the material. When the hot-sheet material is pushed into the mold, the coefficient of friction may prevent it from fully contacting the bottom of the mold, leading to raised corners. A Flowchart 6.17 outlining the troubleshooting process is provided. Excessive heating of the frames can cause stress on the sheet. Insufficient heating in the middle of the sheet may also contribute to this issue, which can be resolved by applying additional heat.

6.6.18 Surface Marking

Surface markings on thermoformed sheets are primarily caused by the mold because the sheet retains the impression of the mold surface. These markings can be eliminated by adjusting the tool temperature or changing the material until cloudy areas disappear. Defects can also be addressed by modifying the sheet temperature, polishing the mold surface, adjusting the tool or plug temperature, altering the plug or tool speed, and inspecting material transport through the machine.

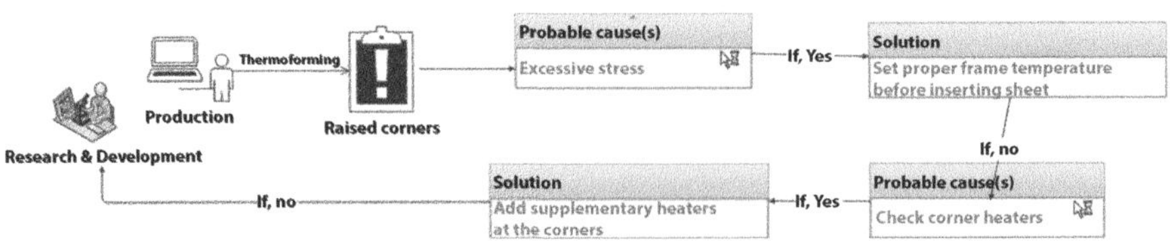

Flowchart 6.17 Troubleshooting process diagram for raised corners.

If the glass transition temperature (Tg) of the thermoforming system is higher than room temperature, it would exhibit excellent scratch and scuff resistance as it would remain in the glassy state. However, if Tg is too high, the system may not possess thermoplastic characteristics at deformation temperatures [29].

Scuffs and blemishes transferred from the mold surface to the forming sheet during processing are known as markoffs. This issue can be frustrating, as it may require scrapping otherwise perfect moldings unless high-cost tooling is utilized [30].

Understanding the presence of surface markings is crucial for identifying the underlying factors that contribute to them during the elastic recovery phase of the material. Flowchart 6.18, depicting the troubleshooting method, is provided for reference.

Cloudy areas or marks on the surface can be caused by the tool-temperature being too low, excessive silicone on the surface, or film defects. Poor transparency, haziness, or blushing of the formed part can result from the sheet or film being too hot (crystallization), the temperature being too low (white break), the mold surface being too rough, the tool or plugs being too cold, the plug or tool speed being too fast, or the material surface being scratched or having abrasions [31].

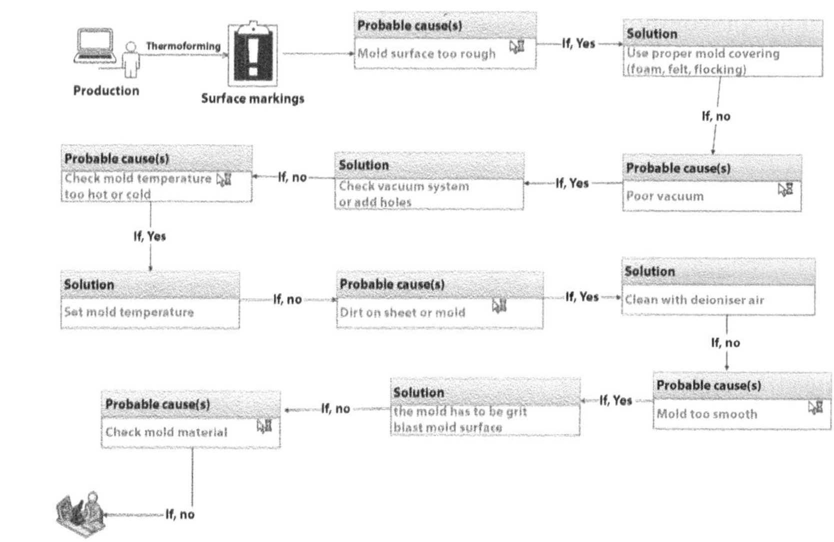

Flowchart 6.18 A systematic troubleshooting process is presented to address issues related to surface markings.

6.6.19 Corners Too Thin

Corners that are excessively thin are primarily affected by factors such as the cooling time, melt temperature, and plug pressure. The plug forcing duration, cooling time, and melt temperature are the most influential factors in the thinning of the corners. In addition, a troubleshooting method is presented in the form of Flowchart 6.19. The application of a vacuum to hot-sheet materials can contribute to thinning. This issue can be resolved by inspecting the radiator or heating plate, increasing the spacing between cavities, adjusting the tool close or plug speed, increasing the material gauge, or re-evaluating the tooling draw ratios. Insufficient spacing between cavities, excessively fast tool closing or plug speed, or an inadequate material gauge for the intended draw can result in cold or hot areas in the heating station and excessively thin walls.

6.6.20 Folds, Webbing, or Wrinkles

In thermoformed PVC foam sheets, the absorption of water can lead to the emergence of defects in the final product, manifesting as wrinkles. This issue primarily arises during the initial introduction of the sheet material into the machine and the subsequent application of temperature. Excessive application of heat can cause the sheet or film to deteriorate, potentially resulting in a modification of its color or premature failure when exposed to outdoor conditions. By following Flowchart 6.20, the occurrence of

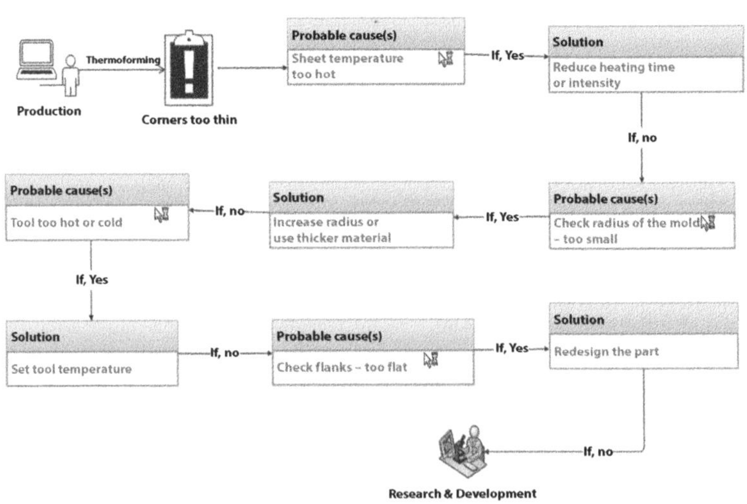

Flowchart 6.19 A troubleshooting process is presents the issue of corners too thin.

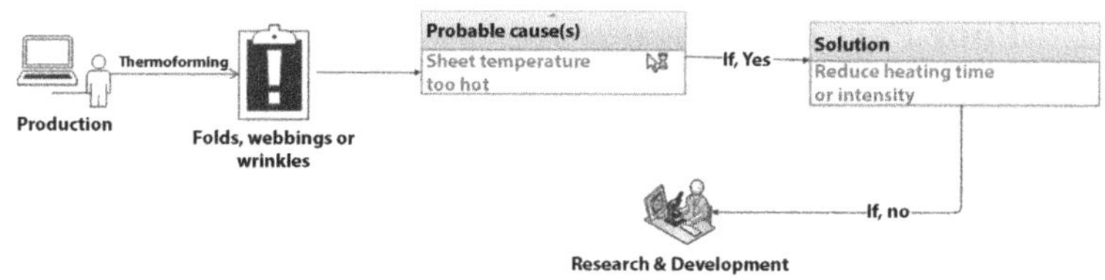

Flowchart 6.20 A troubleshooting process is presented to address issues related to folds, webbings, or wrinkles.

these defects can be mitigated. It is necessary to monitor and regulate the temperature of the radiator (oven) or the heating system. Additionally, any bent or damaged heater should be replaced promptly [32].

6.6.21 Part Deforms During Demolding

During the thermoforming process, the part undergoes deformation, which leads to variations in its shape. Consequently, thermoforming plays a crucial role in determining the performance of the final product. The rate of cooling is a significant factor to consider when a part deforms during demolding. To address this issue, a schematic Flowchart 6.21 is provided to troubleshoot the problem of part deformation during demolding. The deformation of this part primarily arises from the stress exerted on the sheet material during thermoforming. Overheating can contribute to the deformation that occurs after demolding. To mitigate this issue, it is recommended to reduce the temperature and utilize an appropriate sheet material with the correct thickness [33].

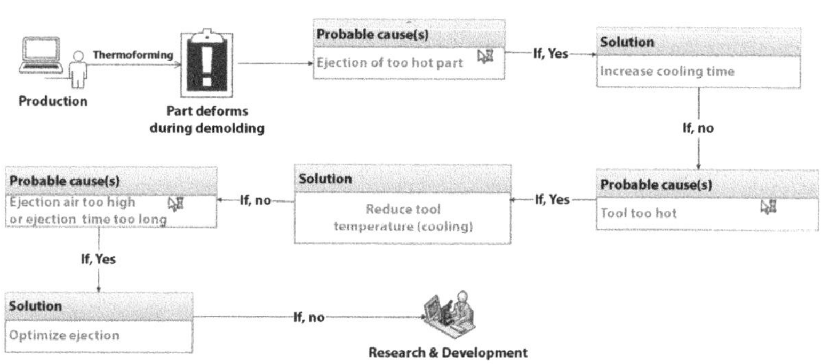

Flowchart 6.21 A troubleshooting process is presented addressing the issue of part deformation during the demolding stage.

6.6.22 Poor Part Detail

The thermomechanical properties of the polymer during thermoforming are influenced by plug velocity, which in turn affects the shape of the final product. Variations in the plug velocity can lead to inconsistent deformation, resulting in poor part detail. Flowchart 6.22 provides a schematic representation of the troubleshooting process. The main cause of poor part detail is the introduction of temperature to the sheet material during thermoforming operations, along with other factors.

If a problem consistently occurs in the same area, increasing the heater temperature may be necessary. Fine details and imperfections in the mold, such as undercuts, can cause the part to become locked to the mold. To prevent this, it is recommended to use generous radii with sufficient draft, ensure that all surfaces are smooth, and avoid using molds that are prone to chipping during thermal cycling at high temperatures. Additionally, it is important to remove this part from the mold as it cools [34].

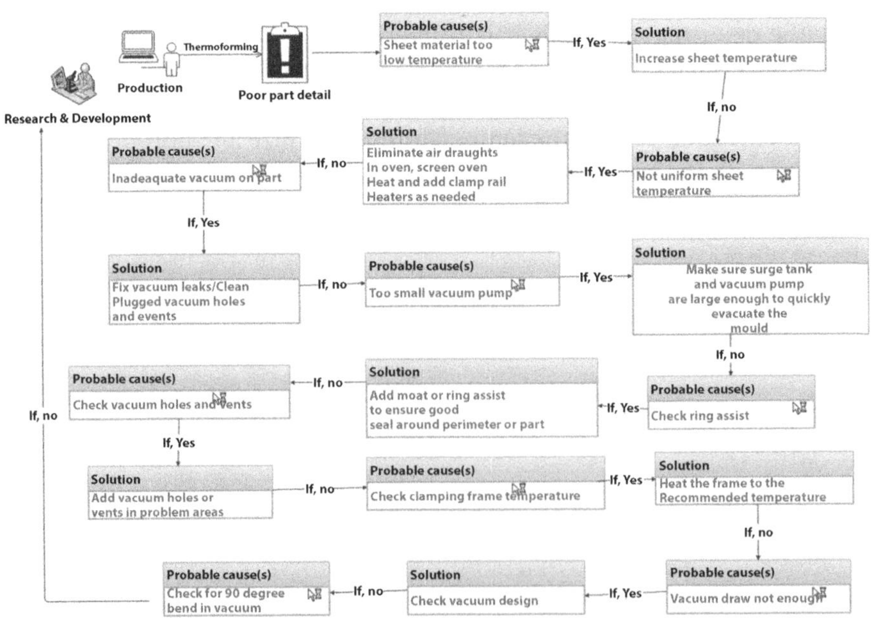

Flowchart 6.22 A troubleshooting process is presented addressing issues related to inadequate part detail.

Uneven sheet temperatures can result in uneven details. It is important to check the temperature of the heaters and their spacing from each other as well as their spacing from the sheet. The sheet should not be placed closer to the heater than the recommended spacing. Any air drafts originating from open doors can prematurely cool the sheet before it can be formed properly. It is also important to identify areas that are too hot and cause excessive stretching [35].

6.6.23 Excessive Post Shrinkage

Shrinkage refers to the dimensional change that occurs over time in a part, often resulting from the mold not being designed with the appropriate shrinkage, as recommended by material data. The phenomenon of post-shrinkage affects both the length and width of the newly produced part. Flowchart 6.23 provides a set of measures to minimize this problem, emphasizing the importance of achieving optimum heating and proper cooling to prevent excessive post shrinkage [36].

The complexity of thermoforming operations can vary, ranging from manual processes to highly automated large-scale operations. Poor product design can result in high rejection rates and poor product quality. Thermoforming costs include various factors such as materials, machinery, and products. To eliminate defects, measures such as checking the compressed air pressure, adjusting part geometry, polishing tool surfaces, changing materials or using release agents, reducing tool temperature, or adjusting heating time/intensity can be taken [37].

Thermoforming is commonly used for parts due to its cost-effective tooling. Maintaining cleanliness in the processing area is crucial for obtaining high-quality parts. Dust and other particles can significantly affect the surface quality of the final product, underscoring the importance of a clean thermoforming process. Each thermoforming process has unique characteristics, and the appropriate parameters for the thermoforming conditions

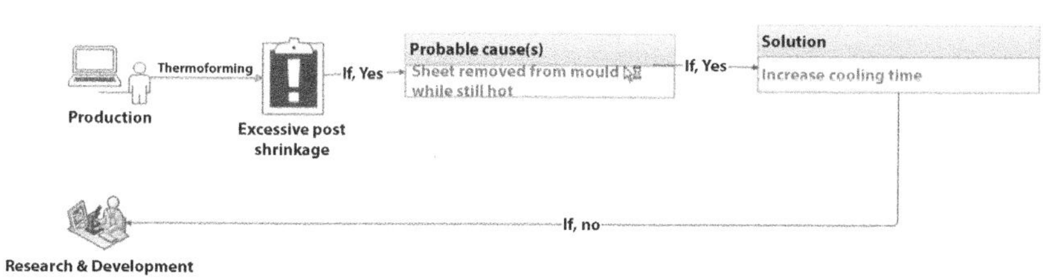

Flowchart 6.23 Troubleshooting process diagram for excessive post shrinkage.

are determined by the specific process. Establishing correct process conditions is essential to ensure the quality of the final product.

In summary, thermoplastic sheet materials have interesting chain molecule properties, all of which are involved in orienting towards the deforming direction in thermoforming. While deforming the sheet materials, they experience the simultaneous influence of external forces such as air pressure, vacuum, and plug movement. In the softening state, deformation, and orientation in the flow direction, the alternative forces supplied by the material's tensile, rheological, and elongational behaviors act against the externally supplied mechanical forces to compensate. These mechanical and alternative forces do not undergo synchronization, and the problems begin with deficiencies in part. However, deficiencies occur, which correspond to solutions in multidimensional configuration such as troubleshooting. Troubleshooting thermoforming leads to optimization of the process to improve productivity. The fundamental ideas for troubleshooting the problem and solutions are provided based on process variables. Troubleshooting and solutions bring together an exact set applicable to product quality.

7
Thermoforming—Optimization

7.1 Introduction

Optimization is a fundamental concept with wide applicability across various disciplines. The primary objective is to identify the optimal solution, set of values, or conditions that maximize a specific objective or performance criterion. This may involve maximizing desired outcomes, minimizing costs or risks, improving efficiency, or achieving predetermined goals.

Optimization serves as a valuable tool for making informed decisions, improving efficiency, and achieving the desired outcomes across diverse fields. The principles of optimization involve identifying the relevant variables, constraints, and an objective function. Subsequently, various algorithms and techniques are utilized to search for the optimal values of the variables that meet the constraints and optimize the objective function. These algorithms can be analytical, heuristic, or metaheuristic in nature depending on the complexity of the problem and available resources [1].

7.2 Thermoforming

Optimization is a critical aspect in the field of thermoforming, particularly when dealing with complex parts or issues, or when aiming to enhance the efficiency and quality of the production process. Thermoforming encompasses a range of operations to convert sheet materials into finished products. The optimization of these operations can yield significant advantages for the industry. By implementing optimization techniques in thermoforming, it is necessary to achieve higher yields, improved product quality, reduced production costs, and increased market competitiveness. Additionally, optimization contributes to the adoption of more sustainable practices by enabling efficient utilization of resources and minimizing waste generation.

Muralisrinivasan Natamai Subramanian. Thermoforming: Processing and Technology, (209–220)

The optimization of thermoforming involves establishing a correlation between the operating variables of each stage and the performance of the part. Because thermoformed parts exhibit non-uniform thickness, which can hinder their performance, the thickness distribution in the final part is commonly used as a variable to evaluate its performance. Furthermore, the effects of processing parameters on the thermoforming of polymeric sheets demonstrate high nonlinearity and full coupling, increasing the complexity of process design [2].

Thermoforming optimization presents significant challenges due to several factors. First, the lack of accurate experimental data hampers the optimization process. While certain aspects, such as sheet temperature, may provide reliable experimental data, other crucial data for various sheet materials are often unavailable. Second, parameter optimization is a time-consuming task that involves comprehensive optimization of numerous parameters at a semi-empirical level. Finally, the parameters involved in thermoforming are highly interdependent. Therefore, any significant alteration in one parameter necessitates corresponding adjustments in multiple other parameters. Optimizing the parameters individually is extremely slow. This interdependency extends not only within the parameters of a single element but also between the parameters of different elements [3].

Thermoforming is a plastic processing method that involves different techniques, such as varying the stretch factors, areas of stretching within the mold, and the application of pre-stretching. Unlike other polymer processing methods, thermoforming requires a separate optimization process due to its distinct characteristics.

The thermoforming process can be used to produce various plastics. The distribution of the wall thickness in thermoforming is influenced by several process parameters, including the coefficient of friction, shape of clamping rings, use of plug assist, temperature distribution on the heated sheet, and design and manufacturing of the mold [4, 5].

7.3 Optimization Process

Optimization techniques, such as modeling and computer simulations, are of great importance in the field of thermoforming. These techniques provide a necessary means to refine processes, select the most favorable parameters, and design configurations that align with specific objectives. The primary goals of thermoforming optimization typically involve improving processing efficiency and product quality. Consequently, modeling and computer simulations are commonly used in thermoforming processes.

The practice of modeling and computer simulation in thermoforming optimization entails conducting controlled experiments to systematically investigate the effects of multiple variables on the process or product. By manipulating the parameters and observing the resulting outcomes, it is necessary to identify the optimal combination of settings that will enable achieving desired objectives. By harnessing these optimization techniques, it is possible to streamline processes, minimize material waste, reduce energy consumption, and ultimately enhance the overall efficiency and sustainability of operations.

7.4 Numerical Modeling

Numerical modeling has emerged as a viable alternative to costly and time-consuming trial-and-error testing methods. In the realm of numerical modeling of thermoforming, previous research has primarily concentrated on the heating and forming stages [6, 7]. Various mathematical approaches, such as implicit finite element modeling, explicit finite element modeling, and computational fluid dynamics modeling, have been used. However, these approaches often involve simplified representations of the processing conditions, such as assuming isothermal conditions or infinite friction. Consequently, there is a growing demand for more accurate representations of actual processing conditions [8, 9].

To address this need, dedicated test rigs have been developed to characterize the deformation response of sheet polymers under thermoforming conditions. The data obtained from these test rigs have facilitated the development of new material models for process simulations, enhancing the treatment of the temperature dependency and material deformation response [10]. Nevertheless, there are still limitations in other areas, such as the treatment of thermal and frictional contacts. Efforts have been made to measure the effects of friction and heat transfer using specialized test rigs. Additionally, the focus is on monitoring the force exerted during the thermoforming process by utilizing laboratory test rigs and force transducers in a plug [11, 12].

Significant advancements have been made in the field of thermoforming process simulation through the utilization of sophisticated material models and the validation of simulation results against experimental data. This development provides researchers with a powerful tool to optimize the thermoforming process and make informed decisions regarding various processing parameters. Ultimately, this advancement leads to improved manufacturing outcomes for lightweight thin-walled products made from thermoplastic materials [13]. However, the progress in simulating and optimizing the thermoforming process using computational tools has been impeded by a lack of knowledge regarding the response of plastic sheet materials under thermoforming conditions and the challenges associated with accurately measuring the associated processing phenomena.

To address these limitations, an experimental program is designed to enhance the understanding of the physics and chemistry underlying the material behavior during the thermoforming process. Therefore, the success of the process hinges on a comprehensive understanding of the behavior of the material and the selection of appropriate constitutive equations to accurately model its response.

The experimental program begins with a preliminary study of several well-known constitutive equations. These equations describe the relationship between the stress and strain in a material and are essential for understanding its mechanical behavior. The objectives of the experimental program are as follows:

1. Gain a thorough understanding of the physics of the material behavior during thermoforming.
2. Investigate and validate different constitutive equations.
3. Use validated constitutive equations to simulate the thermoforming process and optimize the desired outcome.
4. Compare the formabilities of different polymeric materials for thermoforming applications.

By achieving these goals, the experimental program aimed to contribute to the advancement of thermoforming technology, resulting in more efficient and cost-effective manufacturing processes for plastic products.

7.5 Constitutive Model

One of the illustrated flowcharts, labeled Flowchart 7.1, presents a systematic methodology for investigating fluctuations in wall thickness during the thermoforming process. The flowchart outlines the various steps involved in this investigation. The first step involved conducting thermal stability tests to determine the appropriate temperature range for processing the sheet material. Following this, the flowchart depicts the analysis

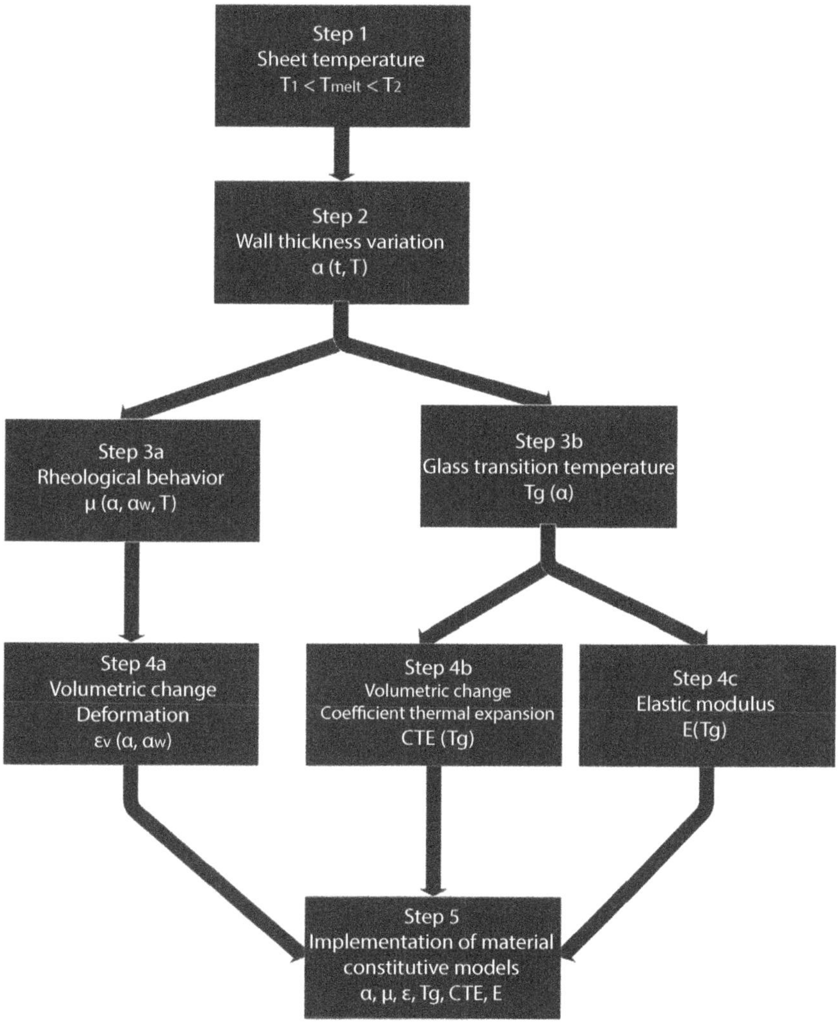

Flowchart 7.1 For the constitutive model for wall thickness variation.

of the temporal evolution and response of the wall thickness variation to temperature changes. Subsequently, step 3a involves the assessment of the rheological properties of the sheet material, whereas step 3b focuses on measuring the glass transition temperature (Tg). These measurements are conducted in relation to the extent of wall thickness variation.

The wall thickness variation (εV) during the forming stage is determined by utilizing the known rheological behavior of the sheet material. This variation is quantified as a function of the degree of wall thickness variation in step 4a. Furthermore, the coefficient of thermal expansion (CTE) (step 4b) and sheet modulus (step 4c) are expressed as functions of the glass transition temperature. These constitutive models of the material can then be incorporated into any finite element software in step 5, allowing for the prediction of the final properties of a composite structure based on the chosen thermoforming cycle [11–14].

In the field of thermoforming simulations, two primary types of constitutive laws are commonly used: hyperelastic constitutive equations and constitutive equations with a memory function. These approaches differ in their fundamental modeling principles. Hyperelastic materials, also referred to as green-elastic materials, exhibit elastic behavior and can be characterized by a strain-energy function. Various models, such as the Mooney–Rivlin, Klosner–Segal, and Ogden models, are derived from the general form of Treolar's equation, which involves elastic constants (Cij) and strain invariants (I1 and I2). A comprehensive understanding of the characterization steps and the development of the material constitutive models can be achieved through experiments and computer software

7.6 Key Findings—Modeling

The primary outcomes of the modeling study underscore the endeavors undertaken, with a particular emphasis on the synthesis of various modeling methodologies. This study highlights the effective application of computational modeling in forecasting wall thicknesses, selecting materials, analyzing rheological behavior, and considering other factors of thermoplastic sheet materials. The study's assumptions have contributed to the attainment of reasonably precise results, which have been communicated effectively. The following section presents the selection of notable modeling highlights.

7.6.1 Selection of Material

Two different grades of ABS have been used in the numerical simulation to evaluate their performance in the thermoforming analysis. The simulation incorporates the material parameters obtained from hot tensile tests. The results of the simulation, which compared the thickness distribution, demonstrated the differences between the two ABS grades in the three-dimensional analysis of the thermoforming process. However, it is important to note that these differences do not necessarily imply the superiority of one grade over another. To achieve a comprehensive understanding and accurate classification of these materials based on thermoforming process parameters, it is essential to employ a more realistic simulation [16].

7.6.2 Wall Thickness

The distribution of wall thickness in plug-assisted thermoformed food containers, which are composed of multiple layers, is influenced by various factors such as the forming temperature, forming air pressure, and heating duration. The position of the wall and side of the container also play a significant role in determining the thickness of the wall [17].

In the process of thermoforming and analysis, it is crucial to consider the variables that affect the properties of the final product, particularly the distribution of the wall thickness. These variables include the temperature of the sheet, depth and timing of the plug, timing of the air, and shape of the plug. The compressive strength of the product is directly related to the distribution of wall thickness and is influenced by the same factors [18].

The objective is to enhance the thermoforming process by minimizing the variations in the wall thickness of finished parts in both the longitudinal and cross sections. This improvement not only enhances the quality of the finished parts but also optimizes the heating time of the sheet, because it directly affects the efficiency of the process. A model is developed based on the relationship between the significant factors affecting the process. The values of these factors are then determined to obtain the most desirable approximated values of the dependent variables. By identifying the optimum values of the independent variables based on the selected dependent variables, a new variable called the total quality is introduced. This variable represents the conversion of the approximated values of multiple dependent factors into a single unit value of total utility [19, 20].

7.7 Mold

The model effectively predicts the distribution of wall thickness in thermoformed products using theoretical analyses involving pressing into conical and truncated molds. The model is based on eight simplifying assumptions that are independent of material properties or forming conditions. The projected wall thickness profiles are aligned with the results obtained from experimental research [21].

7.8 Thermoforming Process

The current study aims to examine the rate at which triangular troughs are manufactured with the objective of identifying analytical solutions for the duration of the forming process. This investigation involves the application of a series expansion and a numerical approach to approximate the time required for thermoforming as well as to assess the sharpness of the trough edges and the presence of frozen-in stress [22, 23].

7.9 Viscoelastic Behavior

The thermoforming process, which involves blowing, stretching, and vacuuming, is simulated using the Mooney–Rivlin model. This model incorporates the viscoplastic behavior and impact of the temperature distribution on the specimen sheet by adjusting the Mooney constant for each element in the finite element analysis. The results obtained from the simulation, which considered distributed temperatures, closely matched the experimental results. This highlights the importance of considering the temperature distribution in the specimen sheet when simulating the thermoforming process. The accuracy of the calculated results is minimally affected by the strain rate of the specimen sheet during thermoforming or the anisotropy of the specimen sheet [24].

7.10 Method of Thermoforming

The objective is to achieve the most favorable outcomes by utilizing both free-forming and plug-assisted forming techniques. While free forming eliminates the need for molds and undesired markings, it does not yield an optimal thickness distribution. Conversely, plug-assisted thermoforming ensures a predominantly uniform thickness but leaves noticeable mold imprints. Consequently, it is advisable to integrate both approaches to optimize their advantages and mitigate their disadvantages [25, 26].

7.11 Heating

Various numerical methods are used in this study, namely the finite element method (FEM), finite difference method, and volume control method [10–15]. Two distinct approaches are commonly observed when dealing with phase changes in materials. The first approach involves independently solving the heat equation for the solid and liquid phases, commonly referred to as Stefan's problem [16]. The second approach involves solving the heat equation concurrently using the enthalpy method [10, 17].

7.12 Cooling

During the cooling stage of the thermoforming process, the temperature distribution within the thermoplastic sheet plays a crucial role in determining the levels of crystallization, shrinkage, and residual stresses in the final thermoformed product. Consequently, a product's quality is heavily influenced by these factors. Accurate simulation of this stage requires precise knowledge of the crystallization and thermophysical properties of the material, which are interconnected through the energy conservation equation.

Various methods have been used to model the cooling stage, which involves infrared heating followed by contact cooling with molds. One approach involves decoupling the energy equation from the crystallization kinetics of the polymer, excluding the term associated with the release of latent heat. This approach is advantageous because of its simplicity of implementation within a numerical model. Another approach involves dividing the physical domain into two regions, solid and liquid, with a dynamic boundary that changes as solidification occurs. However, this method is not well-suited for studying semicrystalline thermoplastics. The final approach allows the simulation of crystallization based on processing conditions and material parameters [27–29].

7.13 Computer Simulation

The primary objective of computer simulations in the field of thermoforming is to accurately determine the distribution of thickness across the entire surface of the final product. This is particularly crucial when dealing with complex three-dimensional shapes, as certain areas, such as corners, edges, protrusions, and indentations, may have thicknesses that fall below an acceptable limit. These defective areas are typically discarded during production. Therefore, computer simulations can aid in the more efficient design of molds, eliminating the need for costly trial-and-error procedures.

By predicting the thickness distribution through computer simulations, mold designers and process engineers can select the optimal choice from a range of alternatives [30].

Thermoforming simulation serves as a valuable tool for enhancing the consistency of thickness distribution in thermoformed products. Various engineering software packages, such as Ansys Polyflow, Abaqus, T-sim, and Pam-form, are based on the principles of Finite Element Analysis. These simulation tools have demonstrated their effectiveness in estimating plastic deformation prior to actual thermoforming processes, resulting in tooling cost savings.

Additionally, the software is capable of simulating heat transfer during the plastic heating phase, considering factors such as positive or negative air pressure and plug assistance. Furthermore, these packages enable the simulation of plastic deformation. With precise boundary conditions, thermoforming simulations can provide accurate results for the final product, including the prediction of thickness distribution [31–33].

Three distinct modeling techniques are used in the simulation of the thermoforming processes [34]. The first technique involves the use of simple mass balance models, that only consider geometric factors and are not influenced by the material behavior. The second technique involves the application of finite element analysis with a membrane approximation, which is suitable for thin-walled parts where the bending resistance is negligible. The third technique involves the use of finite element analysis without membrane approximation, which is used for modeling thick-sheet thermoforming and multilayer thermoforming. Mass balance models have been found to yield inadequate predictions of the wall thickness distribution, even for relatively simple shapes such as truncated cones.

Furthermore, these models are unsuitable for complex three-dimensional shapes. Due to their independence from material behavior, they have limited applicability in material selection or process modification. Although the inclusion of bending resistance and shearing may be crucial in certain cases, most thermoforming problems can be effectively studied using the thin membrane approximation [35, 36]. The following sections provide a brief description of some of the software used in these modeling techniques.

7.14 Polyflow

The focus of the numerical simulation is to study the thermoforming process of three different shapes and examine how various process parameters affect the distribution of thickness and temperature. The use of Polyflow software in numerical simulations helps optimize these parameters and assess their impact on the quality of the end product. By combining simulation results with the knowledge of designers, it is possible to reduce the time required to design a new product. The process parameters analyzed included the initial distribution of temperature, viscosity, forming pressure, and quality of the mold surface. Polyflow software has been used to analyze the influence of these parameters on the distribution of thickness and temperature in the final product. The simulation demonstrates the analytical capabilities of numerical simulations in studying the effects of different parameters on product quality. Integrating these simulation findings with the expertise of designers can lead to a shorter duration of new product designs [37].

7.15 PAM-FORM™ Software

The PAM-FORM™ software has to be used to simulate the thermoforming process of a plastic sheet. It uses the Mooney–Rivlin model, which is a hyper-elastic constitutive law, to simulate bubble inflation and determine the material parameters. The analysis of the thickness distribution and strain during bubble inflation will show a clear connection between the deformation profile and experimental results. Other methods are available for predicting the wall thickness distribution in basic thermoforming molds [38].

7.16 Geometric Element Analysis (GEA)

A commonly used technique in blow molding and thermoforming is the Geometric Element Analysis (GEA) method. This method involves stretching a membrane over a surface with a known geometry without requiring information regarding the polymer properties. When applied to conical thermoforming molds, the GEA method produces equations that can predict the wall thickness distribution for both deep and shallow molds [39].

Validation and verification are two different modeling concepts. A model can fall into one of the following four categories: invalid and unverified, invalid and verified, valid and unverified, or valid and verified. An example of an invalid and verified model is one that accurately incorporates assumptions but deviates significantly from real-world conditions. When different individuals or teams are involved in modeling and programming, the responsibility for validation lies with the modeling person, whereas the programming person focuses on verification [40, 41].

Modeling and computer simulation techniques are valuable tools for optimizing the thermoforming processes. These techniques allow for the examination of various process parameters and variables involved in thermoforming. The primary goal of modeling and simulation is to identify the optimal processing conditions that can enhance the quality of the final product. However, the utilization of diverse modeling techniques and the intricate nature of thermoforming patterns pose challenges for determining the most suitable process variables. Consequently, numerous case studies have been conducted in the field of thermoforming to address this issue.

8

Case Studies

8.1 Introduction

The understanding of thermoforming behavior in different polymers has made significant advancements, driven by a strong emphasis on quality and manufacturing excellence to maintain competitiveness in the field. Thermoforming technology enables the rapid advancement of products to markets and allows intuitive and creative designs [1].

A successful thermoforming process characterizes the properties of materials used as feedstock, typically plastic sheet materials. The properties of materials provide valuable insights for improving processing. These analytical tools are fast and reliable for quality assessment which is important for the successful commercialization of thermoforming technology. These tools ensure that products meet the quality requirements for new applications with the desired standards [2].

Product usage in normal life is of utmost importance. The properties of sheet materials play a significant role in determining the suitability of the final product for their utilization in thermoforming. The mechanical properties of the thermoformed part, such as flexibility and stiffness are influenced by plastic sheet materials [3].

These parts contribute to the development of a response spectrum that encompasses various aspects of the behavior and performance of the material. By studying the product design, the thermoforming process can be optimized to meet consumer standards. It is important to consider the material performance during the thermoforming process. The material ensures that the final product meets the specifications and performance requirements for practical applications.

Case studies are a valuable research method. Thermoforming has been used in various fields. This study involves an in-depth analysis and investigation of processes and products. The research analysis allows for an examination of the design, development, and implementation of innovative projects or products. These studies also provide insights into engineering

Muralisrinivasan Natamai Subramanian. Thermoforming: Processing and Technology, (221–236)

challenges and lessons learned from real-world experiences. Finally, case studies offer knowledge of complex and unique situations. This study provides a rich source of data and contributes to advancements in diverse fields [4].

8.2 Case Studies—Brief Details

A case study in the field of thermoforming involves conducting a comprehensive investigation of a specific problem or issue. The purpose of this study is to gain outstanding knowledge of this subject and to explore the complexities and nuances of this process. The research approach involve the collection and analysis of qualitative data. The data provided detailed and comprehensive outcomes of the thermoforming process [5].

The case study process involved several steps. First, a clear and specific research objective must be defined. This objective serves as a guiding principle throughout the study. A suitable problem is required for the case study. In thermoforming, the wall thickness or geometry of the thermoformed part is a relevant research problem. Data related to the problem must be collected from various sources, such as material parameters, process parameters, and observations, to obtain comprehensive and accurate information.

The collected data must undergo analysis related to the research question. The analysis often involves categorizing the data to identify significant connectivity with thermoforming. The findings, existing literature, and research on the research theme are reviewed. This allows research to relate findings to a broader academic or practical context [6–8].

These findings have been presented in a well-structured report. The interpretations and their implications contribute to the understanding of the research problem and highlight the current limitations [9, 10]. The following section presents two case studies that serve as an initial step in the investigation of the thermoforming process and quality of the resulting products.

8.3 Case Study I

In the food packaging industry, there is a critical juncture at which innovation and necessity intersect. Food packaging is well acquainted with market demand. The market encounters a formidable challenge: the production of food trays that possess three essential qualities: clarity, durability, and cost-effectiveness.

The demand for clear, transparent parts is driven by the desire for an unobstructed view of the enticing content. Achieving a clear and transparent product without compromising durability can be an intimidating task. These parts must be understood in terms of the rigors of transportation, storage, and handling to ensure the freshness and presentation of the delicacies inside.

The search for the optimal solution must encompass not only quality, but also affordability. Cost constraints are a reality that cannot be ignored. The success of a product depends on providing value to customers without increasing production costs. The findings from this study provide an elusive answer that bridges the gaps between clarity, durability, and affordability.

Making prototypes and using thermoforming in production are cost-effective compared to other processing technologies. Experience with a wide range of materials, refining manufacturing processes, scrutinizing details, and conducting trials provides breakthroughs. This provides a material that possesses a rare combination of attributes: thermoforming nucleated polypropylene. This innovative material offers transparency, enabling food trays to showcase their content with unparalleled clarity. Its robustness has rivaled that of the traditional materials. This material provides durability due to its effective stretching and orientation effects during the deformation process of thermoforming. This is indispensable to the packaging industry. However, the true achievement is the affordability of the nucleated polypropylene trays.

This cost-effective option aligns perfectly and provides a solution for overcoming challenges and obstacles. Introducing these thermoformed trays results in extraordinary quality, which serves as a testament to unwavering dedication. These clear trays prominently display the inside of the product, enhance the presentation of products, and elevate the overall experience. The success of a product is attributed to its innovation and excellence. Food trays have set new industry benchmarks.

8.3.1 Challenge

The objective of this case study is to find a solution for producing food trays using nucleated polypropylene sheet material through thermoforming. The food trays should be transparent, durable, and affordable.

8.3.2 Problem Statement

The challenge in the food packaging industry lies in developing a technique to economically produce transparent, durable, and affordable food trays. Consequently, efforts have been made to identify solutions that would meet the requirements for product quality.

8.3.3 In-Depth Analysis

Extensive research and analysis have been conducted on the use of different percentages of nucleating agents in polypropylene, along with calcium stearate in a master batch, to be added to pure PP materials during extrusion. Different sheet materials were produced with proportions of 2%, 3%, and 5% nucleating agents, along with 0.1% calcium stearate in the final sheet material. Calcium stearate was used to reduce the frictional forces during thermoforming.

Of these, nucleated PP sheets, which account for 2% of the total, have proven to be a clear, transparent, and affordable product with a gloss level of 96%. This material has been chosen because of its numerous advantages, such as clarity, durability, and cost-effectiveness, which make it well suited for meeting the industrial requirements of thermoforming production.

The decision to adopt thermoforming PP technology and modify the process demonstrates a commitment to embrace innovation and improve efficiency. To ensure a consistent supply of high-quality thermoforming nucleated PP sheet materials, proactive steps are required to maintain uninterrupted production.

In the case of adopting new materials, there are two challenges during the implementation phase. It is not uncommon to encounter issues related to process optimization. Quality control when integrating new materials and technologies into existing manufacturing processes indicates a proactive and problem-solving approach as necessary adjustments are made.

In this situation, continuous improvement is important. The ability to learn from initial challenges and adapt processes accordingly is likely to result in smoother operations and improved product quality. Overall, the decision to utilize thermoforming PP and the subsequent actions taken reflect a thoughtful and strategic approach to enhancing food tray production and are effective in meeting business objectives.

8.3.4 Quantitative and Qualitative Data

The combination of quantitative and qualitative data supports the utilization of nucleated PP for tray applications with remarkable success. The benefits of this solution are evident in various business aspects. The quantitative data indicate a reduction in production costs compared to previous materials. Evidence from various aspects suggests that nucleated PP in thermoforming allows for easier stretching and deformation within a broad range of thermoforming windows. Improved customer satisfaction provides evidence of its quality and advantages. The product aligns well with performance and needs. Moreover, due to nucleation, the PP material will have improved tensile and impact strength, which enhances the overall efficiency of the supply chain by reducing the risk of product damage during transit.

The highlight of nucleated PP is its perfect balance of properties which makes it suitable for meeting both consumer demands and operational goals. Therefore, the decision to use nucleated PP material and subsequent improvements serve as a testament to the value of data-driven decision-making and strategic material choices. Nucleated PP lead to increased productivity and success, serving as an excellent example for businesses seeking to enhance their products and operations through informed material selection and innovation.

8.3.5 Findings

The strategic decision to adopt thermoforming nucleated polypropylene (PP) in the industry has proven to be advantageous because it allows the production of clear and durable food trays. By utilizing a nucleating agent-grade material, the company has been able to differentiate itself in the market and position its product as a reliable and high-quality food packaging material.

Quantitative data has demonstrated that the use of nucleated PP has streamlined manufacturing processes, resulting in an increase in production efficiency and cost savings. This increased efficiency not only reduces costs, but also enhances productivity, enabling the industry to effectively meet customer demand, the clarity and durability of the food trays reflect the material's ability to meet consumer expectations and satisfaction. This in turn increases the likelihood of repeat buyers and positive recommendations for the industry.

Additionally, the recyclability of nucleated polypropylene (PP) provides potential sustainability benefits. The successful implementation of the material demonstrates a commitment to innovation and to stay ahead in the market. This can enhance trust and credibility among existing and potential markets.

With a strong market position and reputation for quality products, the material is well-positioned for long-term growth and market expansion. The adoption of PP materials with a nucleating agent grade in sheet manufacturing and thermoforming highlights the significance of conducting research, analysis, and making data-driven decisions to achieve business goals. By utilizing this material, the industry has not only enhanced its product offering, but also strengthened its position as a dependable and competitive player in the food packaging industry. The success of this initiative showcases the advantages of making strategic material choices and their potential to drive growth and success in business operations.

8.3.6 Immersive Chronicle

A small food-packaging company faced a critical decision that would have a significant impact on its future direction. In response to the increasing demand for food trays, the company aimed to distinguish itself from its competitors and provide a product that would exceed customer expectations. Driven by the desire to innovate and provide the best possible solution, the company embarked on extensive research and exploration endeavors.

The company entered the materials industry with the goal of achieving a perfect balance between transparency, durability, and cost-effectiveness. Countless hours were dedicated to analyzing data, scrutinizing market trends, and studying the properties of various materials. Through meticulous research and careful consideration, the company ultimately discovered that utilizing polypropylene (PP) with a high-grade nucleating agent to create thermoformed sheets was the optimal solution.

This discovery ignited excitement, as thermoforming nucleated PP has the potential to revolutionize the production of food trays and establish new industrial standards. To implement the findings, the company recognized the necessity of investing in cutting-edge thermoforming polypropylene (PP) technology. This is determined to turn the vision into reality.

As the machines commenced operation and started producing the first thermoformed nucleated polypropylene (PP) food trays, the process eagerly anticipated the realization of the envisioned outcomes. The trays emerged, boasting unparalleled clarity and unwavering sturdiness that surpassed anything already seen.

The initial challenges encountered were unwavering determinations. The company collaborated closely with the production team and fine-tuned the manufacturing processes to achieve optimal efficiency. Each obstacle presents an opportunity for learning and improvement, showcasing the company's unwavering commitment to excellence.

The introduction of trays into the market elicited a highly positive and enthusiastic response. Consumers were greatly impressed by the trays' remarkable clarity, which effectively highlighted the visual attractiveness of the enclosed food. Furthermore, the exceptional durability of the trays has been widely praised, serving as a testament to the unwavering dedication of the industry in manufacturing top-notch products. As each contented customer surfaced, the organization's reputation experienced a significant boost, solidifying its position within the market.

These trays had a positive impact beyond the scope of customer satisfaction. Distributors and retailers also underwent substantial transformations. Durable thermoformed polypropylene (PP) food trays provided efficient protection for the products during transportation, effectively preventing any damage and instilling confidence in every stage of the supply chain.

As the company's success story gained traction, it served as an exemplary model of the potential that could be realized through a visionary approach. The industry was motivated by its eagerness to adopt innovative approaches and its commitment to making decisions based on data, encouraging others to aspire for excellence. They rapidly established themselves as reliable and preferred partners for fulfilling food packaging requirements and embodying trustworthiness.

Today, the previously unassumed food packaging company has emerged as a prominent player in the industry and undergone significant transformation. The trajectory from a pivotal juncture to extraordinary achievement exemplifies the significance of tenacity, importance of data analysis, and willingness to embrace transformative shifts.

Individuals' captivating storyline persists as they progress, overcome obstacles and establish new standards within the industry. With the utilization of thermoforming polypropylene (PP) as a protective covering, these entities exhibit unwavering resilience, consistently maintaining a competitive edge, and making a lasting impact in the realm of food packaging.

8.4 Case Study II

In the highly competitive automotive manufacturing industry, a prominent company has found itself at a critical juncture. With the increasing demand for advanced vehicles, companies have recognized the urgent need to enhance the quality of interior parts, specifically thermoformed components such as dashboards, door panels, and center consoles. These components play a crucial role in defining the aesthetics, performance, and overall safety of vehicles.

The company is motivated by an unwavering commitment to innovation and a relentless pursuit of perfection. The thermoformed interior parts must be revolutionized. Maintaining a competitive edge in the industry requires meeting the ever-growing expectations of discerning automotive consumers.

To achieve this, the complexities of material science, production processes, and design aesthetics must be delved into deeply. Countless hours must be dedicated to exploring new materials, experimenting with innovative manufacturing techniques, and pushing the boundaries of what was previously deemed possible.

The passion and dedication must be fueled by a vision of creating interior parts that not only exceed industry standards but also enhance the driving experience to new levels. Although faced with a significant challenge, their determination remains unwavering.

The results of these efforts must be undeniable: thermoformed interior parts that embody exceptional performance, impeccable safety standards, and a strong commitment to cost-effectiveness. These outcomes should be remarkable. Dashboards must showcase a perfect combination of form and function, seamlessly integrating advanced technologies while capturing the eye with sophisticated design elements. The door panels exuded a sense of strength, providing passengers with a feeling of security and comfort during each journey. The center consoles became the epitome of convenience, offering drivers intuitive control and seamless connectivity.

Automotive interiors must be designed to reflect the unique lifestyles of customers and enhance their enjoyment while driving. In addition to aesthetic appeal, safety enhancements must be prioritized, demonstrating a commitment to protecting all individuals on the road. Each component was carefully engineered to meet rigorous safety standards, providing an additional layer of assurance for both the drivers and passengers.

The positive impact of these advancements has reverberated in the automotive industry. The company's improved thermoformed interior parts have become a benchmark for excellence, setting new standards that others aspire to achieve. Vehicles have gained admiration and recognition, earning customers worldwide trust and loyalty. However, these achievements are only the beginning of the journey. Empowered by their success, the company continues to push the boundaries of innovation, explore new possibilities, and reimagine the future of automotive interior design.

8.4.1 Issue

The assigned responsibility to manufacture essential automotive components for the interior design of vehicles. It must surpass the aesthetic properties of existing products.

8.4.2 Problem Statement

The purpose of this study is to address the challenges faced by a renowned automotive manufacturer in improving the efficiency, safety, and cost-effectiveness of thermoformed interior components. Components such as dashboards, door panels, and center consoles play a crucial role in enhancing the overall interior aesthetics of automobiles. Manufacturers aim to enhance these elements to meet the growing demands of the automotive industry.

8.4.3 In-Depth Analysis

After a thorough examination of the limitations and issues associated with the current thermoformed interior parts, it is evident that the industry has taken a proactive approach to seeking a suitable solution for the customer. The following steps and actions have been implemented to address these challenges:

1. Limitations of the Existing Thermoformed Interior Parts:
   a) Insufficient Durability: The current interior parts are prone to wear and tear, resulting in shorter lifespans and increased replacement costs.
   b) Compromised Safety: The parts may not have met the necessary safety standards, potentially jeopardizing passenger safety in the event of accidents or impacts.
   c) Aesthetic Deficiencies: The visual appearance of the thermoformed parts did not meet customer expectations or align with the desired design aesthetics, compromising the overall appeal of the vehicle's interior.
2. High Manufacturing Costs:
   a) Material Costs: The current manufacturing process may have utilized costly materials, resulting in increased production expenses.

b) Inefficient Process: Inefficiencies in the manufacturing process result in higher labor costs, longer production times, and increased waste.
c) Collaborative Approach: The customer's collaboration demonstrates a commitment to finding a long-term solution to the identified issues. This partnership likely facilitated a deeper understanding of customer's specific needs and challenges.
d) Thorough Evaluation: The team conducted a comprehensive evaluation of the current manufacturing process, material selection, and quality control measures, demonstrating a commitment to understand the underlying causes of the limitations and inefficiencies.
e) Extensive Research: The industry's commitment to extensive research signifies its dedication to finding innovative and effective solutions. This research may have involved studying the latest materials and manufacturing technologies available on the market.
f) Recommendation for Advanced Materials: By proposing the utilization of advanced materials, the team aimed to tackle durability and safety concerns related to interior components. These materials offer improved resistance to wear and tear, enhanced impact resistance, and superior performance characteristics.
g) Performance and Safety Enhancement: The proposed utilization of advanced materials is likely to enhance the overall performance and safety of thermoformed interior parts, meeting or exceeding industry standards.
h) Cost-Effectiveness Consideration: The team's focus on cost-effectiveness demonstrates an understanding of customer profitability challenges. The selected advanced materials were assessed not only for their performance and safety advantages, but also for their impact on overall manufacturing costs.
i) Prototyping and Testing: It is probable that the team developed prototypes using the recommended advanced materials to validate their performance and durability. Rigorous testing may be conducted to ensure that the parts meet all safety requirements.
j) Continuous Support and Improvement: The industry's commitment to ongoing support and continuous

improvement indicates a long-term partnership with customers. Monitoring the performance of newly implemented solutions and making necessary adjustments will help maintain customer satisfaction and profitability.

By addressing the identified limitations and manufacturing challenges through a collaborative and comprehensive approach, the industry aims to provide customers with thermoformed interior parts that are more durable, safer, aesthetically pleasing, and more cost-effective. This initiative highlights the value of proactive problem solving and customer-focused engineering solutions.

8.4.4 Analysis of Quantitative and Qualitative Data

This study incorporated a comprehensive range of both quantitative and qualitative data. The results can be classified into the following categories:

One noteworthy quantitative measure is the observed improvement in the overall performance. This improvement is a result of the successful integration of advanced materials, such as sheet materials from the blends of PC/ABS and ABS/PMMA, into the production of thermoforming parts. Customers reported this improvement, indicating a positive and noticeable impact.

Another quantitative measure is the notable increase in impact resistance due to the presence of the ABS material, which provides the impact strength to the product. This surpasses industry standards and serves as a testament to the heightened safety of the interior parts. This provided a measurable and valuable outcome.

Furthermore, a significant reduction in manufacturing costs is achieved through the optimization of material utilization and the implementation of improved production techniques. This reduction represents a tangible financial advantage resulting from the manufacturing process of thermoformed interior parts.

The customer has provided feedback on the qualitative attributes of the new thermoformed parts, including their superior quality, appearance, and longevity. Although these attributes cannot be precisely measured in numerical terms, the blends provide valuable insights into the customer's perception of the product's overall characteristics and performance.

(a) Polycarbonate/Acrylonitrile–butadiene–styrene (PC/ABS) blend:

The PC/ABS blend sheet exhibited a favorable combination of attributes that make it well-suited for thermoforming applications. By combining the strengths of polycarbonate (PC) and acrylonitrile–butadiene–styrene (ABS), this blend achieved a harmonious array of characteristics.

The advantages of using PC/ABS blend sheets for thermoforming include their robustness and rigidity, resilience and impact resistance, resistance to chemicals and weathering, versatility, thermoformability, and cost-effectiveness. These properties make it a reliable option for a wide range of industries including the automotive, electronics, and medical sectors.

(b) Acrylonitrile–butadiene–styrene/polymethylmethacrylate (ABS/PMMA) blend:

The ABS/PMMA blend is a highly advantageous material for thermoforming because it combines the desirable properties of acrylonitrile–butadiene–styrene (ABS) and polymethyl methacrylate (PMMA). This blend offers many benefits, including exceptional strength and toughness, excellent clarity and transparency, a well-balanced set of properties, resistance to chemicals and weathering, versatility, and the ability to be easily thermoformed. As a result, this material is well suited for a diverse range of applications in industries that require both mechanical performance and optical clarity.

8.4.5 Customer Satisfaction and Appreciation

Qualitative indicators of project success include customer satisfaction and appreciation. The feedback received from customers, expressing contentment with the solution and gratitude for the resulting cost savings and improved profitability, serves as evidence of the positive impact the implemented solution has had on their business and overall experience.

A comprehensive analysis should encompass both quantitative and qualitative data. Quantitative data provide precise measurements and numerical evidence of the improvements achieved, whereas qualitative data offer deeper insights into customer satisfaction, preferences, and the overall perception of the solution. By combining both types of data, a comprehensive and compelling case study can be presented, effectively demonstrating the success of the project and its positive impact on customers' business.

8.4.6 Findings

The implementation of upgraded manufacturing processes, as well as the introduction of PC/ABS and ABS/PMMA blend sheet thermoforming, has proven to be a significant measures for improving the performance, safety, and cost-effectiveness of thermoformed interior parts. By surpassing industry standards, customers have not only enhanced their products, but have also established themselves as frontrunners in automotive interior design. This achievement demonstrates the effectiveness, originality, and positive impact of the solution on customers' business. These outcomes are the result of the team's diligent efforts, unwavering commitment, and expertise in achieving a successful outcome. It is evident that these efforts have had a significant impact on the automotive industry and have strengthened customers' competitive edge.

Continuing to build upon these accomplishments and exploring new avenues for improvement and innovation is of utmost importance. Given the constantly evolving nature of the automotive industry, maintaining and expanding the customer's position as a leader in the field will require staying at the forefront of advancement.

8.4.7 Captivating Account

Maintaining a competitive advantage is of paramount importance in the dynamic and ever-evolving field of automotive design and manufacturing. When a prominent manufacturer of automotive interior parts is approached by a customer facing unique challenges, it presents an opportunity to make a significant impact. Customers have been grappling with obstacles that necessitate innovative solutions. The existing thermoformed interior parts are deficient in terms of performance, safety, and cost-effectiveness, impeding their ability to meet industry standards and satisfy the growing demands of the clientele. Given intensifying competition, it is imperative to find a way to differentiate themselves in the market.

Recognizing the urgency and significance of the task at hand, a comprehensive analysis of the customer's manufacturing process, materials, and product specifications is conducted. Through this meticulous examination, we aimed to identify key areas for improvement and present a comprehensive plan to overhaul the approach.

The core of the solution involves the integration of PC/ABS and ABS/PMMA blend sheet thermoforming, which enhances the quality and durability of the thermoformed interior parts. These materials have been carefully selected to meet the most stringent safety standards and optimize cost-effectiveness. Additionally, a proposal has been made to upgrade the manufacturing process to enhance efficiency and consistency, ensuring that the end products consistently surpass industry benchmarks.

The transformation commenced, and as the upgraded parts rolled off the assembly line, the results have been remarkable. Positive feedback from customers has been pouring in, leaving no doubt that collaborative efforts have had a significant impact. The enhanced parts not only met but also exceeded their expectations.

In addition to satisfying customers, the success of our solution is also demonstrated by tangible results. Improved product performance has increased demand and sales, propelling customers to unprecedented levels of success. However, this story does not end. Customers' achievements did not go unnoticed in the fiercely competitive automotive industry.

The news quickly spread about automotive manufacturers that have revolutionized interior parts, and soon it is acknowledged as a leader in automotive interior design. Their newfound status as trendsetters has attracted the attention of influential industry players and potential partners, opening doors to exciting opportunities and collaborations.

To share this compelling success story with a wider audience, a comprehensive case study is conducted. The objective should showcase that the dedication to providing innovative solutions is not mere rhetoric but rather the driving force behind the customer's success.

By presenting this case study to the automotive community, and the intention is to inspire others to adopt the same approach. It is firmly believed that through continuous exploration of improvements in products and processes, the entire industry can elevate itself, providing customers with superior, safer, and more cost-effective solutions.

As reflected on this remarkable journey, it serves as a reminder of the power of collaboration and expertise. Extraordinary results can be obtained by working closely with customers. It has been a source of pride to know that the solutions not only address challenges, but also create opportunities for growth and success.

With each success story, the industry reaffirms its commitment to delivering innovative solutions that drive success for its customers. The automotive industry is a dynamic and evolving landscape. However, it must be enthusiastic about leading the way, inspiring progress, and shaping mobility for future generations.

By presenting this case study to the automotive community, the intention is to inspire others to adopt the current approach. It is firmly believed that through the continuous exploration of product and process enhancements, the entire industry can elevate itself. In turn, this will provide customers with superior, safer, and more cost-effective solutions.

As reflected in this remarkable journey, it serves as a reminder of the power of collaboration and expertise. By working closely with customers, extraordinary results can be achieved. The solutions not only address challenges, but also create opportunities for growth and success.

With every success story, there is a commitment to delivering innovative solutions that drive customer success. The automotive industry will always be a dynamic and evolving landscape. However, it is about being at the forefront of innovation, inspiring progress, and shaping the future of mobility for generations.

8.5 The Significance of Case Studies in Thermoforming

Case studies are real-world situations that enhance efficiency and product quality. These studies help identify the most suitable materials for specific applications based on their performance. They provide insights into the root causes of defects in products and assist in devising effective solutions. Case studies, which are real-life examples, can be utilized to analyze processes and products. In product and mold design validation, these studies can help validate and identify the necessary areas for improvement. This study helps assess the performance of thermoformed products in various industries, such as automotive, packaging, and aerospace.

Thermoforming involves a wide range of thermoplastic sheets. Every plastic sheet exhibits distinct properties and behaviors during the forming process. Case studies hold considerable value in industry for several reasons. They allow the optimization of the thermoforming process to enhance efficiency and product quality. Modeling and simulation work with assumptions and predict one variable at a time, mainly focusing on property-oriented relationships. Case studies play an important role in thermoforming processes. These case studies have led to several application-oriented problem-solving methods in industries, such as packaging, automotive, and agriculture. In the future, thermoforming will have even more applications, including surprising uses in robotics.

9
Applications

9.1 Introduction

Technological advancements have significantly enhanced the competitive advantage of plastic sheets and thermoformed products over non-plastic materials, such as paper/paperboard, glass, and aluminum. Plastic sheets can be produced through extrusion or casting methods, and are available in a wide range of colors, textures, and levels of transparency. Thermoformed products manufactured from plastic sheets have experienced significant growth, especially in the packaging industry [1]. The popularity of thermoformed products can be attributed to their versatility and cost effectiveness. One of the main advantages of thermoforming is its ability to create complex shapes and design.

Furthermore, thermoforming offers benefits in terms of both cost and time efficiency. Thermoformed products can be produced quickly and at a lower cost compared to other manufacturing methods. Additionally, thermoformed products offer hygiene benefits, making them suitable for industries that require strict cleanliness standards [2].

9.2 World of Thermoformed Products

The versatility of thermoformed products has led to their utilization in various sectors, including food and healthcare packaging, construction, appliances, signage, automotive, and aviation. The selection of materials is based on factors such as cost, working properties, and processability. Thermoform parts can be customized to meet specific requirements, including consideration of mechanical properties in relief geometries. The mechanical properties not only impact the visual appeal of the part but also determine its strength. For instance, part with ridges can affect the overall aesthetics of a product. Incorporating patterns and reliefs on the walls of thermoformed parts not only enhances the value of the parts

Muralisrinivasan Natamai Subramanian. Thermoforming: Processing and Technology, (237–260)

that convey information but also improves the overall consumer experience. This approach effectively combines functionality and design to create packaging that is both practical and visually appealing [3].

9.2.1 Packaging

The demand for transparency in packaging is driving the industry to explore different shapes, materials, and food presentations within packaging. A popular choice for thermoforming applications that require gas and moisture barriers, especially in the production of dairy and drinking cups, fat tubs, fruit and vegetable trays, flowerpots, biscuit trays, meat trays, and microwave food trays, is a combination of polypropylene (PP) and ethylene vinyl alcohol (EVOH) in a multilayer sheet. This combination allows for reliable and durable packaging suitable for retortable applications, making it a preferred option across various industries [4, 5].

Plug-assisted thermoforming and vacuum forming are essential industrial processes that contribute significantly to the production of deep-drawn plastic food packaging and various other products in different sectors. These processes are crucial for meeting the packaging demands of various food products and for meeting the requirements of different industries.

Retortable packaging offers several advantages over traditional metal cans and glass containers, making it the preferred choice for food products that require heat treatment or have diverse applications. It has revolutionized the food packaging industry by offering secure, transparent, and adaptable solutions for a variety of food products. Its ability to withstand heat treatment, compatibility with microwaves, and resistance to high temperatures makes it ideal for packaging fresh and cooked food items, providing consumers with convenient and dependable options [6–8].

The selection of specific thermoplastic materials and thermoforming processes play vital roles in the production of food packaging and other plastic products. Each material possesses unique characteristics that make it suitable for various applications, ensuring safe and effective packaging of food and other items to meet consumer needs [9–11].

9.2.2 Agriculture

Thermoformed high-density polyethylene (HDPE), low-density polyethylene (LDPE), and polypropylene (PP) products are commonly used in horticulture and agriculture to produce seedling trays, plant pots, and protective covers. These materials possess desirable characteristics such as moisture retention and durability, making them suitable for these applications. Additionally, their resistance to moisture and chemicals further enhances their suitability for use in these sectors [12, 13].

Thermoformed polystyrene (PS) trays find applications in nurseries and gardening centers for initiating and nurturing seedlings. On the other hand, thermoformed polyethylene terephthalate (PET) containers are used in the horticulture industry for plant pots and trays. These containers offer an excellent growing environment and can be easily recycled [14].

Thermoformed polycarbonate is commonly used in the construction of greenhouses because of its ability to resist ultraviolet (UV) radiation and excellent light transmission properties. This material allows sunlight to pass through while safeguarding plants from harmful UV rays and extreme weather conditions [15, 16].

Thermoplastic olefin (TPO) materials are commonly used in agricultural equipment, including tractor hoods, fenders, and interior components. The impact resistance and weatherability of TPO materials make them advantageous for use in farm machinery exposed to harsh conditions [17].

Bioplastic pots and planters are used in gardening and horticulture for the cultivation and transportation of plants. These bioplastic alternatives can be directly planted on the ground, providing a more sustainable option than traditional plastic pots. Bioplastic trays are also used in nurseries and greenhouses for the germination and propagation of seedlings. These trays can be planted directly into the soil along with seedlings, reducing transplant shock and minimizing waste [18].

Thermoformed polylactic acid (PLA) or polylactide trays and pots are used in horticulture and gardening for plant growth and transportation. Similar to bioplastic alternatives, these trays and pots can be directly planted into the soil along with seedlings, reducing transplant shock and minimizing waste [19].

9.2.3 Appliances and Consumer Goods

Thermoformed LDPE products can be custom-designed and fabricated to meet specific requirements in various industries ranging from specialty packaging to unique product designs. Similarly, thermoformed LLDPE products can be custom-designed to suit specific requirements in various industries, including specialty packaging and unique product designs. Thermoformed PP is a cost-effective and easily customizable material, making it a popular choice for everyday consumer products, such as storage containers, stationery organizers, and display stands. Thermoformed

PS is commonly used in the manufacturing of consumer goods such as cosmetic packaging, toy components, and stationery organizers [20].

High-impact polystyrene (HIPS) is a versatile material that can be used in a wide range of industries. It is frequently used to create custom enclosures for electronic equipment, power supplies, and specialized devices. Additionally, HIPS is often utilized in the manufacturing of stationery items such as rulers and pencil boxes because of its high impact resistance and ease of manufacturing. It is also used in the production of various household appliances and consumer goods, including refrigerator liners, washing machine components, vacuum cleaner housings, and toys, due to its moldability and impact resistance [21].

Thermoformed ABS parts are commonly used for exterior casings and structural components of household appliances such as refrigerators, washing machines, and vacuum cleaners due to their high impact strength and heat resistance [22]. Thermoformed PVC clamshells are widely used for packaging and displaying small consumer goods such as electronics, toys, and hardware items. These clamshells provide visibility to the product while securely encasing it, reducing the risk of tampering. Thermoformed PET is used in manufacturing a wide range of consumer goods, including storage containers, organizers, and household items [23].

PMMA, which is known for its excellent light transmission properties, is widely used in light fixtures and luminaires. Thermoformed PMMA is specifically used to create diffusers, lenses, and covers for LED and fluorescent lighting systems [24]. Thermoformed TPEs are used in household appliances to provide impact resistance and enhance user comfort. They can be found in refrigerator-door gaskets, dishwasher components, and oven handles. Thermoformed TPOs are used in various consumer products, including storage containers, cases for electronics, and home appliances [25].

PPO, known for its high-quality appearance, chemical resistance, and ease of thermoforming, is suitable for manufacturing consumer goods, such as luggage, briefcases, and protective cases for electronic devices. Thermoformed PPO is commonly used for exterior housings of various appliances, such as coffee makers, toasters, blenders, and vacuum cleaners, because of its high impact strength, heat resistance, and dimensional stability.

PLA can be formed into various office supplies, such as clipboards, document trays, and penholders [26]. Thermoformed PP/EVOH products are commonly used for cosmetic packaging because of their excellent barrier properties to moisture and oxygen, which protect cosmetic products from degradation and contamination. These products are frequently used in packaging makeup, creams, lotions, and other cosmetic products. Thermoformed PC/ABS products can also be utilized in various other applications, including luggage, office furniture, and toys [27].

9.2.4 Analytical and Laboratory Equipment

PEEK is used in the production of sample vials, chromatography columns, and various laboratory apparatus due to its chemical inertness and ability to withstand high pressure and temperature. Thermoformed COC, on the other hand, has applications in the manufacturing of laboratory equipment such as petri dishes, cuvettes, microplates, and other diagnostic devices due to their chemical resistance and convenient sterilization properties. The precision-molding capabilities of COC enable the fabrication of microfluidic devices and lab-on-a-chip systems utilized in medical diagnostics and chemical analysis [28, 29].

9.2.5 Automotive and Aerospace Industry

Thermoformed high-density polyethylene (HDPE) is utilized in the automotive sector to manufacture components, such as truck bed liners, mud flaps, and interior trim pieces. Its advantageous properties include toughness, chemical resistance, and abrasion resistance, making it suitable for these applications. Similarly, thermoformed low-density polyethylene (LDPE) is used in the automotive industry to create various interior components such as door panels, trim pieces, and storage compartments.

LDPE's ability to dampen vibrations and its impact resistance are valuable properties for these applications. In addition, linear low-density polyethylene (LLDPE) is used in the automotive sector to produce components, such as truck bed liners, mud flaps, and protective covers. Durability and resistance to environmental factors are beneficial for these applications. Thermoformed polypropylene (PP) is another material used in the automotive industry to create interior parts, such as door panels, dashboards, and trim components. Its lightweight nature and durability make it the preferred material for automotive applications.

Polystyrene (PS) is used to create interior automotive parts, such as dashboard panels, door panels, and trim components. High-impact polystyrene (HIPS) is also used in the automotive sector to produce interior parts, such as door panels, dashboard components, and trim pieces. Its impact resistance and ease of fabrication make it suitable for such applications. Acrylonitrile butadiene styrene (ABS) is widely used in the automotive sector to manufacture various interior and exterior components, including dashboards, door panels, interior trim, wheel covers, and exterior body parts. This material is selected due to its excellent impact resistance and durability. ABS is also utilized in the aerospace and aviation industries for interior components in aircraft cabins as well as for structural parts where a combination of strength and low weight is required. Polyvinyl chloride (PVC) is thermoformed to manufacture automotive interior components, such as door panels, dashboard covers, and trim pieces. Its durability and ease of customization make it a suitable choice for these applications [30, 31].

Polyethylene terephthalate (PET) is used for manufacturing various automotive parts, such as interior trim, protective covers, and packaging for automotive components. Polymethyl methacrylate (PMMA) is used in the aerospace industry for windows, canopies, and cabin components due to its lightweight nature and high strength-to-weight ratio. Thermoformed PMMA is used in various automotive components, such as light covers, tail light lenses, interior trim, and instrument panels. They offer excellent optical clarity and weather resistance, making them suitable for outdoor applications. Thermoformed polycarbonate is used in aerospace applications, including aircraft windows, canopies, and windshields, due to its high strength-to-weight ratio, ability to withstand extreme temperatures, and resistance to impact. Thermoformed polycarbonates are used in the automotive industry for headlight lenses, sunroofs, windows, and interior components. Its lightweight nature, impact resistance, and optical clarity make it an ideal choice for these applications [32].

Polyether ether ketone (PEEK) has applications in the automotive sector for various components, such as gaskets, seals, connectors, and electrical insulators. Its resistance to chemicals, wear, and heat makes it suitable for demanding automotive applications. PEEK is also used in the aerospace industry for thermoformed parts, such as brackets, clips, and housing, because of its high strength-to-weight ratio, resistance to aerospace fluids, and resistance to high temperatures [33]. Thermoplastic olefin (TPO) materials are extensively used in the automotive industry to manufacture interior and exterior components. They are used in producing bumper fascias, body-side moldings, door panels, interior trim parts, and other structural components. The impact resistance and lightweight nature of TPOs make them popular choices for automotive applications.

Polyphenylene oxide (PPO) is thermoformed to create interior components for aircraft, such as overhead storage bins, seatback trays, and window reveals, due to its good balance of weight savings and performance. Thermoformed PPO is also used in aerospace applications for interior components, seat parts, and structural elements, due to its lightweight nature, flame resistance, and mechanical strength. PPO's high dimensional stability, excellent mechanical properties, and resistance to automotive fluids make it an ideal choice for manufacturing automotive components such as instrument panels, interior trims, door handles, and radiator grilles. Cyclic olefin copolymers (COCs) are suitable for various aerospace applications, including windows, lenses, and interior components, because of their light weight, transparency, and resistance to chemicals and radiation. The high light transmission, resistance to UV degradation, and heat resistance of COCs make them excellent materials for automotive lighting applications, including headlamp lenses and LED light covers [34].

Polyoxymethylene (POM) is suitable for manufacturing automotive components such as gears, bearings, fuel system components, door handles, and interior trim parts, due to its high strength, rigidity, and dimensional stability. Polyvinylidene fluoride (PVDF) products are used in the automotive sector for making fuel and brake lines, as well as other fluid-handling components, due to their chemical resistance and ability to withstand high temperatures. Thermoformed PVDF is also used in aerospace and defense applications including aircraft components, fuel and hydraulic system parts, electrical connectors, and cable insulation. Its lightweight nature, high strength, and resistance to extreme temperatures make it valuable for these industries [35].

Liquid crystalline polymers (LCPs) are used in the aerospace and defense sectors because of their light weight, high strength, and high-temperature resistance. They have applications in the production of aircraft components, satellite parts, and protective covers for electronic devices. Thermoformed thermoplastic composites (TPCs) are used to create interior components for private jets and helicopters, offering lightweight and aesthetically appealing solutions. TPCs are also used in automotive interior components, such as door panels, instrument panels, center consoles, and seat backs, providing weight reduction, improved impact resistance, and design freedom for car manufacturers. In the aerospace industry, TPCs are used for interior components such as overhead bins, seatbacks, and cabin panels, helping to reduce aircraft weight and improve fuel efficiency. Thermoformed polycarbonate/acrylonitrile butadiene styrene (PC/ABS) products are commonly used in the automotive industry for various applications, including dashboards, instrument panels, door panels, and interior trims. These parts were chosen for their strength, impact resistance, and dimensional stability. Thermoplastic composite materials made of continuous fibers (carbon, glass, aramid, etc.) and a polymer matrix are used because of their good mechanical properties associated with a low density. Commercial and military aircraft make great use of composite materials in their airframes and primary structures. These parts, made by thermoforming in the automotive industry, reduce the mass of the components and, consequently, the fuel consumption of vehicles [36–38].

9.2.6 Blister Packaging

Thermoformed polystyrene (PS) is commonly used in blister packaging for small consumer goods, electronics, and pharmaceutical products. This material provides visibility and protection to the packaged items. Thermoformed polyvinyl chloride (PVC) blister packs are widely used in the pharmaceutical industry to securely hold individual tablets, capsules, or medical devices. These packs offer protection, visibility and ensure accurate dosages [39].

9.2.7 Building/Construction Products

High-density polyethylenes (HDPEs) are used in construction for various applications including access panels, wall cladding, and protective covers. Their resistance to moisture and chemicals makes it suitable for outdoor and industrial use. Thermoformed linear low-density polyethylene (LLDPE) products are used in the construction industry as vapor barriers, roofing membranes, and protective covers for construction materials. The weather resistance and flexibility of LLDPE make it advantageous for these purposes [40].

Thermoformed polystyrene (PS) is used to create decorative moldings, trims, and architectural components. Expanded polystyrene (EPS), a type of thermally expanded polystyrene, is commonly used as insulation in construction due to its excellent insulating properties and lightweight characteristics. High-impact polystyrene (HIPS) is used in the construction sector to produce various interior and exterior components such as decorative moldings, false ceilings, and protective coverings [41].

Acrylonitrile butadiene styrene (ABS) has been used in the construction of architectural models, signage, and cladding elements. Its ease of formation and the ability to replicate intricate details make it a preferred choice. Polymethyl methacrylate (PMMA) is used in the manufacture of furniture and home decor items, including chairs, tables, and decorative pieces. Its versatility in terms of color and shape makes it an attractive option for designers [42].

Thermoformed polycarbonate is utilized to create light diffusers, lenses, and covers for LED lighting fixtures. It also offers light dispersion, impact resistance, and thermal stability. Thermoformed thermoplastic elastomers (TPEs) are used in furniture and upholstery due to their ability to create soft and comfortable surfaces. They are commonly used in chair pads, armrests, and seating cushions [43].

Thermoplastic olefin (TPO) materials are used in the construction industry to produce roofing membranes, wall cladding, and other building components. In particular, TPO roofing membranes are popular due to their weather resistance and energy efficiency. Thermoformed polyoxymethylene (POM) is used in furniture hardware, such as drawer slides, locks, handles, and hinges, due to its smooth surface, low friction, and durability [44].

Thermoplastic composite materials (TCPs) are utilized for architectural elements, such as cladding, façade panels, and roofing materials due to their lightweight nature and weather-resistant properties. Thermoformed ABS/PMMA products are also used in interior design for various applications such as countertops, shelves, and light diffusers. These products are favored for their durability, ease of cleaning, and the ability to be produced in a range of colors and finishes [45, 46].

9.2.8 Electrical and Electronics

Low-density polyethylene (LDPE) is used in the electronics industry to produce protective covers and packaging for small electronic devices such as smartphones, tablets, and computers. Thermoformed linear low-density polyethylene (LLDPE) is also used in the electronics industry to create protective covers and packaging materials for electronic devices. These materials have advantageous electrical insulating properties for these applications [47].

Thermoformed polypropylene (PP) is used in the electronics industry to produce protective cases, covers, and trays for electronic devices and components. Thermoformed polystyrene (PS) is used to manufacture protective cases and packaging for various electronic devices including small appliances, cell phones, and accessories. It is also used to create protective packaging for delicate or fragile items, such as electronics, consumer goods, and medical devices [48].

High-impact polystyrene (HIPS) is used to create protective casings for electronic devices, such as remote controls, gaming consoles, and computer peripherals. Acrylonitrile butadiene styrene (ABS) is commonly used in the production of consumer electronics due to its electrically insulating properties, toughness, and lightweight nature. It is frequently used to manufacture items such as computer and laptop casings, TV and monitor frames, remote controls, and other electronic device housings [49].

Thermoformed polyethylene terephthalate (PET) is used to create protective packaging for electronic devices during shipping and handling. PET is also used in the electronics industry for packaging electronic components and accessories such as mobile phone cases and protective covers. Thermoformed polymethyl methacrylate (PMMA) is used in electronics, including display screens, touch panels, and light-guide plates for LED displays. Its optical properties, including high light transmittance, make it an ideal choice for such applications [50].

Thermoformed polycarbonate is used to manufacture protective covers for electronic devices, such as smartphones, tablets, and laptops. Its transparency allows users to see the screen clearly, while protecting the device from scratches and impacts. Polyether ether ketone (PEEK) is used in electrical connectors, insulators, and switches due to its excellent dielectric properties, high-temperature resistance, and inherent flame resistance [51, 52].

Thermoplastic olefin (TPO) materials are used in the electronics industry to create protective covers and housing for electronic devices and electrical components. Due to its inherent flame resistance, low moisture absorption, and good dielectric properties, thermoformed polyphenylene oxide (PPO) is widely used in electrical and electronic applications such as connectors, switches, sockets, circuit breakers, and motor housings [53].

Cyclic olefin copolymers (COCs) are utilized in the production of electronic device components, such as smartphones, tablet display windows, camera lenses, and protective covers, where optical clarity and scratch resistance are crucial. The high optical clarity, low birefringence, and good dimensional stability of COCs make them a preferred material for producing optical lenses, light guides, and electronic displays [54].

Polyoxymethylene (POM) is preferred for electrical connectors, switches, insulators, and other components used in electrical and electronic equipment due to its low moisture absorption and good electrical insulating properties [55, 56]. Bioplastic materials can be used for the protective packaging of electronics and other fragile products, reducing the environmental impact and waste generated from traditional packaging materials. Polylactic acid (PLA) is used in the manufacture of protective packaging for electronic devices and components [57, 58].

Thermoformed foam plastics are used to create protective packaging for electronics, such as smartphones, tablets, laptops, and other electronic devices, safeguarding them from damage during transit. TPCs are used to produce casings and enclosures for electronic devices, providing protection and improved thermal management. TPCs are used in electronic device casings, covers, and accessories due to their lightweight and customizable properties [59].

Thermoformed PP/ethylene vinyl alcohol copolymer (EVOH) products are also used for electronic packaging because they have excellent barrier properties to moisture and other gases. This helps protect the electronic components from corrosion and moisture damage. Thermoformed PP/EVOH products are commonly used in packaging circuit boards, memory cards, and other electronic components [60]. Thermoformed polycarbonate/acrylonitrile butadiene styrene (PC/ABS) products are also used in the electronics industry for various applications, including housings, enclosures, and circuit boards. These parts are chosen for their strength, lightweight nature, and good electrical properties [61].

9.2.9 Health-Care Packaging

Foam plastic products are used in disposable food containers, plates, cups, and trays, providing insulation for hot and cold items. Thermoformed LDPE products are extensively used in the food service industry to create disposable cups, plates, bowls, and utensils. They are preferred for their hygienic properties, cost-effectiveness, and ease of disposal, making them ideal for fast-food restaurants, catering services, and other food-related businesses [62].

Thermoformed polypropylene (PP) is used to manufacture disposable cutlery, plates, cups, and food containers for catering and food service industries. Thermoformed polystyrene (PS) is used to create disposable cutlery such as spoons, forks, and knives, as well as single-use plates and bowls for parties, events, and food service applications. Thermoformed PS is commonly used for packaging food items, particularly takeaway containers, cups, plates, and trays. Its insulation properties help maintain the desired temperature in hot or cold foods [63].

Thermoformed polyvinyl chloride (PVC) is used to produce disposable cups, containers, and lids commonly used in the food and beverage industries [64]. Thermoformed polyethylene terephthalate (PET) is used in the food service industry as disposable containers and trays to serve and store food. These containers are suitable for takeaway meals, ready-to-eat salads, bakery products, etc.

The excellent gas barrier properties and transparency of COC make it suitable for food packaging applications, particularly for products that require an extended shelf life and protection against external factors [65].

PLA-based thermoformed products are commonly used in single-use cutlery, plates, and bowls at events and gatherings. PLA cups are widely used for cold beverages, and PLA lids offer a sustainable alternative to traditional plastic lids. PLA can be thermoformed into disposable cutlery such as spoons, forks, and knives. These items are used in various food-service establishments and events. PLA is frequently used to produce trays, clamshells, cups, and food-packaging containers. It is an excellent material for takeaway containers, salad boxes, and baking items.

Thermoformed PP/EVOH products are ideal for food packaging because of their excellent barrier properties to oxygen, moisture, and other gases. This helps extend the shelf life of food products and protects them from spoilage. Thermoformed PP/EVOH products are commonly used for packaging meat, cheese, fruits, vegetables, and other perishable foods [66]. Thermoformed polycarbonate/acrylonitrile butadiene styrene (PC/ABS) products are used in the food service industry for a variety of parts, including food containers, trays, and lids. These parts are chosen for their strength, food safety, and heat resistance. The food-processing industry utilizes thermoformed PEEK to manufacture machinery. This material is preferred because of its ability to withstand repeated exposure to hot water, steam, and cleaning chemicals, making it suitable for use in the food industry.

9.2.10 Medical Applications

In the medical field, thermoformed HDPE is used to produce a range of equipment components, including instrument trays, medical device housings, and protective covers. The sterilizability and chemical resistance of the material are crucial in medical settings. LDPE, known for its chemical resistance and non-reactivity, has been applied in the medical and pharmaceutical contexts. Thermoformed LDPE products are used to create medical trays, vials, specimen containers, and protective packaging for medical devices [67].

Thermoformed LLDPE products are used in the medical field to manufacture items, such as medical packaging, disposable trays, and protective covers. The cleanliness, chemical resistance, and non-reactive properties of LLDPE are essential for these applications. Due to its resistance to chemicals and moisture, thermoformed PP is used for the packaging of medical instruments, devices, and pharmaceutical products. It is also used in the production of medical device components such as trays, housings, and packaging for sterilized medical products due to its hygienic and sterilizable nature [68].

Thermoformed PS is utilized in the healthcare industry to create sterile packaging and trays for medical devices and pharmaceutical products [69]. HIPS is used to manufacture various medical equipment components, including sample trays, petri dishes, test tube holders, and disposable medical instruments. ABS has applications in the medical industry for the creation of equipment casings, enclosures, and instrument panels. Its ability to withstand repeated sterilization processes and resistance to chemicals makes it suitable for medical applications [70].

Thermoformed PVC is used in the production of medical products such as trays for medical devices, packaging for diagnostic kits, and other disposable medical equipment. Thermoformed PET is used to create custom trays and protective packaging for medical devices, ensuring sterilization and safe transportation. PET is also used to manufacture blister packs for medicines, medical device packaging, and diagnostic test trays due to its ability to protect products from contamination and provide visibility [71].

PMMA has applications in medical settings such as dental products (e.g., denture baseplates), surgical instrument components, and protective covers for medical equipment. Their biocompatibility and clarity are advantageous in these contexts. Thermoformed polycarbonates are commonly used in the medical industry for items such as face shields, medical equipment covers, and surgical instrument components. Its durability, optical clarity, and sterilization ability make it suitable for medical applications [72].

PEEK is widely used in the medical industry to produce surgical instruments, orthopedic implants, and spinal implants. Its biocompatibility, radiolucency (allowing X-rays to pass through), and excellent mechanical properties make it an ideal material for critical applications. TPO is used in the medical industry for manufacturing certain medical device components as well as equipment housing and enclosures, due to its chemical resistance and ease of sterilization [73].

The biocompatibility and resistance of PPO to gamma radiation make it suitable for use in medical device applications, such as IV connectors, surgical instruments, and various components of medical equipment. COC's excellent barrier properties, high transparency, and resistance to moisture and chemicals make it ideal for packaging sensitive medical and pharmaceutical products, such as syringes, vials, blister packs, and inhalers [74].

The biocompatibility and resistance of POM to chemicals make it suitable for certain medical device applications including syringes, catheter components, and surgical instruments [75]. LCPs are used in the medical industry to manufacture surgical instruments, implantable devices, drug delivery systems, and medical tubing. Their biocompatibility, chemical resistance, and ability to withstand sterilization processes make them suitable for medical applications [76].

Thermoformed bioplastics are used in medical device packaging such as sterile trays and clamshell containers. The biodegradable nature of these materials is advantageous in medical settings, where disposables are common. Thermoformed PLA is used in the packaging of medical and pharmaceutical products due to its biocompatibility and biodegradability. Thermoformed foam plastics are used for medical device packaging, providing protection to sensitive medical equipment, and ensuring sterility. They are also utilized in orthopedic and prosthetic devices for cushioning and support. TPCs have applications in the medical sector for creating lightweight and strong components in imaging equipment, patient beds, and mobility aids [77].

Thermoformed PP/EVOH products are also used in pharmaceutical packaging due to their excellent barrier properties to moisture and oxygen. This helps to protect pharmaceutical products from degradation and contamination. Thermoformed PP/EVOH products are commonly used as packaging tablets, capsules, liquids, and other pharmaceutical products. Thermoformed PC/ABS products are used in the medical industry for various purposes, including surgical trays, patient monitors, and IV bags. These parts are chosen for their strength, sterility, and good chemical resistance [78].

Thermoformed ABS/PMMA products are also utilized in the medical industry for a variety of applications such as trays, surgical instruments, and patient monitors. These products are preferred due to their durability, sterility, and ability to be manufactured in various shapes and sizes.

9.2.11 Packaging

Low-density polyethylenes (LDPEs) are extensively used in the packaging industry to manufacture lightweight and durable containers, trays, clamshells, and blister packs. These thermoformed packaging solutions offer excellent protection for a wide range of products, including food, electronics, pharmaceuticals, and consumer goods.

Linear low-density polyethylenes (LLDPEs) are widely used in the packaging industry to create flexible and durable packaging solutions. They are commonly used to produce food containers, pouches, lids, and wraps. The high strength and puncture resistance of LLDPE make it ideal for safeguarding and preserving various products, including food, industrial goods, and consumer products.

Thermoformed polypropylene (PP) is commonly used for packaging in food, pharmaceutical, and consumer goods industries. It is used to create trays, clamshells, blister packs, and containers that provide protection and display products effectively. High-impact polystyrene (HIPS) is commonly used in packaging applications due to its excellent impact resistance and lightweight properties. It is used to manufacture disposable cups, plates, trays, containers, clamshells, and protective packaging for electronics and fragile items.

Thermoformed acrylonitrile butadiene styrene (ABS) is used as a protective packaging solution, custom tray, and transport case to ensure the safety of delicate and valuable items during shipping [79]. Thermoformed polyvinyl chloride (PVC) is commonly used to produce trays for packaging and display various products such as electronics, cosmetics, pharmaceuticals, and food items. These trays provide protection and organization for products during transport and retail displays.

While polyethylene terephthalate (PET) bottles are typically blow-molded, thermoforming can be utilized to create specialized or custom-shaped bottles for specific applications. Thermoformed PET is widely used in packaging applications due to its clarity, rigidity, and barrier properties. It is commonly used in food trays, blister packs, clamshell packaging for electronics and consumer goods, and cosmetic packaging. Thermoformed polymethyl methacrylate (PMMA) can be used in packaging applications where transparency, impact resistance, and aesthetics are essential. It is commonly used in cosmetic packaging, food displays, and protective covers for electronics.

Thermoformed polycarbonate is used in packaging applications where product visibility is desired. It is commonly used in blister packaging and clamshell packaging for consumer goods and electronics. Thermoformed thermoplastic olefins (TPOs) are used in the packaging industry to create protective and durable packaging solutions. They are commonly used in protective cases, shipping containers, and trays for various products.

Thermoformed cyclic olefin copolymers (COCs) are used to create protective packaging solutions for sensitive electronic components, medical devices, and high-end consumer products. It provides both physical protection and a clear view of the contents [80]. Liquid crystal polymers (LCPs) are utilized in specialized packaging applications, where barrier properties, temperature resistance, and dimensional stability are crucial. They are used to package sensitive electronic devices, pharmaceuticals, and food products.

Bioplastic trays, containers, and blister packs are commonly used in food packaging and consumer products to reduce the environmental impact of single-use plastics in the packaging industry. Thermoformed polylactic acid (PLA) is used to create attractive and eco-friendly retail packaging materials for various consumer products. Thermoformed foam plastics are widely used as protective packaging materials. They are often used to create custom-fit cushioning and inserts for delicate and fragile items during shipping and transportation. Thermoformed polypropylene/ethylene vinyl alcohol copolymer (PP/EVOH) products can also be used for various other applications, such as packaging toys, household products, and industrial components [81].

9.2.12 Semiconductor Industry

Polyether ether ketone (PEEK) is used in the semiconductor industry for thermoformed parts, including wafer carriers, sockets, and handling trays. Their purity, chemical resistance, and high-temperature stability are advantageous in cleanroom environments. Polyvinylidene fluoride (PVDF) is used in the semiconductor industry for manufacturing components, such as wafer carriers, tubing, and fittings. Its high purity and low extractability make it suitable for critical applications in semiconductor manufacturing processes [82].

9.2.13 Signs/Displays

Thermoformed plastics are commonly used to create signs and displays due to their ability to be molded into various shapes and visual appeal. Thermoformed low-density polyethylene (LDPE) is commonly used to create attractive point-of-purchase displays in retail settings. These displays are used to showcase products and attract customer attention. Thermoformed polypropylene (PP) is utilized in retail settings to create attractive and custom-designed product displays to attract customers' attention. Thermoformed polystyrene (PS) is utilized to create eye-catching displays and products for retail stores and promotional events.

High-impact polystyrene (HIPS) sheets are used for advertising displays, point-of-sale signage, and exhibition panels due to their ease of printing and cutting. Acrylonitrile butadiene styrene (ABS) is often used to create eye-catching and durable point-of-purchase displays, retail packaging, and product stands. Its ease of formation allows for creative and intricate designs [83].

Thermoformed polyvinyl chloride (PVC) sheets are used to create outdoor signage, displays, and promotional materials due to their weather resistance and ability to hold vibrant graphics. PVC is also used to create eye-catching point-of-purchase displays to promote products in retail stores, supermarkets, and other commercial settings. Thermoformed PVC is used to create various types of display stands, holders, and racks for retail and trade show applications. These displays help showcase products effectively and attract customer attention.

Polyethylene terephthalate (PET) sheets can be thermoformed to create visually appealing displays and signs. The transparency and excellent printability of PET makes it suitable for retail displays, point-of-sale advertising, and other promotional materials. Polymethyl methacrylate (PMMA) is a popular material used for creating signs, point-of-purchase displays, and retail fixtures. Thermoforming allows for the creation of complex shapes and enhances the visual appeal of advertising and marketing materials.

Thermoformed polycarbonate is used to create signage, point-of-sale displays, and advertising materials due to its optical clarity and weather resistance [84].

Thermoformed cyclic olefin copolymer (COC) is often used to create clear and visually appealing displays for products in retail settings, as its transparency enhances product visibility and aesthetics [85]. Thermoformed acrylonitrile butadiene styrene/polymethyl methacrylate (ABS/PMMA) products are commonly used for signage because they are durable, weather resistant, and can be easily customized. They are often used to create signs that are exposed to elements, such as outdoor and traffic signs.

9.2.14 Sports and Recreation

High-density polyethylene (HDPE) is used in the manufacturing of various recreational products, including kayaks, canoes, sleds, and playground equipment. Their lightweight and durable properties make them suitable for these applications. Low-density polyethylene (LDPE) is suitable for creating toys and recreational products due to its safety, flexibility, and impact resistance. Examples of such products include inflatable items, pool toys, and children's play equipment.

Linear low-density polyethylene (LLDPE) is used in the production of various sporting goods such as protective gear, water sports equipment, and recreational items. Their impact resistance and lightweight nature are valuable properties for these applications. LLDPE is also used in the manufacturing of toys and recreational items, such as inflatable structures, pool floats, and play equipment, due to its flexibility and toughness.

High-impact polystyrene (HIPS) is widely used in the education sector to create model kits, project displays, and educational aids. This is due to its ease of shaping and affordability. Acrylonitrile butadiene styrene (ABS) is a popular material for making toys and games because of its ability to be molded into various shapes and child-safe properties. Thermoformed ABS is used to manufacture various sports and recreational equipment, such as protective gear, helmets, sporting equipment enclosures, and recreational vehicle components [86, 87].

Thermoformed polymethyl methacrylate (PMMA) has been used in the creation of aquariums, terrariums, and other transparent enclosures for pets and exhibits. Its optical clarity allows a clear view of its contents. In sports and recreation, thermoformed polycarbonate is used in items such as face shields for helmets, protective goggles, and visors for masks. Impact resistance helps safeguard athletes during various activities. Thermoformed polyether ether ketone (PEEK) is used in the production of high-performance sporting goods such as tennis racquets, snowboard bindings, and bicycle components, where its lightweight and strong properties are advantageous.

Thermoplastic elastomers (TPEs) are used in the footwear industry to produce comfortable shoe insoles, midsoles, and outsoles. They offer excellent shock absorption and cushioning properties, making them suitable for use in athletic and casual shoes. TPEs are also popular materials for children's toys and childcare products, as they are free from phthalates and BPA, and provide a soft, safe, and durable surface. TPEs are used in the production of sporting goods, such as shoe soles and handle grips for bicycles and sports equipment, due to their elasticity, cushioning, and excellent grip characteristics.

Thermoplastic olefin (TPO) materials are used in the manufacturing of outdoor and recreational equipment because of their resistance to UV radiation, chemicals, and impact. Examples include kayak hulls, playground equipment, and outdoor furniture. Polyphenylene oxide (PPO) is suitable for producing sports equipment, such as helmets, protective gear, and equipment casings due to its impact resistance and lightweight nature.

Polyoxymethylene (POM) is used in various sports and recreational equipment, such as ski bindings, skateboard wheels, and kayak fittings, due to its toughness and resistance to impact. Thermoplastic composite materials (TPCs) are used in the sports and recreation industry to produce lightweight and high-performance equipment such as bicycle frames, snowboards, skis, and helmets.

9.2.15 Transportation

HDPE is utilized to create protective packaging solutions such as custom-molded inserts or foam replacements, that help safeguard delicate or sensitive items during transportation and storage. Thermoformed TPOs are used in the production of transportation-related parts, including truck bed liners, trailer skirts, and vehicle interior panels. Their durability and resistance to harsh environmental conditions make them suitable for these applications. TPCs are utilized in various transportation sectors, such as trains, buses, and marine vessels, to create lightweight and durable interior components.

9.2.16 Waste Management

Thermoplastics transform the current linear patterns of production and consumption in circular systems. Product value is increasingly seen as a key challenge for achieving sustainability goals and ensuring future competitiveness. HDPE is widely used for creating waste management products such as recycling bins, trash cans, and waste containers. Their durability and resistance to chemicals and UV radiation make them ideal for outdoor use [88, 90].

9.2.17 Water Filtration Systems

The resistance of PPO to hot water and steam, along with its ability to withstand chemical exposure, makes it an excellent material for components used in water filtration systems. Thermoplastic composite materials (TPCs) are utilized in the defense sector to manufacture lightweight armor panels, vehicle components, and aerospace applications. Polylactic acid (PLA) is used in certain biomedical applications due to its biocompatibility. It is commonly used in tissue engineering scaffolds, drug delivery systems, and biodegradable medical implants.

9.2.18 Industrial Applications

In industrial applications, HDPE is commonly used to produce containers, such as drums, barrels, and bulk containers. Thermoformed HDPE containers offer excellent resistance to chemicals and are suitable for storing and transporting various substances including chemicals, food ingredients, and hazardous materials. Additionally, thermoformed HDPE trays are used in automotive, electronics, and pharmaceutical industries for material handling and storage purposes.

LDPE products are used in various industrial applications including protective covers, trays, and bins for material handling and storage. Thermoformed LLDPE products serve as liners in industrial applications such as ponds, reservoirs, and chemical storage tanks. The chemical resistance and water impermeability of LLDPE make it suitable for such applications. Thermoformed PP is utilized in the production of various industrial components, including trays, handling trays, and material-handling bins.

Thermoformed PVC is used to manufacture various industrial components and protective covers for machinery and equipment. PMMA is used to create protective covers, machine guards, and safety barriers due to its impact resistance and ability to withstand different environmental conditions. Thermoformed polycarbonate is used in various components of industrial equipment, such as machine guards, protective covers, and safety barriers. Its impact resistance and durability contribute to improving workplace safety.

Thermoformed PEEK is used to manufacture pump and valve components for industrial applications requiring chemical resistance and dimensional stability. Thermoformed TPEs are used in various industrial applications, including gaskets, seals, and vibration-dampening components, due to their weather resistance and durability. Thermoformed PPO is used in various industrial applications including pump housings, impellers, and other components that require chemical resistance, strength, and dimensional stability.

The low friction and resistance to wear and chemicals of POM make it suitable for manufacturing seals, O-rings, and other sealing components in various industries. Its high dimensional stability and machinability make it a preferred material for precision engineering components, such as gears, bushings, and cams. The resistance of POM to chemicals and its low water absorption make it suitable for plumbing applications, including faucet components, valves, and pipe fittings. In addition, the self-lubricating properties and high wear resistance of POM make it an excellent material for gears, bushings, and bearings, particularly in applications that require low friction and high precision. Thermoformed POM is utilized in conveyor systems and material-handling equipment, including rollers, guides, and chain links, due to its low friction and wear resistance.

TPCs are used in the renewable energy sector to create lightweight components for wind-turbine blades and solar-panel structures. They also find applications in industrial equipment for producing lightweight and robust components for machinery and manufacturing processes. Thermoformed

ABS/PMMA products are also utilized in industrial applications such as covers, guards, and enclosures. These products are chosen for their durability, chemical resistance, and ability to be manufactured in various sizes and thicknesses.

9.2.19 Marine and Watercraft Industry

The marine and watercraft industries utilize HDPE to produce various components, including buoys, floats, boat seats, and storage boxes. The material's resistance to water and its ability to withstand harsh environments make it well suited for these applications. Similarly, thermoformed TPO is used in the marine industry for manufacturing boat components such as hulls, seating, and instrument panels. Its resistance to water and UV radiation makes it a suitable choice for marine environments.

9.3 Market Trends

It is important to note that market trends and dynamics can change over time, and external factors such as technological advancements, environmental concerns, and economic fluctuations can influence the trajectory of the thermoforming polymer material market. Therefore, continuous monitoring and adaptation to evolving conditions are crucial in the industry.

9.3.1 Current Market

The plastic sheet market, particularly in the US, is experiencing significant growth and has great potential. Polymers, which account for 85% of the global plastic usage, play a crucial role in this market. In the US, the demand for plastic sheets is projected to increase by approximately 3% annually from 2011 to 2015, reaching 7.7 billion pounds. The estimated cost of the final plastic sheet product is $18.5 billion, with resins contributing 39% of this cost. The popularity of plastic sheets in food packaging is a major driver of demand, as they offer versatility in creating various packaging formats such as trays, containers, and films. Plastic sheets are also used in construction, automotive, and electronics industries. However, the maturity of certain applications and efforts to reduce packaging waste may pose challenges for the growth of the plastic sheet market. Despite these challenges, there are emerging applications of plastic sheet materials that can drive future growth.

Overall, the plastic sheet industry shows promise because of its versatility, cost-effectiveness, and wide range of applications. However, the industry must address certain challenges to sustain and expand its growth in the future [91, 92]. Over the years, there has been a significant increase in the consumption of thermoforming polymers. The expected growth from 600 metric tons in 2015 to 840 metric tons in 2023 indicates a substantial increase in demand. The primary reason for this increase is the growth of the food packaging industry. Thermoforming is a versatile and efficient process for creating various food packaging formats, and has become highly sought-after as the demand for packaged food continues to rise. Its ability to manufacture trays, containers, and films using thermoforming techniques makes it an attractive option for food packaging companies.

The growth in other industries, such as automotive, electronics, and medical also contributes to the increased demand for thermoforming polymer materials. Considering these factors, the overall outlook for the thermoforming polymer material market appears positive. The anticipated growth in the food packaging industry and the rising popularity of thermoformed products in other sectors are expected to sustain the demand for these materials in the future.

9.3.2 Future Market

The thermoforming plastic materials industry is expected to experience a steady growth of approximately 3% per year over the next 12 years, indicating a positive outlook. Advancements in thermoforming technologies have expanded the applications of thermoformed products, further driving the demand for thermoforming plastic materials.

The versatility of thermoforming, which allows for the creation of various food packaging formats, such as trays, containers, and films, is a key factor in boosting demand in the food packaging industry. Moreover, the benefits of thermoformed products, such as being lighter and more durable than traditional molded products, make them increasingly attractive to other sectors, such as automotive, electronics, and medicine. With new and innovative thermoforming technologies enabling the creation of products with more complex shapes and features, the market is expanding into new territories, driving further demand for thermoforming plastic materials. The projected growth in the consumption of thermoforming plastic materials, as shown in the table, supports a positive forecast for the industry. By 2035, the market is expected to reach 1,278 metric tons, reflecting a steady and sustained growth from 2023 to 2035 [93–96].

In summary, plastic sheets and thermoformed products have experienced significant growth, particularly in the packaging industry. Thermoformed products are versatile and used in various sectors, including food and healthcare packaging, construction, appliances, signage, automotive, and aviation. Thermoformed products offer hygiene benefits, making them suitable for industries with strict cleanliness standards. The selection of specific thermoplastic materials and thermoforming processes play vital roles in the production of food packaging and other plastic products.

The growth in industries such as automotive, electronics, and medicine has also contributed to the increased demand for thermoforming polymer materials. The overall outlook for the thermoforming polymer material market appears positive, with steady growth expected. The consumption of thermoforming polymer materials is increasing, primarily due to the growth in the food packaging industry. Advancements in thermoforming technologies have expanded the applications of thermoformed products.

10
Conclusion

Thermoforming is an important process for manufacturing thin-walled parts that has attracted the attention of industries. It is an industrial process that uses thermoplastic sheet materials, have the potential to expand to the commercial market. Furthermore, significant advancements in products with thin walls can create new market opportunities. Thin-walled products are a fascinating concept for inexpensive applications.

The growth of new materials and applications in development creates a perfect opportunity to advance thermoforming process technology. Better growth for thermoforming has been sustained in the coming years. Currently, the thermoforming industry has the necessary equipment, technology, and materials to sustain strong growth. Thermoformed parts are a part of everyday life for both urban and rural populations. Their fields of use extend to almost all areas. About 30%–35% of plastics are used for packaging applications, with an annual increase in demand of 25%–30% in this area due to their disposable nature.

Thermoforming offers processing advantages, and the products produced using this technique are used in packaging, automotive, and other fields. The selection of materials begins with an extrusion process. The properties of the products, even those with thin walls, exceed those of other processing techniques with the same wall thickness. Even with variations in the sheet materials, the thermoforming process has become increasingly challenging and demanding.

Materials with varying morphological properties were then thermoformed. Optimizing and increasing productivity require efforts to refine the processing technology for large-volume products that will be used in various fields. The productivity of thermoforming is determined by the ratio of the parts produced to the amount of sheet material used. Being able to create value from sheet materials, parts of large-volume products contribute to cost optimization. In thermoforming, sheet materials provide the maximum output when their thickness is efficiently utilized. Thickness

Muralisrinivasan Natamai Subramanian. Thermoforming: Processing and Technology, (261–264)

reduces product costs and enhances productivity and efficiency. Modeling and simulation help minimize variations in thermoforming operations.

Improving productivity is a common objective in manufacturing thermoformed products, particularly in a dynamic business environment. With the aim of minimizing waste and achieving higher levels of productivity, the quantity and quality of products are closely related to the resources used in thermoforming operations. With comprehensive understanding, productivity is important for the development and implementation of effective planning and control systems.

Careful consideration of the economics involved in thermoforming comprise various factors, including sheet material costs, part design, tooling expenses, regrind costs, setup charges, line time charges, quality control measures, finished part quality, market demand, and product demand. The basic raw material for thermoforming is plastic sheet material, which has a direct impact on production expenses.

The thermoforming process influences revenue generation and profitability. This process plays an important role in preserving product quality. Product optimization is important to meet consumer expectations. Enhancing product appeal and reducing costs is possible with thermoforming when compared to other processing techniques. This is because thermoforming allows the construction of prototype molds and the production of parts using cheaper materials such as wood, aluminum, and plastic molds.

The influence of market demand and pricing strategies in thermoforming on revenue generation and profitability are significant factors to consider. Thermoforming plays a crucial role in preserving the product quality. The selection and optimization of an appropriate product are important for various reasons, including cost reduction, enhancing product appeal, facilitating transportation, meeting consumer expectations, and addressing environmental concerns. The prototype and its production are flexible in thermoforming and can greatly benefit the design and development cycle, ultimately leading to an improved consumer-oriented final product.

Thermoforming plays a critical role in the decision-making process for product design, especially for small- and medium-sized entities operating in dynamic markets. Thermoforming is emphasized as a processing technique to support the development of small industries. Thermoforming is a desirable economic process that is used for prototyping molds before moving on to the final part of the design. Small-scale production is not possible with other processing techniques, which are costly compared to thermoforming.

The study of thermoplastic materials examines how changes in product characteristics, such as improvements or reductions in quality, affect the demand for quantity. Thermoformed products can respond to changes in price based on demand. An economic framework has to be developed using thermoforming products including market access.

With current global economic challenges, thermoforming is currently one of the most suitable production technologies for processors in various industries. This enables them to create new products in a shorter timeframe and with lower investments in molds and equipment. Thermoformed products are typically classified as permanent, industrial, or disposable products. Thermoforming, which is the art and engineering of fabricating functional plastic parts from sheets, is becoming a mature and competitive technology in the packaging and structural parts industries.

Thermoforming offers reliable technology and economical operation, making it a sensible investment for the future. Thermoforming remains a preferred process for manufacturing thin-wall packaging. Thin-wall packaging offers material savings on high-volume parts at a unit cost that is unachievable with injection molding. Thermoforming can offer high-end in-mold graphics using thermoforming tools. This facility offers commercial-scale processing of thermoformed containers with in-mold labels.

Thermoforming has evolved technically to become the preferred process for many applications in a wide range of products. The expansion of thermoforming into a wide range of markets is driving innovation in both equipment and materials as applications become more diverse and intricate.

The basics of thermoforming, with special emphasis, were placed on thermoplastic sheet materials and processes. These materials have great potential for use in parts with high specific strengths and stiffness. Thermoforming allows for the production of a wide variety of parts with different properties. The properties of the parts depend on factors such as the type of polymer used and the specific thermoforming process. The practical application of thermoplastic materials in various fields shows that they have become an important factor in the design of thin, lightweight parts, often with specific applications. Projections regarding the future of thermoforming are positive, indicating promising trends for future products. This also includes the use of various thermoplastic materials such as PBT and PLA. This thermoforming process has opened new perspectives.

References

Chapter 1

1. Ebewele, R.O., *Polymer science and technology*, p. 1, CRC Press, Boca Raton, FL, USA, 1996.
2. McKelvey, J.M., *Polymer processing*, p. 1, John Wiley, New York, NY, USA, 1962.
3. Xanthos, M. and Todd, D.B., *Kirk-Othmer encyclopedia of chemical technology*, 4th Edition, vol. 19, J.I. Kroschwitz (Ed.), p. 290, John Wiley & Sons, New York, NY, USA, 1996.
4. Xanthos, M., Modification of polymer mechanical and rheological properties with functional fillers, in: *Functional Fillers for Plastics*, pp. 17–38, 2005.
5. Kazmer, D., Simulation of the blow molding and thermoforming processes, in: *The International Industrial Engineering Conference. Chicago. IL*, pp. 269–275, 1992.
6. Muralisrinivasan, N.S., *Update on troubleshooting in thermoforming*, iSmithers Shawbury, SY4 4NR, UK, 2010.

Chapter 2

1. Williamson, J.B., Lewis, S.E., Johnson, R.R., Manning, I.M., Leibfarth, F.A., C-H functionalization of commodity polymers. *Angew. Chem.*, 58, 26, 8654–8668, 2019.
2. Lutz, J.-F., Ouchi, M., Liu, D.R., Sawamoto, M., Sequence-controlled polymers. *Science*, 341, 2013, 341, 1238149.
3. Ojumu, T.V., Yu, J., Solomon, B.O., Production of polyhydroxyalkanoates, a bacterial biodegradable polymer. *Afr. J. Biotechnol.*, 3, 18–24, 2004.
4. Gross, R.A. and Kalra, B., Biodegradable polymers for the environment. *Science*, 297, 5582, 803–807, 2002.
5. Wencel-Delord, J. and Glorius, F., C-H bond activation enables the rapid construction and late-stage diversification of functional molecules. *Nat. Chem.*, 5, 5, 369–75, 2013.
6. Brown, H.R., A molecular interpretation of the toughness of glassy polymers. *Macromolecules*, 24, 10, 2752–2756, 1991.

7. Brown, H.R., Char, K., Deline, V.R., Green, P.F., Effects of a diblock copolymer on adhesion between immiscible polymers. 1. Polystyrene (PS)-PMMA copolymer between PS and PMMA. *Macromolecules*, 26, 16, 4155–4163, 1993.
8. Van Krevelen, D.W. and Hoftyzer, P.J., *Properties of polymers, their estimation and correlation with chemical structure*, Elsevier Scientific, New York, 1976.
9. Imre, B., García, L., Puglia, D., Vilaplana, F., Reactive compatibilization of plant polysaccharides and biobased polymers: Review on current strategies, expectations and reality. *Carbohydr. Polym.*, 209, 20–37, 2019.
10. ISO-International Organization for Standardization, Standard ISO 16620-1:2015, Plastics—Biobased content—Part 1: General principles, International Organization for Standardization, Geneva, Switzerland, 2015.
11. Johnson, P.R., A general correlation of the flammability of natural and synthetic polymers. *J. Appl. Polym. Sci.*, 18, 2, 491–504, 1974.
12. Siracusa, V. and Lotti, N., Biobased plastics for food packaging, in: *Reference Module in Food Science*, pp. 1–19, Elsevier, Amsterdam, The Netherlands, 2018.
13. Chen, H., Lu, Q., Cao, X., Wang, N., Wang, Z.L., Natural polymers based triboelectric nanogenerator for harvesting biomechanical energy and monitoring human motion. *Nano Res.*, 15, 3, 2505–2511, 2022.
14. Brody, A.L., Bugusu, B.A., Han, J.H., Sand, C.K., McHugh, T.H., Innovative Food Packaging Solutions. *J. Food Sci.*, 73, 107–116, 2008.
15. Page, B., Rigid metal packaging, in: *Packaging Technology*, A. Emblem, and H. Emblem, (Eds.), pp. 122–162, Elsevier, Amsterdam, The Netherlands, 2012.
16. Jia, L., Evans, S., van der Linden, S., Motivating actions to mitigate plastic pollution. *Nat. Commun.*, 10, 1, 4582, 2019.
17. Price, C.C., Some aspects of the mechanism of addition polymerization. *Ann. N. Y. Acad. Sci.*, 44, 4, 351–370, 1943.
18. Hamielec., A.E., MacGregor, J.F., Penlidis, A., Multicomponent free-radical polymerizations in batch, semi-batch, and continuous reactors. *Proceedings IUPAC Symposium*, Genoa, Italy, 1987.
19. Gerstenberg, H., Sckuhr, P., Steiner, R., Stirred tank reactors for polymerization. *Ger. Chem. Eng.(Engl. Transl.);(Germany, Federal Republic)*, 6, 3, 129–141, 1983.
20. Holt, T. and Simpson, W.C., Observations on intramolecular reaction in addition polymerization systems. *Proc. R. Soc. Lond. Ser. A. Math. Phys. Sci.*, 238, 154–174, 1956.
21. Christiansen, R.L., Kumar, A., Ray, W.H., Reactor design for addition polymerization under conditions of reversible reaction. *Preprint AlChE Annual Meeting*, Chicago, IL, November 1985.

22. Arriola, D.J. and Ray, W.H., A framework for modelling complex addition polymerization kinetics. *Preprint AIChE Annual Meeting*, New York, November 1987.
23. Reichert, K.-H. and Geisler, H. (Eds.), *Polymer Reaction Engineering*, VCH Publishers, New York, 1989.
24. Secor, R.M., The kinetics of condensation polymerization. *AIChE J.*, 15, 6, 861–865, 1969.
25. Nomura, N., Tsurugi, K., RajanBabu, T.V., Kondo, T., Homogeneous two-component polycondensation without strict stoichiometric balance via the tsuji–trost reaction: Remote control of two reaction sites by catalysis. *J. Am. Chem. Soc.*, 126, 17, 5354–5355, 2004.
26. Miller, I.K. and Zimmerman, J., Condensation polymerization and polymerization mechanisms. Chapter 8, pp. 159–173, 1985.
27. Moad, G., Rizzardo, E., Thang, S.H., Living radical polymerization by the RAFT process–a third update. *Aust. J. Chem.*, 65, 8, 985–1076, 2012.
28. Pelegri-O'Day, E.M. and Maynard, H.D., Controlled radical polymerization as an enabling approach for the next generation of protein–polymer conjugates. *Acc. Chem. Res.*, 49, 9, 1777–1785, 2016.
29. Maszewski, K., Atom transfer radical polymerization (ATRP): Current status and future perspectives. *Macromolecules*, 45, 4015–4039, 2012.
30. Seguela, R., Critical review of the molecular topology of semicrystal-line polymers: The origin and assessment of intercrystalline tie molecules and chain entanglements. *J. Polym. Sci. Part B: Polym. Phys.*, 43, 14, 1729–1748, 2005.
31. Fernández, J., Amestoy, H., Sardón, H., Aguirre, M., Varga, A.L., Sarasua, J.-R., Effect of molecular weight on the physical properties of poly(ethylene brassylate) homopolymers. *J. Mech. Behav. Biomed. Mater.*, 64, 209–19, 2016.
32. Carmesin, I. and Kremer, K., The bond fluctuation method: A new effective algorithm for the dynamics of polymers in all spatial dimensions. *Macromolecules*, 21, 9, 2819–2823, 1988.
33. Takada, K., Ito, T., Kitano, K., Tsuchida, S., Takagi, Y., Chen, Y., Satoh, T., Kakuchi, T., Synthesis of homopolymers, diblock copolymers, and multiblock polymers by organocatalyzed group transfer polymerization of various acrylate monomers. *Macromolecules*, 48, 3, 511–519, 2015.
34. Hawker, C.J. and Wooley, K.L., The convergence of synthetic organic and polymer chemistries. *Science*, 309, 1200–1205, 2005.
35. Bercea, M., Cazacu, M., Wolf, B.A., Chain connectivity and conformational variability of polymers: Clues to an adequate thermodynamic description of their solutions, 1. *Macromol. Chem. Phys.*, 204, 11, 1371–1380, 2003.
36. Barth, C., Horst, R., Wolf, B.A., (Vapour + liquid) equilibria of (water + dimethylformamide): Application of the headspace-gas chromatography for the determination of thermodynamic interactions. *J. Chem. Thermodyn.*, 30, 641–652, 1998.

37. Wang, J., Müller, M., Wang, Z.-G., Nucleation in A/B/ AB blends: Interplay between microphase assembly and macrophase separation. *J. Chem. Phys.*, 130, 15, 154902, 2009.
38. Du, Z., Jiang, Z., Li, C., Thermodynamic optimization of the Ru–Sc system. *J. Alloys Compd.*, 427, 1-2, 148–152, 2007.
39. Petri, H.-M., Schuld, N., Wolf, B.A., Hitherto ignored effects of chain length on the Flory-Huggins interaction parameters in concentrated polymer solutions. *Macromolecules*, 28, 14, 4975–4980, 1995.
40. Zhang, J., Jin, J., Cooney, R., Fu, Q., Qiao, G.G., Thomas, S., Merkel, T.C., Synthesis of perfectly alternating copolymers for polymers of intrinsic microporosity. *Polym. Chem.*, 6, 28, 5003–5008, 2015.
41. Huang, W., Luo, C., Li, B., Han, Y., Solvent-induced crystallization of spherical micelles formed by copolymer blends. *Macromolecules*, 39, 23, 8075–8082, 2006.
42. Shull, K.R., Kramer, E.J., Hadziioannou, G., Tang, W., Segregation of block copolymers to interfaces between immiscible homopolymers. *Macromolecules*, 23, 22, 4780–4787, 1990.
43. Fredrickson, G.H., The theory of polymer dynamics. *Curr. Opin. Solid State Mater. Sci.*, 1, 6, 812–816, 1996.
44. Bates, F.S. and Fredrickson, G.H., Block copolymer thermodynamics: Theory and experiment. *Annu. Rev. Phys. Chem.*, 41, 1, 525–557, 1990.
45. Bates, F.S. and Fredrickson, G.H., Block copolymers—Designer soft materials. *Phys. Today*, 52, 2, 32–38, 1999.
46. Shull, K.R., Interfacial phase transitions in block copolymer/homopolymer blends. *Macromolecules*, 26, 9, 2346–2360, 1993.
47. Ryan, M.E., Stephenson, M.J., Grosser, K., Karadin, L.J., Kaknes, P., Statistical analysis of product variability associated with continuous and cut sheet thermoforming operations. *Polym. Eng. Sci.*, 36, 19, 2432–2442, 1996.
48. Binder, K., Phase transitions of polymer blends and block copolymer melts in thin films. *Adv. Polym. Sci.*, 138, 1–89, 1999.
49. Balbo Block, M.A. and Hecht, S., Poly (propylene oxide)– poly (phenylene ethynylene) block and graft copolymers. *Macromolecules*, 41, 9, 3219–3227, 2008.
50. Ko, M.J., Kim, S.H., Jo, W.H., The effects of copolymer architecture on phase separation dynamics of immiscible homopolymer blends in the presence of copolymer: A Monte Carlo simulation. *Polymer*, 41, 16, 6387–6394, 2000.
51. Anastasiadis, S.H., Gancarz, I., Koberstein, J.T., Compatibilizing effect of block copolymers added to the polymer/polymer interface. *Macromolecules*, 22, 3, 1449–1453, 1989.
52. Brown, H.R., Deline, V.R., Green, P.F., Evidence for cleavage of polymer chains by crack propagation. *Nature*, 341, 6239, 221–222, 1989.
53. Chang, K., Macosko, C.W., Morse, D.C., Ultralow interfacial tensions of polymer/polymer interfaces with diblock copolymer surfactants. *Macromolecules*, 40, 10, 3819–3830, 2007.

54. Retsos, H., Margiolaki, I., Messaritaki, A., Anastasiadis, S.H., Interfacial tension in binary polymer blends in the presence of block copolymers: Effects of additive MW. *Macromolecules*, 34, 15, 5295–5305, 2001.
55. Touris, A., Lee, S., Hillmyer, M.A., Bates, F.S., Synthesis of tri-and multiblock polymers with asymmetric poly (ethylene oxide) end blocks. *ACS Macro Lett.*, 1, 6, 768–771, 2012.
56. Bates, F.S., Hillmyer, M.A., Lodge, T.P., Bates, C.M., Delaney, K.T., Fredrickson, G.H., Multiblock polymers: Panacea or Pandora's box? *Science*, 336, 6080, 434–440, 2012.
57. Retsos, H., Anastasiadis, S.H., Pispas, S., Mays, J.W., Hadjichristidis, N., Interfacial tension in binary polymer blends in the presence of block copolymers. 2. Effects of additive architecture and composition. *Macromolecules*, 37, 2, 524–537, 2004.
58. Giacomelli, C., Schmidt, V., Aissou, K., Borsali, R., Block copolymer systems: From single chain to self-assembled nanostructures. *Langmuir : The ACS Journal of Surfaces and Colloids*, 26, 20, 15734–15744, 2010.
59. Saegusa, T., Chujo, Y., Aoi, K., Miyamoto, M., One-shot block copolymerization, in: *Makromolekulare Chemie. Macromolecular Symposia*, vol. 32, pp. 1–10, 1990.
60. Huang, J. and Turner, S.R., Recent advances in alternating copolymers: The synthesis, modification, and applications of precision polymers. *Polymer*, 116, 572–586, 2017.
61. Kobayashi, S. and Saegusa, T., Block and graft copolymers of 2-oxazolines. *Makromol. Chem.: Macromol. Chem. Phys.*, 12, S19851, 11–24, 1985.
62. Goswami, K.G., Saha, B., De, P., Alternating copolymers with glycyl-glycine and alanyl-alanine side-chain pendants: Synthesis, characterization and solution properties. *J. Macromol. Sci., Part A*, 57, 9, 675–683, 2020.
63. Feldman, D., *Developments in block copolymers-2*, I. Goodman (Ed.), p. 229, Elsevier Applied Science Publishers, London and New York, 1985.
64. Kennedy, J.P., *Developments in block copolymers-1*, I. Goodman (Ed.), p. 358, Applied Science (Elsevier Science, New York, 1982.
65. Zhang, X., Wang, L., Jiang, T., Lin, J., Phase behaviors of supramolecular graft copolymers with reversible bonding. *J. Chem. Phys.*, 139, 18, 184901, 2013.
66. de Gennes, P.G. and Witten, T.A., Scaling concepts in polymer. *Phys. Today*, 33, 6, 51–54, 1980.
67. Yablon, D., Chakraborty, I., Passino, H., Iyer, K., Doufas, A., Shivokhin, M., Deep learning to establish structure property relationships of impact copolymers from AFM phase images. *MRS Commun.*, 11, 962–968, 2021.
68. Mncwabe, S., Luruli, N., Marantos, E., Nhlapo, P., Botha, L., Fractionation of polypropylene impact copolymers using temperature rising elution fractionation (TREF), in: *Macromolecular Symposia*, vol. 313(1), WILEY-VCH Verlag, Weinheim, pp. 33–42, 2012.

69. Song, S., Wu, P., Feng, J., Ye, M., Yang, Y., Influence of pre-shearing on the crystallization of an impact-resistant polypropylene copolymer. *Polymer*, 50, 1, 286–295, 2009.
70. Parrish, M. and Brown, N., Crazing of polymers at low temperatures in He, Ar, N2 and at reduced pressure. *Nat. Phys. Sci.*, 237, 77, 122–123, 1972.
71. Magagula, S.I., Ndiripo, A., van Reenen, A.J., Heterophasic ethylene-propylene copolymers: New insights on complex microstructure by combined molar mass fractionation and high temperature liquid chromatography. *Polym. Degrad. Stab.*, 171, 109022, 2020.
72. Gent, A.N., Relaxation processes in vulcanized rubber. II. Secondary relaxation due to network breakdown. *J. Appl. Polym. Sci.*, 6, 22, 442–448, 1962.
73. Alkonis, J.J., MacKnight, W.J., Shen, M. (eds.), *Introduction to polymer viscoelasticity*, Second Edition, p. 249, Wiley, New York, 1983.
74. Yousefi, A., Lafleur, P.G., Gauvin, R., Kinetic studies of thermoset cure reactions: A review. *Polym. Compos.*, 18, 2, 157–168, 1997.
75. Duan, Y., Saigal, A., Greif, R., Zimmerman, M.A., Impact behavior and modeling of engineering polymers. *Polym. Eng. Sci.*, 43, 1, 112–124, 2003.
76. Jehanno, C., Alty, J.W., Roosen, M., De Meester, S., Dove, A.P., Chen, E.Y.-X., Leibfarth, F.A., Sardon, H., Critical advances and future opportunities in upcycling commodity polymers. *Nature*, 603, 7903, 803–814, 2022.
77. Stokes, V.K. and Nied, H.F., Solid phase sheet forming of thermoplastics—Part I: Mechanical behavior of thermoplastics to yield. *J. Eng. Mater. Technol. Trans.*, 108, 107–112, 1986.
78. Kurdi, A., Kan, W.H., Chang, L., Tribological behaviour of high performance polymers and polymer composites at elevated temperature. *Tribol. Int.*, 130, 94–105, 2019.
79. Bin Ahmad, Z. and Ashby, M.F., Failure-mechanism maps for engineering polymers. *J. Mater. Sci.*, 23, 6, 2037–2050, 1988.

Chapter 3

1. Geyer, R., Jambeck, J.R., Law, K.L., Production, use, and fate of all plastics ever made. *Sci. Adv.*, 3, e1700782, 2017.
2. Duc-Toan, N., Seung-Han, Y., Dong-Won, J., Tae-Hoon, C., Young-Suk, K., Incremental sheet metal forming: Numerical simulation and rapid prototyping process to make an automobile white-body. *Steel Res. Int.*, 82, 7, 795–805, 2011.
3. Faust, K., Denifl, P., Hapke, M., Recent advances in catalytic chemical recycling of polyolefins. *ChemCatChem*, 15, 1, 2023.
4. Günay, K.A., Theato, P., Klok, H.-A., Standing on the shoulders of Hermann Staudinger: Post-polymerization modification from past to present. *J. Polym. Sci. Part A: Polym. Chem.*, 51, 1, 1–28, 2013.
5. Boenig, H.V., *Polyolefins: Structure and properties*, Elsevier, New York, 1966.

6. Galli, P., The future of science has begun—Chemical, biochemical and cellular topology. *Proceedings of the 2nd International Conference*, Milan, Italy, p. 43, 1992.
7. Sperling, L.H., *13. Multicomponent polymeric materials. Introduction to polymer science*, pp. 687–756, John Wiley & Son Inc, NJ, USA, 2006.
8. Spalding, M.A. and Chatterjee, A., *Handbook of industrial polyethylene and technology: Definitive guide to manufacturing, properties, processing, applications and markets*, Wiley Scrivener, Beverly, MA, USA, 2018.
9. Karnoub, A., Antypas, I.R., Dyachenko A, G., Savostina, T.P., Buckling resistance of cylinders made of textile composite material. *IOP Conf. Ser.: Earth Environ. Sci.*, 403, 1, 012162, 2019.
10. Furushima, Y., Nakada, M., Murakami, M., Yamane, T., Toda, A., Schick, C., Method for calculation of the lamellar thickness distribution of not-reorganized linear polyethylene using fast scanning calorimetry in heating. *Macromolecules*, 48, 8831–8837, 2015.
11. Keith, H.D., Padden Jr., F.J., Vadimsky, R.G., Intercrystalline links in polyethylene crystallized from the melt. *J. Polym. Sci. Part A-2: Polym. Phys.*, 4, 2, 267–281, 1966.
12. La Mantia, F.P. and Acierno, D., Influence of the molecular structure on the melt strength and extensibility of polyethylenes. *Polym. Eng. Sci.*, 25, 279–283, 1985.
13. Agrawal, P., Silva, M.H.A., Cavalcanti, S.N., Freitas, D.M.G., de Araújo, J.P., Oliveira, A.D.B., Mélo, T.J.A., Rheological properties of high-density polyethylene/linear low-density polyethylene and high-density polyethylene/low-density polyethylene blends. *Polym. Bull.*, 79, 2321–2343, 2021.
14. Chen, Y., Zou, H., Liang, M., Liu, P., Rheological, thermal, and morphological properties of low-density polyethylene/ultra-high-molecular-weight polyethylene and linear low-density polyethylene/ultra-high-molecular-weight polyethylene blends. *J. Appl. Polym. Sci.*, 129, 945–953, 2013.
15. Xu, J. and Guo, B., Poly(butylene succinate) and its copolymers: Research, development and industrialization. *Biotechnol. J.*, 5, 1149–1163, 2010.
16. Madera-Santana, T.J., Robledo, D., Azamar, J.A., Ríos-Soberanis, C.R., Freile-Pelegrín, Y., Preparation and characterization of low density polyethylene-agar biocomposites: Torque-rheological, mechanical, thermal and morphological properties. *Polym. Eng. Sci.*, 50, 585–591, 2010.
17. Berdyshev, B.V. and Borisov, A.A., Operating regimes of production equipment for thermal shaping of polymer articles. *Chem. Pet. Eng.*, 49, 648–651, 2014.
18. Yamaguchi, M., Todd, D.B., Gogos, C.G., Effect of processing history on rheological properties and processability of LDPE. *Proceeding of ACS National Meeting*, Chicago, IL, Aug, 2001.
19. Popli, R., Glotin, M., Mandelkern, L., Benson, R.S., Dynamic mechanical studies of α and β relaxations of polyethylenes. *J. Polym. Sci.: Polym. Phys. Ed.*, 22, 3, 407–448, 1984.

20. Yamaguchi, M., Rheological properties of linear and crosslinked polymer blends: Relation between crosslink density and enhancement of elongational viscosity. *J. Polym. Sci. Part B*, 39, 228–235, 2001.
21. Bremner, T. and Rudin, A., Persistence of regions with high segment density in polyethylene melts. *J. Polym. Sci. Part B*, 30, 1247–1260, 1992.
22. Gulmine, J.V., Janissek, P.R., Heise, H.M., Akcelrud, L., Polyethylene characterization by FTIR. *Polym. Test.*, 21, 5, 557–563, 2002.
23. Crewe, A.V. and Wall, J., Contrast in a high resolution scanning transmission electron microscope. *Optik*, 30, 461–474, 1970.
24. Lu, Y.C. and Shinozaki, D.M., Temperature dependent viscoelastic properties of polymers investigated by small-scale dynamic mechanical analysis. *Exp. Mech.*, 50, 71–775, 2010.
25. Lampman, S., *Characterization and failure analysis of plastics*, ASM International, OH, USA, 2003.
26. Fotheringham, D. and Parker, B., Spherulitic cracking in high density polyethylene. *J. Mater. Sci.*, 11, 979–980, 1976.
27. Colak, O. and Dusunceli, N., Modeling Viscoelastic and viscoplastic behavior of high density polyethylene (HDPE). *J. Eng. Mater. Technol.-Trans. ASME*, 128, 572–578, 2006.
28. Oliveira, A.D.B., Freitas, D.M.G., de Araújo, J.P., Cavalcanti, S.N., Câmara, D.S., Agrawal, P., Mélo, T.J.A., HDPE/LLDPE blends: Rheological, thermal, and mechanical properties. *Mater. Res. Innovations*, 24, 289–294, 2020.
29. Jiang, Z., Tang, Y., Rieger, J., Enderle, H.-F., Lilge, D., Roth, S.V., Gehrke, R., Wu, Z., Li, Z., Men, Y., Structural evolution of tensile deformed high-density polyethylene at elevated temperatures: Scanning synchrotron small- and wide-angle X-ray scattering studies. *Polymer*, 50, 4101–4111, 2009.
30. Addiego, F., Dahoun, A., G'Sell, C. *et al.*, Characterization of volume strain at large deformation under uniaxial tension in highdensity polyethylene. *Polymer*, 47, 4387–4399, 2006.
31. Breedon, J.E., Jackson, J.F., Marcinkowski, M.J., Taylor, M.E., Study of polyethylene spherulites using scanning electron microscopy. *J. Mater. Sci.*, 8, 1071–1082, 1973b.
32. Fukuda, T., Technological progress in high - voltage XLPE power cables in Japan - Part I. *IEEE Electr. Insul. Mag.*, 4, 9–16, 1988.
33. Hedvig, P., *Dielectric spectroscopy of polymers*, Academia Kiado, Budapest, 1977.
34. Hussein, I.A. and Williams, M.C., Rheological evidence for high-temperature phase transitions in melts of high-density polyethylene. *Macromol. Rapid Commun.*, 19, 6, 323–325, 1998.
35. Popelar, C.F., Popelar, C.H., Kenner, V.H., Viscoelastic material characterization and modeling for polyethylene. *Polym. Eng. Sci.*, 30, 577–86, 1990.
36. Romanini, D., Synthesis technology, molecular structure, and rheological behavior of polyethylene. *Polym.-Plast. Technol. Eng.*, 19, 2, 201–226, 1982.

37. Scer, F., Rasoul, F., Kosior, E., Micro-morphology of modified and nonmodified blends of polypropylene with linear low density polyethylene. *Polym. Eng. Sci.*, 39, 6, 1100–1108, 1999.
38. van Krevelen, D.W. and Hoftyzer, P.J., *Properties of polymers, their estimation and correlation with chemical structure*, Elsevier Scientific, New York, 1976.
39. Martins, P.M., Ribeiro, J.M., Teixeira, S., Petrovykh, D., Cuniberti, G., Pereira, L., Lanceros-Méndez, S., Photocatalytic microporous membrane against the increasing problem of water emerging pollutants. *Materials*, 12, 1649, 2019.
40. Yamaguchi, M. and Abe, S., LLDPE/LDPE blends. I. Rheological, thermal, and mechanical properties. *J. Appl. Polym. Sci.*, 74, 13, 3153–3159, 1999.
41. Kukaleva, N., Cser, F., Jollands, M., Kosior, E., Comparison of structure and properties of conventional and "high-crystallinity" isotactic polypropylenes and their blends with metallocene-catalyzed linear low-density polyethylene. I. Relationships between rheological behavior and thermal and physical properties. *J. Appl. Polym. Sci.*, 77, 1591–1599, 2000.
42. Harrison, I.R., Runt, J., Stanislow, L.J., Bell, D.A., The heat of fusion of polymer crystals: Polyethylene crystals in suspension. *J. Polym. Sci.: Polym. Phys. Ed.*, 17, 1, 63–71, 1979.
43. Kim, Y.C., Ahn, W., Kim, C.Y., A study on multiple melting of isotactic polypropylene. *Polym. Eng. Sci.*, 37, 6, 1003–1011, 1997.
44. Hoffman, J.D. and Weeks, J.J., X-ray study of isothermal thickening of lamellae in bulk polyethylene at the crystallization temperature. *J. Chem. Phys.*, 42, 12, 4301–4302, 1965.
45. Raimo, M., Cascone, E., Martuscelli, E., Review melt crystallisation of polymer materials: The role of the thermal conductivity and its influence on the microstructure. *J. Mater. Sci.*, 36, 3591–3598, 2001.
46. Duffo, P., Monasse, B., Haudin, J.M., G'sell, C., Maxime Dahoun, A., Rheology of polypropylene in the solid state. *J. Mater. Sci.*, 30, 701–711, 1995.
47. Song, S., Feng, J., Wu, P., Yang, Y., Shear-enhanced crystallization in impact-resistant polypropylene copolymer: Influence of compositional heterogeneity and phase structure. *Macromolecules*, 42, 18, 7067–7078, 2009.
48. Natta, G. and Corradini, P., *Proceedings of the Conference Nuovo Cimento XV*, p. 40, 1960.
49. Wunderlich, B., *Macromolecular physics*, vol. 3, p. 61, Academic Press, New York, NY, USA, 1980.
50. Brandrup, J. and Immergut, E.H., *Polymer handbook*, 3rd Edition, p. 5, John Wiley & Sons, New York, NY, USA, 1989.
51. Järvelä, P., Shucai, L., Järvelä, P., Dynamic mechanical properties and morphology of polypropylene/maleated polypropylene blends. *J. Appl. Polym. Sci.*, 62, 5, 813–826, 1996.
52. Mezghani, K., Anderson Campbell, R., Phillips, P.J., Lamellar thickening and the equilibrium melting point of polypropylene. *Macromolecules*, 27, 4, 997–1002, 1994.

53. Kumar, P., Kumar, G.K., Kommoji, S., Banerjee, R., Ghosh, A.K., The effect of material characteristics and mould parameters on the thermo-forming of thick polypropylene sheets. *J. Plast. Film Sheeting*, 30, 162–180, 2014.
54. Lee, N.J. and Jang, J., The effect of fiber-content gradient on the mechanical properties of glass-fiber-mat/polypropylene composite. *J. Compos. Sci. Technol.*, 60, 209–17, 2000.
55. Stange, J., Uhl, C., Münstedt, H., Rheological behavior of blends from a linear and a long-chain branched polypropylene. *J. Rheol.*, 49, 5, 1059–1079, 2005.
56. Macauley, N.J., Harkin-Jones, E.M.A., Murphy, W.R., The influence of extrusion parameters on the mechanical properties of polypropylene sheet. *Polym. Eng. Sci.*, 38, 4, 662–673, 1998.
57. Macauley, N.J., Harkin-Jones, E.M.A., Murphy, W.R., Method of assessing thermoformability of extruded polypropylene sheet. *Plast. Rubber Compos. Process. Appl.*, 26, 4, 165–171, 1997.
58. Ojeda Sersun, D.J., Swafford, K.A., Orosco, H., Rodgers, C., Davidson, M., Bains, G.S., Becerra, B., The effects of cooling time on the dimensional stability of thermoforming polypropylene. *J. Prosthet. Orthot.*, 35, 180–185, 2022.
59. Duarte, F.M., Barroso, V.C., Maia, J.M., Covas, J.A., A computational study on the influence of the rheological behavior of polystyrene and its blends on their thermoforming ability. *J. Polym. Eng.*, 25, 115–148, 2005.
60. Mathur, V. and Sharma, K., Thermal response of polystyrene/polymethylmethacrylate (PS/PMMA) polymeric blends. *Heat Mass Transfer*, 52, 2901–2911, 2016.
61. Sasimowski, E. and Filipek, P., Thermoforming of a polystyrene sheet with a vibrating male mold. *Adv. Sci. Technol. Res. J.*, 13, 246–254, 2019.
62. Hockey, J.B., In search of useful data on medical plastics. *Med. Device Technol.*, 19, 6, 20, 22-3, 2008.
63. Brandrup, J. and Immergut, E.H., *Polymer handbook*, 3rd Edition, Wiley-Interscience, New York, NY, USA, 1989.
64. Boyer, R.F., *Encyclopedia of polymer science and technology*, H.F. Mark, N.M. Bikales, C.G. Overberger, G. Menges (Eds.), p. 13, John Wiley & Sons, New York, NY, USA, 1970.
65. Van Krevelen, D.W. and Hoftyzer, P.J., *Properties of polymers: Correlations with chemical structure*, Elsevier Publishing Company, Amsterdam, The Netherlands, 1972.
66. Buffel, B., Amerijckx, M., Hamblok, M., van Mieghem, B., Desplentere, F., Van Bael, A., Experimental and computational analysis of the heating step during thermoforming of thermoplastics. *Key Eng. Mater.*, 651-653, 1003–1008, 2015.
67. Muralisrinivasan, N.S., *Update on troubleshooting in thermoforming*, iSmithers Shawbury, SY4 4NR, UK, 2010.

68. Grassi, V.G., Forte, M.M.C., Dal Pizzol, M.F., Aspectos morfológicos e relação estrutura-propriedades de poliestireno de alto impacto. *Polímeros*, 11, 158–168, 2001.
69. Bucknall, C.B., *Toughened plastics*, Applied Science, London, 1977.
70. Goh, S.H., Siow, K.S., Yap, K.S., Miscibility of poly (styrene-co-acrylonitrile) and poly (α-methyl styrene-co-acrylonitrile) with copolymers of methyl methacrylate. *J. Appl. Polym. Sci.*, 29, 1, 99–103, 1984.
71. Brinke, T., Gerrit, F.E.K., MacKnight, W.J., Phase behavior in copolymer blends: Poly (2, 6-dimethyl-1, 4-phenylene oxide) and halogen-substituted styrene copolymers. *Macromolecules*, 16, 12, 1827–1832, 1983.
72. Ghobadnam, M., Mosaddegh, P., Rezaei Rejani, M. *et al.*, Numerical and experimental analysis of HIPS sheets in thermoforming process. *Int. J. Adv. Manuf. Technol.*, 76, 1079–1089, 2015.
73. Lee, S.J., Jeoung, H.G., Ahn, K.H., Influence of sol- vent contents on the rubber-phase particle size distribution of high-impact polystyrene. *J. Appl. Polym. Sci.*, 89, 13, 3672–3679, 2003.
74. Echte, A., *Rubber-toughened plastics*, C.K. Riew (Ed.), p. 15, American Chemical Society, Washington, 1989.
75. Bourgin, P., Cormeau, I.C., Saint-Matin, T., A first step towards the modelling of the thermoforming of plastic sheets. *J. Mater. Process. Technol.*, 54, 1–11, 1995.
76. Pezzin, G., Rheology and plasticization of polyvinyl chloride. *Pure Appl. Chem.*, 26, 2, 241–256, 1971.
77. Sieglaff, C.L., Rheological properties of polyvinyl chloride. Parts I–general flow properties. *Polym. Eng. Sci.*, 4, 2, 129–133, 1964.
78. Daniels, V.D. and Rees, N.H., Analysis of the ultraviolet/visible spectrum of degraded poly (vinyl chloride) to determine polyene concentrations. *J. Polym. Sci.: Polym. Chem. Ed.*, 12, 9, 2115–2122, 1974.
79. Muralisrinivasan, N.S., *Update on troubleshooting in thermoforming*, iSmithers Shawbury, UK, 2010.
80. Rangel-nafaile, C., García-rejón, A., García de León, A., Residence time effects during capillary flow of a rigid PVC formulation. *J. Vinyl Addit. Technol.*, 6, 63–72, 1984.
81. Termine, E.J., Heat distortion temperature modification of rigid PVC and chlorinated PVC: A method to produce low smoke compositions. *J. Vinyl Addit. Technol.*, 12, 204–207, 1990.
82. Ahn, S.J., Kim, B.K., Lee, K.H., Jeong, H.M., Morphology and physical properties of ABS/NBR: The effect of melt viscosity of SAN and the content of NBR. *J. Macromol. Sci., Part B*, 39, 691–700, 2000.
83. Xu, X.F., Wang, R., Tan, Z.Y., Yang, H.D., Zhang, M.Y., Zhang, H.X., Effects of polybutadiene-g-SAN impact modifiers on the morphology and mechanical behaviors of ABS blends. *Eur. Polym. J.*, 41, 8, 1919–1926, 2005.

84. Castellani, L., Frassine, R., Pavan, A., Rink, M., Rate and temperature dependence of fracture toughness in ABS resins in relation to dispersed-phase structure. *Polymer*, 37, 1329–1338, 1996.
85. Liang, J.-Z., Effects of extrusion conditions on rheological behavior of acrylonitrile-butadienestyrene terpolymer melt. *J. Appl. Polym. Sci.*, 85, 606–611, 2002.
86. Aoki, Y., Rheological properties of ABS polymer melts having a good dispersion of rubber particles. *J. Non-Newton. Fluid Mech.*, 22, 91–99, 1986.
87. Throne, J.L., *Thermoforming*, Hanser Publishers, Munich, 1987.
88. Bucknall, C.B., *Toughened plastics*, Applied Science, London, 1977.
89. Lee, J.K., Scott, C.E., Virkler, T.L., Effects of ABS rubber particles on rheology, melt failure, and thermoforming. *Polym. Eng. Sci.*, 42, 7, 1541–1557, 2002.
90. Lee, J.K., Virkler, T.L., Scott, C.E., Effects of rheological properties and processing parameters on ABS thermoforming. *Polym. Eng. Sci.*, 41, 2, 240–261, 2001.
91. Donald, A.M. and Kramer, E.J., Plastic deformation mechanisms in poly(acrylonitrile-butadiene styrene) [ABS]. *J. Mater. Sci.*, 17, 1765–1772, 1982.
92. Lau, H.C., Bhattacharya, S.N., Jo Field, G., Influence of rheological properties on the sagging of polypropylene and ABS sheet for thermoforming applications. *Polym. Eng. Sci.*, 40, 7, 1564–1570, 2000.
93. Busico, V. and Cipullo, R., Microstructure of polypropylene. *Prog. Polym. Sci.*, 26, 3, 443–533, 2001.
94. Olivera, S., Muralidhara, H.B., Venkatesh, K., Gopalakrishna, K., Vivek, C.S., Plating on acrylonitrile–butadiene–styrene (ABS) plastic: A review. *J. Mater. Sci.*, 51, 3657–3674, 2016.
95. Moore, J.D., Acrylonitrile-butadiene-styrene (ABS)-a review. *Composites*, 4, 3, 118–130, 1973.
96. Saheb, D.N. and Jog, J.P., Crystallization and equilibrium melting behavior of PBT/PETG blends. *J. Polym. Sci. Part B: Polym. Phys.*, 37, 17, 2439–2444, 1999.
97. Avramova, N., Fakirov, S., Avramov, I., High performance polyamide-6 and blends of poly (ethylene terephthalate) and poly (butylene terephthalate) through ultraquenching. *Die Angew. Makromol. Chem.: Appl. Macromol. Chem. Phys.*, 199, 1, 129–135, 1992.
98. Boyce, M.C., Socrate, S., Llana, P.G., Constitutive model for the finite deformation stress-strain behavior of poly_ethylene terephthalate_above the glass transition. *Polymer*, 41, 2183–2201, 2000.
99. Bodor, G., *Structural investigation of polymers*, 1st ed., p. 143, Ellis Harwood Limited, England, 1991.
100. Stern, C., Frick, A., Weickert, G., Relationship between the structure and mechanical properties of polypropylene: Effects of the molecular weight and shear-induced structure. *J. Appl. Polym. Sci.*, 103, 1, 519–533, 2007.

101. Jia, J. and Raabe, D., Evolution of crystallinity and of crystallographic orientation in isotactic polypropylene during rolling and heat treatment. *Eur. Polym. J.*, 42, 8, 1755–1766, 2006.
102. Saidi, A., Akmal, M., Hassan, A., Wahit, M.U., Choy, L.J., Anuar, H., Thermal dynamic mechanical analysis and mechanical properties of polybutylene terephthalate/polyethylene terephthalate blends. *Jurnal Teknologi*, 82, 5, 73–83, 2020.
103. Avramova, N., Amorphous poly(ethylene terephthalate)/poly(butyleneterephthalate) blends: Miscibility and properties. *Polymer*, 36, 801–808, 1995.
104. Asano, T., Balta′ Calleja, F.J., Flores, A., Tanigaki, M., Mina, M.F., Sawatari, C., Itagaki, H., Takahashi, H., Hatta, I., Crystallization of oriented amorphous poly(ethylene terephthalate) as revealed by X-ray diffraction and microhardness. *Polymer*, 40, 23, 6475–6484, 1999.
105. Choy, C.-L., Ong, E.L., Chen, F.C., Thermal diffusivity and conductivity of crystalline polymers. *J. Appl. Polym. Sci.*, 26, 2325–2335, 1981.
106. Painter, P.C. and Coleman, M.M., *Essentials of polymer science and engineering*, DES Tech Publications, Lancaster, 2009.
107. Zhang, A.M., Zhao, G.Q., Guan, Y.J., Effect of surface modifiers and surface modification methods on properties of acrylonitrile -butadiene -styrene/poly (methyl methacrylate)/nano-calcium carbonate composites. *J. Appl. Polym. Sci.*, 127, 2520, 2013.
108. Gedemer, T.J., Technical papers. *30th Annual Technical Conference*, p. 361, Society of Plastics Engineers, Chicago, May, 1972.
109. *Materials for microlithography: Radiation-sensitive polymers*, Thompson, L.F., Willson, C.G., Frechet, J.M.J. (Eds.), p. 266, American Chemical Society, Washington, DC, USA, 1984.
110. Federico, C.E., Bouvard, J.L., Combeaud, C., Billon, N., Modelling strain rate and temperature dependent mechanical response of PMMAs at large deformation from below to above Tg. *Polymer*, 202, 122710, 2020.
111. Silva, G.G., de Freitas Rocha, P.M., de Oliveira, P.M.S., Neves, B.R.A., Domain size effects on the thermal properties of PS/PMMA blends. *Appl. Surf. Sci.*, 238, 64–72, 2004.
112. Li, Z. and Lambros, J., Strain rate effects on the thermomechanical behavior of polymers. *Int. J. Solids Struct.*, 38, 3549–3562, 2001.
113. Marshall, G.P., Coutts, L.H., Williams, J.G., *J. Mater. Sci.*, 9, 1409–1419, 1974.
114. Bousmina, M.M., Muller, R.M., *J. Rheol.*, 37, 663–679, 1993.
115. Richeton, J., Ahzi, S., Daridon, L., Remond, Y., A formulation of the cooperative model for the yield stress of amorphous polymers for a wide range of strain rates and temperatures. *Polymer*, 46, 6035–6043, 2005.
116. Blumenthal, W.R., Cady, C.M., Lopez, M.F., Gray III, G.T., Idar, D.J., Influence of temperature and strain rate on the compressive behavior of PMMA and polycarbonate polymers. in: *AIP Conference Proceedings*, vol. 620, no. 1, pp. 665–668, American Institute of Physics, 2002.

117. White, T.R., Melting behaviour of crystalline polymer fibres. *Nature*, 175, 895–896, 1953.
118. Ke, B. and Sisko, A.W., Differential thermal analysis of high polymers. III. Polyamides. *J. Polym. Sci.*, *50*, 153, 87–98, 1961.
119. Watanabe, M. and Yamaguchi, H., The friction and wear properties of nylon. *Wear*, 110, 379–388, 1986.
120. de P. Daubeny, R., Bunn, C.W., Brown, C.J., The crystal structure of polyethylene terephthalate. *Proc. R. Soc. Lond. Ser. A. Math. Phys. Sci.*, 226, 531–542, 1954.
121. Stoklásek, P., Mizera, A., Maňas, M., Influence of beta rays on nylon 6 mechanical properties. *Key Eng. Mater.*, 756, 27–34, 2017.
122. Okada, A., Kawasumi, M., Usuki, A., Kojima, Y., Kurauchi, T., Kamigaito, O., Nylon 6–clay hybrid. *MRS Proc.*, 171, 45, 1989.
123. Deanin, R.D., *Polymer structure, properties and applications*, pp. 4–6 and 5-4, Cahners Books, Boston, MA, 1972.
124. Margolis, M. (Ed.), *Engineering Thermoplastics: Properties and Applications*, p. 393, Wiley, New York, 1985.
125. Palabiyik, M. and Bahadur, S., Mechanical and tribological properties of polyamide 6 and high density polyethylene polyblends with and without compatibilizer. *Wear*, 246, 1–2, 149–58, 2000 Nov 1.
126. Ramesh, C., Keller, A., Eltink, S.J.E.A., Studies on the crystallization and melting of nylon 66: 3. Melting behaviour of negative spherulites by calorimetry. *Polymer*, 35, 24, 5300–5308, 1994.
127. Ishikawa, T., Nagai, S., Kasai, N., Thermal behavior of α nylon-12. *J. Polym. Sci.: Polym. Phys. Ed.*, 18, 6, 1413–1419, 1980.
128. Watanabe, M. and Yamaguchi, H., The friction and wear properties of nylon. *Wear*, 110, 379–388, 1986.
129. Fu, J.H., Chen, X.D., Xu, Q.J., Wang, Y.J., Wang, X.J., The compatibilization and toughening modification of PPO/PA alloy. *China Plast. Ind.*, 42, 36–39, 2014.
130. Starkweather, H.W., Whitney, J.F., Johnson, D.R., Crystalline order in nylon 66. *J. Polym. Sci. Part A*, 1, 715–723, 1963.
131. Starkweather, H.W. and Brooks, R.E., Effect of spherulites on the mechanical properties of nylon 66. *J. Appl. Polym. Sci.*, 1, 236–239, 1959.
132. Bashford, D., Polyoxymethylene; Polyacetals (POM), in: *Thermoplastics: Directory and Databook*, D. Bashford, (Ed.), pp. 255–267, Springer, Dordrecht, The Netherlands, 1997.
133. Cruz-Pinto, J.J.C., Martins, J.A., Jovita Oliveira, M., The isothermal crystallization of engineering polymers — POM and PEEK. *Colloid Polym. Sci.*, 272, 1–16, 1994.
134. Khalatbari, H. and Lazoglu, I., Friction stir incremental forming of polyoxymethylene: Process outputs, force and temperature. *Mater. Manuf. Process.*, 36, 94–105, 2021.

135. He, J.Q., Wang, Q., Yao, B.Y., Ho, J., Mechanical properties of high strength POM-FRCC and its performance under elevated temperatures. *Constr. Build. Mater.*, 290, 15, 2021.
136. Wang, Z.H., Guo, R.X., Liu, G.S., Guo, L.X., Yan, Y., Study on flexural fatigue properties of POM fiber airport pavement concrete. *Polymers*, 14, 2979, 2022.
137. Tatarka, P.D., Improved properties and cost efficiencies of cyclic olefin copolymer enhanced forming films. *Proceedings of the SPE ANTEC Conference*, Cincinnati, OH, USA, p. 1149, 2007.
138. Yun, S.-S., Shin, D.H., Jang, K.-S., Influence of ionomer and cyanuric acid on antistatic, mechanical, thermal, and rheological properties of extruded carbon nanotube (CNT)/polyoxymethylene (POM) nanocomposites. *Polymers*, 14, 1849–1871, 2022.
139. Riga, A.T., Heat distortion and mechanical properties of polymers by thermal-mechanical analysis. *Polym. Eng. Sci.*, 14, 764–767, 1974.
140. Li, X. and Mason, J.P., New perspective on heat deflection temperature of glassy polycarbonate. *Macromol. Mater. Eng.*, 294, 306–314, 2009.
141. Zhang, X., Hendro, W., Fujii, M. *et al.*, Measurements of the thermal conductivity and thermal diffusivity of polymer melts with the short-hot-wire method. *Int. J. Thermophys.*, 23, 1077–1090, 2002.
142. Curran, D.R., Shockey, D.A., Seaman, L., Dynamic fracture criteria for a polycarbonate. *J. Appl. Phys.*, 44, 4025–4038, 1973.
143. Legrand, D.G., Crazing, yielding, and fracture of polymers. I. Ductile brittle transition in polycarbonate. *J. Appl. Polym. Sci.*, 13, 2129–2147, 1969.
144. Mills, N.J., The mechanism of brittle fracture in notched impact tests on polycarbonate. *J. Mater. Sci.*, 11, 363–375, 1976.
145. Bauwens-Crowet, C., Bauwens, J.C., Homes, G., The temperature dependence of yield of polycarbonate in uniaxial compression and tensile tests. *J. Mater. Sci.*, 7, 176–183, 1972.
146. Turner, J.A., Menary, G.H., Martin, P.J., Yan, S., Modelling the temperature dependent biaxial response of poly(ether-ether-ketone) above and below the glass transition for thermoforming applications. *Polymers*, 11, 1042, 1–2, 2019.
147. Velisaris, C.N. and Seferis, J.C., Crystallization kinetics of polyetheretherketone (PEEK) matrices. *Polym. Eng. Sci.*, 26, 1574–1581, 1986.
148. Sarasua, J.-R., Remiro, P.M., Pouyet, J.M., The mechanical behaviour of PEEK short fibre composites. *J. Mater. Sci.*, 30, 3501–3508, 1995.
149. Searle, O. and Pfeiffer, R., Victrex® poly(ethersulfone) (PES) and Victrex® poly(etheretherketone) (PEEK). *Polym. Eng. Sci.*, 25, 474–476, 1985.
150. Lu, Z.J. and Friedrich, K., On sliding friction and wear of PEEK and its composites. *Wear*, 183, 624–631, 1995.
151. Yuan, M., Galloway, J.A., Hoffman, R.J., Bhatt, S.M., Influence of molecular weight on rheological, thermal, and mechanical properties of PEEK. *Polym. Eng. Sci.*, 51, 94–102, 2011.

152. Rae, P.J., Brown, E.N., Orler, E.B., The mechanical properties of poly (ether-ether-ketone)(PEEK) with emphasis on the large compressive strain response. *Polymer*, 48, 2, 598–615, 2007.
153. Stoelting, J., Karasz, F.E., MacKnight, W.J., Dynamic mechanical properties of poly (2, 6-dimethyl-1, 4-phenylene ether)-polystyrene blends. *Polym. Eng. Sci.*, 10, 3, 133–138, 1970.
154. Prest Jr., W.M. and Porter, R.S., Rheological properties of poly (2, 6-dimethylphenylene oxide)—Polystyrene blends. *J. Polym. Sci. Part A-2: Polym. Phys.*, 10, 9, 1639–1655, 1972.
155. Sato, Y., Takikawa, T., Takishima, S., Masuoka, H., Solubilities and diffusion coefficients of carbon dioxide in poly (vinyl acetate) and polystyrene. *J. Supercrit. Fluids*, 19, 2, 187–198, 2001.
156. Sada, E., Kumazawa, H., Yakushiji, H., Bamba, Y., Sakata, K., Wang, S.T., Sorption and diffusion of gases in glassy polymers. *Ind. Eng. Chem. Res.*, 26, 3, 433–438, 1987.
157. Zhang, Z., Yang, M., Cai, K., Chen, Y., Liu, S., Liu, W., Liu, J.-L., Effect of the flame retardants and glass fiber on the polyamide 66/polyphenylene oxide composites. *Materials*, 15, 813–835, 2022.
158. Sato, Y., Takikawa, T., Yamane, M., Takishima, S., Masuoka, H., Solubility of carbon dioxide in PPO and PPO/PS blends. *Fluid Phase Equilib.*, 194, 847–858, 2002.
159. Robertson, J., Shenian, P., Punkar, J., Properties, processing and potential automotive applications of PPO, polyphenylene oxide resins. SAE Technical Paper 660002, 1966.
160. Kim, S.Y., Jang, H.G., Yang, C.-M., Yang, B., Multiscale prediction of thermal conductivity for nanocomposites containing crumpled carbon nanofillers with interfacial characteristics. *Compos. Sci. Technol.*, 155, 169–176, 2018.
161. Yamamoto, B.E., Trimble, A.Z., Minei, B., Nejhad, M.N.G., Development of multifunctional nanocomposites with 3-D printing additive manufacturing and low graphene loading. *J. Thermoplast. Compos. Mater.*, 32, 383–408, 2018.
162. Nofar, M. and Oğuz, H., Development of PBT/recycled PET blends and the influence of using chain extender. *J. Polym. Environ.*, 27, 1404–1417, 2019.
163. Yang, W., Hu, Y., Tai, Q., Lu, H., Song, L., Yuen, R.K., Fire and mechanical performance of nanoclay reinforced glass-fiber/PBT composites containing aluminum hypophosphite particles. *Compos. Part A Appl. Sci. Manuf.*, 42, 794–800, 2011.
164. Roenner, N., Yuan, H., Krämer, R.H., Rein, G., Computational study of how inert additives affect the flammability of a polymer. *Fire Saf. J.*, 106, 189–196, 2019.
165. Han, M.S., Lim, B.H., Jung, H.C., Hyun, J.C., Kim, S.R., Kim, W.N., Reactive blends of poly(butylene terephthalate)/polyamide-6 with ethylene glycidyl methacrylate. *Korea-Aust. Rheol. J.*, 13, 169–177, 2001.

166. Sun, T., Datta, A., de Souza, J.P., Baird, D.G., Thermoforming of *in situ* reinforced thermoplastic composites. *ANTEC'91*, p. 933, 1991.
167. Trombley, K.F., Cyclic olefin copolymer: An alternative plastic film for pharmaceutical blister packages, pp. 28–38, 2001.
168. Shin, J.Y., Park, J.Y., Liu, C., He, J.-S., Kim, S.C., Chemical structure and physical properties of cyclic olefin copolymers (IUPAC technical report). *Pure Appl. Chem.*, 77, 801–814, 2005.
169. Khanarian, G., Rubber toughened and optically transparent blends of cyclic olefin copolymers. *Polym. Eng. Sci.*, 40, 2590–2601, 2000.
170. Throne, J.L., Polystyrene foam sheet expansion during heating. *J. Polym. Eng.*, 6, 313–344, 1986.
171. Auras, R., Harte, B., Selke, S., An overview of polylactides as packaging materials. *Macromol. Biosci.*, 4, 835, 2004.
172. Kosior, E., Braganca, R.M., Fowler, P., Lightweight compostable packaging: Literature review. *Retailer Information Milestone Report for The Waste & Resources Action Programme*, Banbury, UK, June 2006.
173. Lee, R.E., Guo, Y., Tamber, H., Planeta, M., Leung, S.N., Thermoforming of polylactic acid foam sheets: Crystallization behaviors and thermal stability. *Ind. Eng. Chem. Res.*, 55, 560–567, 2016.
174. Costa, F.R., Dutta, N.K., Choudhury, N.R., Bhowmick, A.K., Thermoplastic elastomers, in: *Current Topics in Elastomers Research*, A.K. Bhowmick (Ed.), pp. 101–164, Taylor & Francis, Boca Raton, FL, 2008.
175. Litvinov, V.M., EPDM/PP thermoplastic vulcanizates as studied by proton NMR relaxation: Phase composition, molecular mobility, network structure in the rubbery phase, and network heterogeneity. *Macromolecules*, 39, 25, 8727–8741, 2006.
176. Locke, D., Chaney, J., Bubak, J., Parks, M., Head impact compliance and optimization with modified TPO resins. *SPE Automotive TPO Global Conference '99*, Troy, MI, September 20-22, 1999.
177. Shah, S. and Kakarala, N., *An overview of advances in TPO skin material and process technology" by TPO in automotive '98*, Novi, MI, October 12-14, 1998.
178. Ryntz, R.A., The effects of thermoplastic poly(olefin) (TPO) morphology on subsequent paintability and thermal shock performance. *Prog. Org. Coat.*, 27, 241–254, 1996.
179. Wang, W.-J., Ye, Z., Fan, H., Li, B.-G., Zhu, S., Dynamic mechanical and rheological properties of metallocene-catalyzed long-chain-branched ethylene/propylene copolymers. *Polymer*, 45, 16, 5497–5504, 2004.
180. Arriola, D.J., Carnahan, E.M., Hustad, P.D., Kuhlman, R.L., Wenzel, T.T., Catalytic production of olefin block copolymers via chain shuttling polymerization. *Science*, 312, 5774, 714–719, 2006.
181. Estes, G.M., Cooper, S.L., Tobolsky, A.V., Block polymers and related heterophase elastomers. *J. Macromol. Sci.-Rev. Macromol. Chem.*, C4, 2, 313, 1970.

182. Huh, D.S. and Cooper, S.L., Dynamic mechanical properties of polyurethane block polymers. *Polym. Eng. Sci.*, 11, 5, 369–376, 1971.
183. Ferguson, J., Hourston, D.J., Meredith, R., Patsavoudis, D., Mechanical relaxations in a series of polyurethanes with varying hard to soft segment ratio. *Eur. Polym. J.*, 8, 3, 369–383, 1972.
184. Sorimpuk, N.P., Choong, W.H., Chua, B.L., Thermoforming characteristics of PLA/TPU multi-material specimens fabricated with fused deposition modelling under different temperatures. *Polymers*, 14, 4304–4318, 2022.
185. Hufenbach, W., Böhma, R., Thieme, M., Winkler, A., Mäder, E., Rausch, J. *et al.*, Polypropylene/glass fibre 3D-textile reinforced composites for automotive applications. *Mater. Des.*, 32, 3, 1468–1476, 2011.
186. Yanagimoto, J. and Ikeuchi, K., Sheet forming process of carbon fiber reinforced plastics for lightweight parts. *CIRP Ann.*, 61, 1, 247–250, 2012.
187. Delorenzi, H.G. and Nied, H.F., Blow molding and thermoforming of plastics. Finite elem. model. *Comput. Struct.*, 26, 1-2, 197–206, 1987.
188. Campbell, F.C., *Manfacturing processes for advanced composites*, Elsevier, New York, 2004.
189. Okine, R.K., Analysis of forming parts from advanced thermoplastic composite sheet materials. *J. Thermoplast. Compos. Mater.*, 2, 50–76, 1989.
190. Akkerman, R. and Haanappel, S.P., Thermoplastic composites manufacturing by thermoforming, in: *Advances in Composites Manufacturing and Process Design*, p. 111–129, Woodhead Publishing, 2015.
191. Brachos, V., Borbone, V., Robblee, G., The emergence of thermoplastic composite materials through cost-effective molding techniques, in: *Proceedings of SAMPE'08—Material and Process Innovations: Changing our World*, vol. 52, pp. 18–22, Long Beach, California, May 2008.
192. McCool, R., Murphy, A., Wilson, R., Jiang, Z., Price, M., Butterfield, J., Hornsby, P.R., Thermoforming carbon fibre-reinforced thermoplastic composites. *Proc. Inst. Mech. Eng. Part L: J. Mater. Des. Appl.*, 226, 102–91, 2012.
193. Dufour, C., Wang, P., Boussu, F., Soulat, D., Experimental investigation about stamping behaviour of 3D warp interlock composite preforms. *Appl. Compos. Mater.*, 21, 5, 725–738, 2014.
194. Yin, H., Peng, X., Du, T., Chen, J., Forming of thermoplastic plain woven carbon composites: An experimental investigation. *J. Thermoplast. Compos. Mater.*, 28, 5, 730–742, 2013.
195. Nurul Fazita, M.R., Jayaraman, K., Bhattacharyya, D., Formability analysis of bamboo fabric reinforced poly (lactic) acid composites. *Materials*, 9, 7, 539, 2016.
196. Omrani, F., Wang, P., Soulat, D., Ferreira, M., Ouagne, P., Analysis of the deformability of flax-fibre nonwoven fabrics during manufacturing. *Compos. B Eng.*, 116, 471–485, 2017.
197. Gere, D. and Czigany, T., Rheological and mechanical properties of recycled polyethylene films contaminated by biopolymer. *Waste Manage.*, 76, 190–198, 2018.

198. Jo, Y.K. and Lee, D., Biopolymer microparticles prepared by microfluidics for biomedical applications. *Small*, 16, 1903736, 2019.
199. Valdes, A., Mellinas, A.C., Ramos, M., Garrigos, M.C., Jimenez, A., Natural additives and agricultural wastes in biopolymer formulations for food packaging. *Front. Chem.*, 2, 6, 2014.
200. Giesa, T. and Buehler, M.J., Nanoconfinement and the strength of biopolymers. *Annu. Rev. Biophys.*, 42, 651–673, 2013.
201. Jem, K.J. and Tan, B., The development and challenges of poly (lactic acid) and poly (glycolic acid). *Adv. Ind. Eng. Polym. Res.*, 3, 60–70, 2020.
202. Zuzanna, C., Singh, M., Seshadri, S., Additives for improving processing and properties of polylactic acid. *Proceedings of the SPE ANTEC Conference*, Cincinnati, OH, USA, p. 3071, 20072007.
203. Sorimpuk, N.P., Choong, W.H., Chua, B.L., Thermoforming characteristics of PLA/TPU multi-material specimens fabricated with fused deposition modelling under different temperatures. *Polymers*, 14, 4304–4318, 2022.
204. Bi, S., Tan, B., Soule, J.L., Sobkowicz, M.J., Enzymatic degradation of poly(butylene succinate-co-hexamethylene succinate). *Polym. Degrad. Stab.*, 155, 9–14, 2018.
205. Di Lorenzo, M.L., Androsch, R., Righetti, M.C., Low-temperature crystallization of poly(butylene succinate). *Eur. Polym. J.*, 94, 384–391, 2017.

Chapter 4

1. Nam, G.J., Lee, J.W., Ahn, K.H., Three-dimensional simulation of thermoforming process and its comparison with experiments. *Polym. Eng. Sci.*, 40, 10, 2232–2240, 2000.
2. Keating, M.Y., High glass transitions of high-performance thermoplastics. *Thermochim. Acta*, 319, 1–2, 201–212, 1998.
3. Cleaver, E.A., Schellenberg, T., Sakuma, Y., Liquid crystal polymers, specialty plastics, 580.4200A, in: *Chemical Economics Handbook*, SRI International, Menlo Park, CA, May 1996.
4. Wunderlich, B., Motion in the solid state of high polymers. *J. Polym. Sci. Part C Polym. Symp.*, 1, 1, 41–64, 1963.
5. Dainton, F.S., Evans, D.M., Hoare, F.E., Melia, T.P., Thermodynamic functions of linear high polymers: Part I—Polyoxymethylene. *Polymer*, 3, 263–321, 1962.
6. Ayhan, Z. and Zhang, Q.H., Wall thickness distribution in thermoformed food containers produced by a Benco aseptic packaging machine. *Polym. Eng. Sci.*, 40, 1, 1–10, 2000.
7. Argon, A.S. and Bessonov, M., II, Plastic flow in glassy polymers. *Polym. Eng. Sci.*, 17, 3, 174–182, 1977.
8. Richeton, J., Ahzi, S., Vecchio, K.S., Jiang, F.C., Makradi, A., Modeling and validation of the large deformation inelastic response of amorphous polymers

over a wide range of temperatures and strain rates. *Int. J. Solids Struct.*, 44, 24, 7938–7954, 2007.
9. Throne, J., *Thermoforming*, Hanser Publishers, New York, 1986.
10. Schmidt, F.M., Le Maoult, Y., Monteix, S., Modelling of infrared heating of thermoplastic sheet used in thermoforming process. *J. Mater. Process. Technol.*, 143, 225–2315, 2003.
11. Van Dommelen, J.A.W., Parks, D.M., Boyce, M.C., Brekelmans, W.A.M., Baaijens, F.P.T., Micromechanical modeling of the elasto-viscoplastic behavior of semi-crystalline polymers. *J. Mech. Phys. Solids*, 51, 3, 519–541, 2003.
12. Makradi, A., Ahzi, S., Belouettar, S., Ruch, D., Thermoforming process of semicrystalline polymeric sheets: Modeling and finite element simulations. *Polym. Sci. Ser. A*, 50, 550–5575, 2008.
13. Yamaguchi, M. and Suzuki, K.-I., Rheological properties and foam processability for blends of linear and crosslinked polyethylenes. *J. Polym. Sci. Part B Polym. Phys.*, 39, 18, 2159–2167, 2001.
14. Osaki, K., Linear viscoelastic relation concerning shear stresses at the strat and cessation of shear flow. *Nihon Reoroji Gakkaishi (J. Soc Rheol. Jpn.)*, 4, 166–168, 1976.
15. Schneider, K., Zafeiropoulos, N.E., Häußler, L., Stamm, M., High-throughput screening of the influence of thermal treatment on the mechanical properties of semicrystalline polymers: A case study for iPP. *Macromol. Rapid Commun.*, 25, 1, 355–359, 2004.
16. Rault, J., Memory effects in semicrystalline polymers. *Crit. Rev. Solid State Mater. Sci.*, 13, 1, 57–95, 1986.
17. Pezzin, A.P.T., Alberda Van Ekenstein, G.O.R., Zavaglia, C.A.C., Brinke, G.T., Duek, E.A.R., Poly (para-dioxanone) and poly (L-lactic acid) blends: Thermal, mechanical, and morphological properties. *J. Appl. Polym. Sci.*, 88, 12, 2744–2755, 2003.
18. Oya, N., Saitoh, S., Furuhashi, Y., Yoshie, N., Mechanical property tuning of semicrystalline network polymers by controlling rates of crystallization and crosslinking. *J. Polym. Sci. Part A Polym. Chem.*, 50, 10, 1926–1932, 2012.
19. Natta, G., Pino, P., Corradini, P., Danusso, F., Mantica, E., Mazzanti, G., Moraglio, G., Crystalline high polymers of α-olefins. *J. Am. Chem. Soc.*, 77, 6, 1708–1710, 1955.
20. Diez, F.J., Alvariño, C., Lopez, J., Ramirez, C., Abad, M.J., Cano, J., Garabal, S.G., Barral, L., Influence of the stretching in the crystallinity of biaxially oriented polypropylene (BOPP) films. *J. Therm. Anal. Calorim.*, 81, 21–255, 2005.
21. Rosato, D.V. and Rosato, D.V. (Eds.), *Blow Molding Handbook*, Hanser, New York, 1989.
22. Kalyon, D.M. and Kamal, M.R., An experimental investigation of capillary extrudate swell in relation to parison swell behavior in blow molding. *Polym. Eng. Sci.*, 26, 7, 508–516, 1986.
23. Charrier, J.M., *Polymeric Materials and Processing*, Hanser, New York, 1990.

24. La Mantia, F.P. and Acierno, D., Influence of the molecular structure on the melt strength and extensibility of polyethylenes. *Polym. Eng. Sci.*, 25, 5, 279–283, 1985.
25. Dealy, J.M. and Wissbrun, K.F., *Melt Rheology and Its Role in Plastics Processing*, Reinhold, New York, 1995.
26. Cheng, S., Johnson, L., Wang, S.-Q., Crazing and strain localization of polycarbonate glass in creep. *Polymer*, 54, 13, 3363–3369, 2013.
27. Zartman, G.D., Cheng, S., Li, X., Lin, F., Becker, M.L., Wang, S.-Q., How melt-stretching affects mechanical behavior of polymer glasses. *Macromolecules*, 45, 16, 6719–6732, 2012.
28. Ma, Z., Balzano, L., Peters, G.W.M., Dissolution and re-emergence of flow-induced shish in polyethylene with a broad molecular weight distribution. *Macromolecules*, 49, 7, 2724–2730, 2016.
29. Alvarez, G.A. and Cantow, H.-J., Effect of molecular weight heterodispersity on linear viscoelastic properties of entangled polymers. *Polym. Bull.*, 7, 2-3, 137–144, 1982.
30. De Rosa, C., Auriemma, F., de Ballesteros, O.R., Resconi, L., Fait, A., Ciaccia, E., Camurati, I., Synthesis and characterization of high-molecular-weight syndiotactic amorphous polypropylene. *J. Am. Chem. Soc.*, 125, 36, 10913–10920, 2003.
31. Lynd, N.A., Meuler, A.J., Hillmyer, M.A., Polydispersity and block copolymer self-assembly. *Prog. Polym. Sci.*, 33, 9, 875–893, 2008.
32. Leibler, L., Theory of microphase separation in block copolymers. *Macromolecules*, 13, 6, 1602–1617.
33. Stürzel, M., Mihan, S., Mülhaupt, R., From multisite polymerization catalysis to sustainable materials and all-polyolefin composites. *Chem. Rev.*, 116, 3, 1398–1433, 2016.
34. Koopmans, R.J., Extrudate swell of high density polyethylene. Part I: Aspects of molecular structure and rheological characterization methods. *Polym. Eng. Sci.*, 32, 23, 1741–1749, 1992.
35. Koopmans, R.J., Extrudate swell of high density polyethylene. Part I: Aspects of molecular structure and rheological characterization methods. *Polym. Eng. Sci.*, 32, 23, 1755, 1992.
36. Varshouee, G.H., Heydarinasab, A., Vaziri, A., Roozbahani, B., Predicting molecular weight distribution, melt flow index, and bulk density in a polypropylene reactor via a validated mathematical model. *Theor. Found. Chem. Eng.*, 55, 153-166, 2021.
37. Bourgin, P., Cormeau, I.C., Saint-Matin, T., A first step towards the modelling of the thermoforming of plastic sheets. *J. Mater. Process. Technol.*, 54, 1–115, 1995.
38. Gilormini, P., Chevalier, L., Régnier, G., Thermoforming of a PMMA transparency near glass transition temperature. *Polym. Eng. Sci.*, 50, 10, 2004–2012, 2010.

39. Kovacs, A.J., Glass transition in amorphous polymers: A phenomenological study. *Adv. Polym. Sci.*, 3, 3, 394–507, 1963.
40. Xanthos, M. and Todd, D.B., *Kirk-Othmer Encyclopedia of Chemical Technology*, 4th Edition, vol. 19, J.I. Kroschwitz (Ed.), p. 290, John Wiley & Sons, New York, NY, USA, 1996.
41. Elias, M., Machado, R., Canevarolo. Thermal, S., and dynamic-mechanical characterization of uni-and biaxially oriented polypropylene films. *J. Therm. Anal. Calorim.*, 59, 1-2, 143–155, 2000.
42. Van Krevelen, D.W., Chap. 19, in: *Properties of Polymers*, 3rd ed, Elsevier, Amsterdam, Oxford, New York, Tokyo, 1990.
43. Chang, Y.-W. and Cheng, J.-H., Material characterization of polycarbonate near glass transition temperature. *J. Chin. Inst. Eng.*, 35, 8, 967–978, 2012.
44. Williams, J.G., A method of calculation for thermoforming plastics sheets. *J. Strain Anal.*, 5, 1, 49–57, 1970.
45. Suzuki, H., Grebowicz, J., Wunderlich, B., Glass transition of poly (oxymethylene). *Br. Polym. J.*, 17, 1, 1–3, 1985.
46. Riga, A.T., Heat distortion and mechanical properties of polymers by thermal-mechanical analysis. *Polym. Eng. Sci.*, 14, 11, 764–767, 1974.
47. Moynihan, C.T. and Macedo, P.B., Dependence of the glass transition temperature on heating rate and thermal history. *J. Phys. Chem.*, 75, 21, 3379–3381, 1971.
48. Wolpert, S.M., Weitz, A., Wunderlich, B., Time dependent heat capacity in the glass transition region. *J. Polym. Sci. A-2 Polym. Phys.*, 9, 10, 1887–1905, 1971.
49. Gotze, W. and Sjogrenm, L., Relaxation processes in supercooled liquids. *Rep. Prog. Phys.*, 55, 3, 241, 1992.
50. Gilormini, P., Chevalier, L., Régnier, G., Modeling the nonlinear PMMA behavior near glass transition temperature: Application to its thermoforming. *AIP Conf. Proc.*, 1315, 1, 179–184, January 17, 2011.
51. Lee, J.K., Virkler, T.L., Scott, C.E., Influence of initial sheet temperature on ABS thermoforming. *Polym. Eng. Sci.*, 41, 1830–1844, 2001.
52. Hieber, C.A. and Chiang, H.H., Shear-rate-dependence modeling of polymer melt viscosity. *Polym. Eng. Sci.*, 32, 14, 931–938, 1992.
53. Lau, H.C., Bhattacharya, S.N., Field, G.J., Melt strength of polypropylene: Its relevance to thermoforming. *Polym. Eng. Sci.*, 38, 11, 1915–1923, 1998.
54. Takemori, M.T., Towards an understanding of the heat distortion temperature of thermoplastics. *Polym. Eng. Sci.*, 19, 15, 1104–1109, 1979.
55. Lai, M.O. and Holt, D.L., Thickness variation in the thermoforming of poly (methyl methacrylate) and high-impact polystyrene sheets. *J. Appl. Polym. Sci.*, 19, no. 7, 1805–1814, 1975.
56. Siegmann, A., Narkis, M., Rosenzweig, N., Softening temperature of glassy polymers as affected by residual stresses. *Polym. Eng. Sci.*, 19, 223–225, 1979.
57. Balasko, M.M., Summers, J.W., Dickerson, J.L., Rodriguez, E.L., A comparison of oven sag temperature and heat deflection temperature for measuring heat resistance. *J. Vinyl Addit. Technol.*, 11, 184–189, 1989.

58. Gilman, J.J., *The Art and Science of Growing Crystals*, John Wiley & Sons, New York, London, 1963.
59. Azzurri, F. and Alfonso, G.C., Insights into formation and relaxation of shear-induced nucleation precursors in isotactic polystyrene. *Macromolecules*, 41, 1377–1383, 2008.
60. Valcárcel, J.P., Palacios, M.J., Alvarado-Gil, J.J., Determination of the thermophysical properties of polymers (PET) using photoacoustic spectroscopy. *J. Mater. Sci.*, 34, 2113, 1999.
61. Furushima, Y., Schick, C., Toda, A., Crystallization, recrystallization, and melting of polymer crystals on heating and cooling examined with fast scanning calorimetry. *Adv. Polym. Crystallization, Part 1*, 1, 2, e10005, August 2018.
62. Kovalski, L., Duszczyk, J., Katgerman, L., Thermal conductivity of metal powder-polymer feedstock for powder injection moulding. *J. Mater. Sci.*, 34, 1, 1999.
63. Hill, J.W. and Cuculo, J.A., Elongational flow behavior of polymeric fluids. *J. Macromol. Sci. Rev.*, C14, 107, 1976.
64. Cogswell, F.N., Polymer melt rheology during elongational flow. *Appl. Polym. Symp.*, 27, 1, 1975.
65. Ghijsels, A. and De Clippeleir, J., Melt strength behaviour of polypropylenes. *Int. Polym. Process.*, 9, 3, 252–257, 1994.
66. Wissbrun, K.F., Interpretation of the "melt strength" test. *Polym. Eng. Sci.*, 13, 342, 1973.
67. Worn, R., Untersuchung des Rheotensversuchs hinsichtlich seiner Eignung als Wareneingangskontrollverfahren. *Polym. Process Eng.*, 4, 1, 97–123, 1986. 97, 1986.
68. Wong, P.P. and Bollampally, R.S., Thermal conductivity, elastic modulus, and coefficient of thermal expansion of polymer composites filled with ceramic particles for electronic packaging. *J. Appl. Polym. Sci.*, 74, 3396, 1999.
69. Alsleben, M. and Schick, C., The melting of polymers—A three-phase approach. *Thermochim. Acta*, 238, 203, 1994.
70. Romanini, D., Savadori, A., Gianotti, C., Long chain branching in low density polyethylene: 2. Rheological behaviour of the polymers. *Polymer*, 21, 1092, 1980.
71. Busse, W.F., Mechanical structures in polymer melts. I. Measurements of melt strength and elasticity. *J. Polym. Sci. A-2 Polym. Phys.*, 5, 1249–1259, 1967.
72. Koyama, K. and Ishizuka, O., Nonlinearity in uniaxial elongational viscosity at a constant strain rate. *Polym. Proc. Eng.*, 1, 55, 1983.
73. Dusunceli, N. and Colak, O., The effects of manufacturing techniques on viscoelastic and viscoplastic behavior of high density polyethylene (HDPE). *Mater. Des.*, 29, 1117–1124, 2008.
74. Brinson, H.F. and Brinson, L.C., *Polymer Engineering Science and Viscoelasticity: An Introduction*, pp. 99–157, Springer Science, 2008.

75. Kouba, K. and Vlachopoulos, J., The role of rheology in thermoforming process: Experimental and theoretical consideration. *SPE Tech. Pap.*, 50, 114, 1992.
76. Colak, O.U., Modeling deformation behavior of polymers with viscoplasticity theory based on overstress. *Int. J. Plast.*, 21, 145–60, 2005.
77. Sternstein, S.S. and Ho, T.C., Biaxial stress relaxation in glassy polymers: polymethylmethacrylate. *J. Appl. Phys.*, 43, 4370–4383, 1972.
78. Meissner, J., Rheometry of polymer melts. *Annu. Rev. Fluid Mech.*, 17, 45–64, 1985.
79. Sperling, L.H., *An Introduction to Physical Polymer Science*, Fourth ed, Wiley-Interscience, 2006.
80. Sudduth, R.D., Comparison of the viscous and elastic components of two ABS materials with creep, stress relaxation and constant strain rate measurements using the universal viscoelastic model. *J. Appl. Polym. Sci.*, 90, 1298–1318, 2003.
81. Mınstedt, H., Rheology of rubber-modified polymer melts. *Polym. Eng. Sci.*, 21, 259–270, 1981.
82. Shaw, M.T. and MacKnight, W.J., *Introduction to Polymer Viscoelasticity*, Wiley Interscience, 2005.
83. Langdon, T.G., The mechanical properties of superplastic materials. *Metall. Trans. A*, 13, 5, 689–701, 1982.
84. Gregory, D.R., Rheological properties of molten poly (ethylene terephthalate). *J. Appl. Polym. Sci.*, 16, 1479, 1972.
85. Agrawal, P., Silva, M.H.A., Cavalcanti, S.N., Freitas, D.M.G., de Araújo, J.P., Oliveira, A.D.B., Mélo, T.J.A., *Polym. Bull.*, 79, 2321–2343, 2021.
86. Daneshi, S. and Porter, R.S., Blends of isotactic polypropylene and ethylene-propylene rubbers: rheology, morphology and mechanics. *Polymer*, 19, 448–457, 1978.
87. Han, C.D., Yu, T.C., Kim, K.U., Rheological properties of molten polymers. I. Homopolymer systems. *J. Appl. Polym. Sci.*, 15, 1149–1162, 1971.
88. You, L., Pan, Y., Sun, Z.Z., Sheng, X., Shi, W., Li, J., Huang, H., *Adv. Mater. Res.*, 631–632, 100–103, 2013.
89. Utracki, L.A., Viscoelastic behavior of polymer blends. *Polym. Eng. Sci.*, 28, 1401–1404, 1988.
90. Gasco, M.C., Rodriguez, F., Long, T., Polymer–polymer friction as a function of test speed. *J. Appl. Polym. Sci.*, 67, 11, 1831–1836, 1998.
91. Erner, A. and Billon, N., Thermal and friction effects during plug-assisted thermoforming, experimental approach. *Int. J. Form. Process.*, 8, 265–281, 2005.
92. Cha, J., Song, H.Y., Hyun, K., Go, J.S., Rheological measurement of the nonlinear viscoelasticity of the ABS polymer and numerical simulation of thermoforming process. *Int. J. Adv. Manuf. Technol.*, 107, 2449–2464, 2020.
93. Lancaster, J.K., *Encyclopedia of Polymer Science and Engineering*, J. Kroschwitz II (Ed.), p. 1, Wiley, New York, 1985.

94. Lobo, H. and Cohen, C., Measurement of thermal conductivity of polymer melts by the line-source method. *Polym. Eng. Sci.*, 30, 2, 65, 1990.
95. Dessain, B., Moulaert, O., Keunings, R., Bunsell, A.R., Solid phase change controlling the tensile and creep behaviour of gel-spun high-modulus polyethylene fibres. *J. Mater. Sci.*, 27, 4515, 1992.
96. Lopes, C.M.A. and Felisberti, M., II, Thermal conductivity of PET/(LDPE/AI) composites determined by MDSC. *Polym. Test.*, 23, 637–643, 2004.
97. Ross, R.G., Anderson, P., Sundqvist, B., Backstrom, G., Thermal conductivity of solids and liquids under pressure. *Rep. Prog. Phys.*, 47, 1347, 1984.
98. Lee, H.M. and Ok Park, O., Rheology and dynamics of immiscible polymer blends. *J. Rheology*, 38, 5, 1405–1425, 1994.
99. Choy, C.L., Thermal conductivity of polymers. *Polymer*, 18, 984, 1977.
100. dos Santos, W.N., Thermal properties of melt polymers by the hot wire technique. *Polym. Test.*, 24, 932–94, 2005.
101. Eiermann, K., Thermal conductivity of high polymers, in: *Journal of Polymer Science Part C: Polymer Symposia*, vol. 6, no. 1, pp. 157–165. New York: Wiley Subscription Services, Inc., A Wiley Company, 1964.
102. Kern, D.Q., *Process Heat Transfer*, p. 643, McGraw-Hill, New York, 1950.
103. Zhang, X., Hendro, W., Fujii, M. *et al.*, Measurements of the thermal conductivity and thermal diffusivity of polymer melts with the shorthot-wire method. *Int. J. Thermophys.*, 23, 1077–1090, 2002.
104. D'Angelo, G., Tripodo, G., Carini, G., Bartolotta, A., Di Marco, G., Salvato, G.N., Low-temperature excess specific heat and fragility in polymers: Crystallinity dependence. *J. Chem. Phys.*, 109, 7625–7631, 1998.
105. Carini, G., d'Angelo, G., Tripodo, G., Bartolotta, A., DiMarco, G., Low-temperature excess specific heat and fragility in semicrystalline polymers. *Phys. Rev. B Condens. Matter*, 54, 21, 15056–15063, 1996.
106. Hıllmansen, S., Hobeika, S., Haward, R.N., Aleevers, S., The effect of strain rate, temperature and molecular mass on the tensile deformation of polyethylene. *Polym. Eng. Sci.*, 40, 481–9, 2000.
107. Tang, Z., Fujimoto, K., Okazaki, S., All-atom molecular dynamics study of impact fracture of glassy polymers. II: Microscopic origins of stresses in elasticity, yielding, and strain hardening. *Polymer*, 207, 122908, 2020.
108. Dergachev, M.V. and Berdyshev, B.V., Rheological behavior of elastoviscous polymers under a complex shearing load. *Chem. Pet. Eng.*, 35, 358–363, 1999.
109. Kitagawa, M., Onoda, T., Mizutani, K., Stress–strain behavior at finite strains for various strain paths in polyethylene. *J. Mater. Sci.*, 27, 13–23, 1992.
110. Haward, R.N., Strain hardening of thermoplastics. *Macromolecules*, 26, 5860–5869, 1993.
111. Ward, I.M., Ch. 11, in: *Mechanical Properties of Solid Polymers*, 2nd Edn, Wiley, New York, 1985.
112. Tsai, J.T., Some rheological aspects of fracture in thermoforming. *Polym. Eng. Sci.*, 22, 265–2685, 1982.

113. Haward, R.N., The derivation of a strain hardening modulus from true stress-strain curves for thermoplastics. *Polymer*, 35, 3858–3862, 1994.
114. Laun, H.M. and Münstedt, H., Elongational behaviour of a low density polyethylene melt: I. Strain rate and stress dependence of viscosity and recoverable strain in the steady-state. Comparison with shear data. Influence of interfacial tension. *Rheol. Acta*, 17, 415, 1978.
115. Smith, S.B., Finzi, L., Bustamante, C., *Sci. (Washington, D. C.)*, 258, 1122, 1992.
116. Boyce, M.C., Parks, D.M., Argon, A.S., Plastic flow in oriented glassy polymers. *Int. J. Plast.*, 5, 593–615, 1989.
117. Lin, Y.H. and Das, A.K., Monte Carlo simulations of stress relaxation of entanglement-free Fraenkel chains. I. Linear polymer viscoelasticity. *J. Chem. Phys.*, 126, 074902, 2007.
118. DalMaso, G., De Simone, A., Mora, M.G., Quasistatic evolution problems for linearly elastic-perfectly plastic materials. *Arch. Ration. Mech. Anal. (Germany)*, 180, 237–291, 2006.
119. Mathot, V.B.F., *Calorimetry and Thermal Analysis of Polymers*, pp. 369, Hanser Gardner Publishers, Munich, New York, 1994.
120. Haward, R.N., Strain hardening of thermoplastics. *Macromolecules*, 26, 5860–5869, 1993.
121. Hong, K. and Strobl, G., Network stretching during tensile drawing of polyethylene: A study using X-ray scattering and microscopy. *Macromolecules*, 39, 268–273, 2006.
122. Peterlin, A., Molecular model of drawing polyethylene and polypropylene. *J. Mater. Sci.*, 6, 490–508, 1971.
123. Na, B., Zhang, Q., Fu, Q., Men, Y., Hong, K., Strobl, G.R., Viscous-force-dominated tensile deformation behavior of oriented polyethylene. *Macromolecules*, 39, 2584–2591, 2006.
124. De Rosa, C. and Auriemma, F., Structural–mechanical phase diagram of isotactic polypropylene. *J. Am. Chem. Soc.*, 128, 11024–11025, 2006.
125. Bartczak, Z., Effect of chain entanglements on plastic deformation behavior of linear polyethylene. *Macromolecules*, 38, 7702–7713, 2005.
126. Chang, C.-J., Chiu, W., Hsieh, K.-H., Ma, C.-C.M., The molecular orientation and mechanical properties of poly (ethylene terephthalate) under uniaxial extension. *J. Appl. Polym. Sci.*, 50, 855–862, 1993.
127. Krempl, E. and Khan, F., Rate (time) dependent deformation behavior an overview of some properties of metals and solid polymers. *Int. J. Plast.*, 19, 1069–95, 2003.
128. Malkin, A.Y., Arinstein, A.E., Kulichikhin, V.G., Polymer extension flows and instabilities. *Prog. Polym. Sci.*, 39, 959–978, 2014.
129. Boyce, M.C., Parks, D.M., Argon, A.S., Large inelastic deformation of glassy polymers, part I: Rate dependent constitutive model. *Mech. Mater.*, 7, 15–33, 1988.

130. Shinozaki, D.M. and Grovers, G.W., The plastic deformation of oriented polypropylene and polyethylene: deformation mechanisms. *J. Mater. Sci.*, 8, 1012, 1973.
131. Gent, A.N. and Jeong, J., Plastic deformation of crystalline polymers. *Polym. Eng. Sci.*, 26, 285–289, 1986.
132. Zanegin, V.D. and Gerasimov, V., II, Kink bands in low-density oriented polyethylene. *Polym. Sci. U.S.S.R.*, 23, 1279–1293, 1981.
133. Li, J., Nie, Y., Ma, Y., Hu, W., Stress-induced polymer deformation in shear flows. *Chin. J. Polym. Sci.*, 31, 1590–1598, 2013.
134. Arruda, E.M. and Boyce, M.C., Evolution of plastic anisotropy in amorphous polymers during finite straining. *Int. J. Plast.*, 9, 697–720, 1993.
135. Tervoort, T.A. and Govaert, L.E., Strain-hardening behavior of polycarbonate in the glassy state. *J. Rheol.*, 44, 1263–1277, 2000.
136. Boyce, M.C. and Haward, R.N., Chapter 5, in: *The Physics of Glassy Polymers*, 2nd ed, R.N. Haward, and R.J. Young, (Eds.), p. 288, Chapman and Hall, London, 1997.
137. Erk, K.A., Rhein, M., Krafcik, M.J., Ydstie, S., Demonstrating the effects of processing on the structure and physical properties of plastic using disposable PETE cups. *J. Chem. Educ.*, 92, 1876–1881, 2015.
138. Pegoretti, A., Marchi, A., Riccò, T., Determination of the fracture toughness of thermoformed polypropylene cups by the essential work method. *Polym. Eng. Sci.*, 37, 1045–1052, 1997.
139. Bensason, S., Minick, J., Moet, A., Chum, S., Hiltner, A., Baer, E., Classification of homogeneous ethylene-octene copolymers based on comonomer content. *J. Polym.Sci. Part B: Polym. Phys.*, 34, 7, 1301–1315, 1996.
140. Rezayat, M., Numerical computation of cooling-induced residual stress and deformed shape for injection-molded thermoplastics. *ANTEC'89: Plastics Create a World of Difference*, pp. 341–343, 1989.
141. Douven, L.L., Baaijens, F.F., Meijer, H.H., The computation of properties of injection-moulded products. *Prog. Polym. Sci.*, 20, 403–457, 1995.
142. Poh, L., Li, B., Yu, W., Narimissa, E., Wagner, M.H., Modeling of nonlinear extensional and shear rheology of low-viscosity polymer melts. *Polym. Eng. Sci.*, 61, 4, 1077–1086, 2021.
143. Schmidt, F.M., Agassant, J.-F., Bellet, M., Experimental study and numerical simulation of the injection stretch/blow molding process. *Polym. Eng. Sci.*, 38, 9, 1399–1412, 1998.
144. Zhan-Song, Y., Huang, H.-X., Liu, J.-H., Effect of sequence of stretch and blowing on preform growth in PET injection stretch blow molding. *ANTEC, 2005*, Society of Plastics Engineers, Brookfield, CT, pp. 36–40, 2005.
145. Malkin, A.Y., Non-linearity in rheology—An essay of classification. *Rheol. Acta*, 34, 27–39, 1995.
146. Yaoita, T., Isaki, T., Masubuchi, Y., Watanabe, H., Ianniruberto, G., Marrucci, G., Primitive chain network simulation of elongational flows of entangled linear chains: Stretch/orientation-induced reduction of monomeric friction. *Macromolecules*, 45, 2773–2782, 2012.

147. Koyama, K. and Ishizuka, O., Birefringence of polyethylene melt in transient elongational flow at constant strain rate. *J. Polym. Sci. Part B Polym. Phys.*, 27, 297–306, 1989.
148. White, J.L. and Ide, Y., Fiber and yarn processing. *Appl. Polym. Symp.*, 27, 61, 1975.
149. Florian, J., *Practical Thermoforming—Principles and Applications*, Marcel Dekker, New York, 1996.
150. Cormont, J.J.M., Differences between amorphous and crystalline plastics with respect to thermoforming. *Adv. Polym. Technol.*, 5, 209–218, 2003.
151. Kempthorn, J.T., The use of oven heat sag to determine heat resistance in polypropylenes. *Polym. Eng. Sci.*, 33, 248–252, 1993.
152. Leevers, P., Horsfall, I.M., Rager, A., Major, Z., Moore, D.R., Pavan, A., Williams, J.G., High rate fracture toughness testing of thermoplastics. *Polym. Test.*, 33, 79–87, 2014.
153. Kinloch, A.J. and Young, R.J. (Eds.), *The Fracture Behaviour of Polymers*, Elsevier, London, 1985.
154. Seidler, S. and Grellmann, W., Application of the instrumented impact test to the toughness characterization of high impact thermoplastics. *Polym. Test.*, 14, 453–469, 1995.
155. Wittmann, L.-M., Kurth, K.H., Drummer, D., Elongation behaviour of filled semi-crystalline polymers in thermoforming. *Adv. Mater. Process. Technol.*, 6, 647–660, 2020.
156. Suh, C.H. and White, J.L., Particle and polymer chain orientation development in thermoformed talc-filled thermoplastics. *Polym. Eng. Sci.*, 36, 2188–2197, 1996.
157. Zorin, V., Baurova, N., II, Shakurova, A.M., Control of microstructure and properties of filled polymer compositions. *Polym. Sci. Ser. D*, 6, 36–40, 2013.
158. Marko, J.F. and Siggia, E.D., Stretching data. *Macromolecules*, 28, 8759–8770, 1995.

Chapter 5

1. Lau, H.C., Bhattacharya, S.N., Field, G.J., Influence of rheological properties on the sagging of polypropylene and ABS sheet for thermoforming applications. *Polym. Eng. Sci.*, 40, 1564, 2000.
2. Fabrizio, R., Salmaso, L., De Dominicis, M., Illert, J., Krajewski, K., Link, R. and Monti, L., Optimal designs to develop and support an experimental strategy on innovation of thermoforming production process. *Statistica*, 77, 2, 109–131, 2017.
3. Debbaut, B., Goublomme, A., Homerin, O., Koopmans, R., Liebman, D., Meissner, J., Schroeter, B., Reckmann, B., Daponte, T., Verschaeren, P., Agassant, J.F., Vergnes, B., Venet, C., Development of high quality LLDPE and optimised processing for film blowing. *Int. Polym. Process.*, 13, 262, 1998.

4. Parisi, M., Ryan, M., Charrier, J.M., Experimental study on the dynamics of thermoforming of polystyrene. *Polym. Eng. Sci.*, 34, 2, 102–108, 1994.
5. Laun, H.M. and Münstedt, H., Elongational behaviour of a low density polyethylene melt: I. Strain rate and stress dependence of viscosity and recoverable strain in the steady-state. Comparison with shear data. Influence of interfacial tension. *Rheol. Acta*, 17, 4, 415–425, 1978.
6. Ogden, R.W., *Non-Linear Elastic Deformations*, Halsted Press, New York, NY, 1984.
7. Lima, L.-T., Auras, R., Rubino, M., Processing technologies for poly (lactic acid). *Prog. Polym. Sci.*, 33, 8, 820–852, 2008.
8. Akkerman, R. and Pronk, R., Vacuum forming of thermoplastic foam. *Polym. Eng. Sci.*, 39, 2064–2074, 1999.
9. Yang, C. and Hung, S.W., Optimising the thermoforming process of polymeric foams: an approach by using the Taguchi method and the utility concept. *Int. J. Adv. Manuf. Technol.*, 24, 5, 353–360, 2004.
10. Morales, R.A., Candal, M.V., Santana, O., Gordillo, A., Salazar, R., Effect of the thermoforming process variables on the sheet friction coefficient. *Mater. Des.*, 53, 1097–1103, 2014.
11. Bae, H., Lee, H.-J., Park, K., Effect of vibration transmission direction in ultrasonic thermoforming on the formability of micro-corrugations. *Int. J. Precis. Eng. Manuf.*, 18, 697–703, 2017.
12. Levy, S. and DuBois, J., *Plastics Product Design Engineering*, Van Nostrand Reinhold Company, USA, 1977.
13. Tomasoni, D., Colosio, S., Giorleo, L., Ceretti, E., Design for additive manufacturing: thermoforming mold optimization via conformal cooling channel technology. *Procedia Manuf.*, 47, 1117–1122, 2020.
14. Ghobadnam, M., Mosaddegh, P., Rejani, M.R., Amirabadi, H., Ghaei, A., Numerical and experimental analysis of HIPS sheets in thermoforming process. *Int. J. Adv. Manuf. Technol.*, 76, 1079–1089, 2015.
15. Novotny, P., Saha, P., Kouba, K., Optimization of thermoforming. *Proc. Int. Conf. Antec'99*, SPE, New York, USA, pp. 841–843, May 1999.
16. Collins, P., Harkin-Jones, E.M.A., Martin, P.J., The role of tool/sheet contact in plug-assisted thermoforming. *Int. Polym. Process.*, 17, 4, 361–369, 2002.
17. Martin, P.J. and Duncan, P., The role of plug design in determining wall thickness distribution in thermoforming. *Polym. Eng. Sci.*, 47, 6, 804–813, 2007.
18. Chang, Y.-W. and Cheng, J.-H., Numerical and experimental inves tigation of polycarbonate vacuum-forming process. *J. Chin. Inst. Eng.*, 36, 7, 831–841, 2013.
19. Macauley, N.J., Harkin-Jones, E.M.A., Murphy, W.R., Method of assessing thermoformability of extruded polypropylene sheet. *Plastics, Rubber Composites Process. Appl.*, 26, 4, 165–171, 1997.
20. Wriggers, P., Finite element methods for contact problems with friction. *Tribol. Int.*, 29, 8, 651–658, 1996.

21. Weber, G.G., Lush, A.M., Zavaliangos, A., Anand, L., An objective time-integration procedure for isotropic rate-independent and rate-dependent elastic-plastic constitutive equations. *Int. J. Plast.*, 6, 6, 701–744, 1990.
22. Boyce, M.C., Socrate, S., Llana, P.G., Constitutive model for the finite deformation stress–strain behavior of poly (ethylene terephthalate) above the glass transition. *Polymer*, 41, 6, 2183–2201, 2000.
23. Makradi, A., Belouettar, S., Ahzi, S., Puissant, S., Thermoforming process of amorphous polymeric sheets: Modeling and finite element simulations. *J. Appl. Polym. Sci.*, 106, 3, 1718–1724, 2007.
24. Vantal, M.H., Monasse, B., Bellet, M., Numerical simulation of thermoforming of multi-layer polymer sheets, in: *Simulation of Materials Processing: Theory, Methods and Applications*, Shen, S.-F. and P. Dawson, (Eds.), pp. 1089–1095, Balkema, Rotterdam, 1995.
25. Lee, J.K., Scott, C.E., Virkler, T.L., Effects of rheological properties and processing parameters on ABS thermoforming. *Polym. Eng. Sci.*, 41, 2, 240–261, 2001.
26. Stephenson, M.J. and Ryan, M.E., Experimental study of the thermoforming of a blend of styrene-butadiene copolymer with polystyrene. *Polym. Eng. Sci.*, 37, 2, 450–459, 1997.
27. DeLorenzi, H.G. and Nied, H.F., Blow molding and thermoforming of plastics: finite element modeling. *Comput. Struct.*, 26, 1-2, 197–206, 1987.
28. Kim, M.S., Hwang, Y.N., Kwon, S.T., Lee, S.G., Cho, H.H., Structure and physical properties of hydrophilic polyamide copolymers fiber based on nylon 6 and nylon 4, 6. *Fibers Polym.*, 21, 2173–2178, 2020.
29. Harron, G.W., Harkin-Jones, E., Martin, P.J., Plug force monitoring for the control and optimization of the thermoforming process. *Proc. Inst. Mech. Eng. E J. Process Mech. Eng.*, 217, 181–188, 2003.
30. Li, Z., Heo, K., Seol, S., Time-dependent optimal heater control in thermoforming preheating using dual optimization steps. *Int. J. Precis. Eng. Manuf.*, 9, 4, 51–56, 2008.
31. Monteix, S., Schmidt, F., Le Maoult, Y., Ben Yedder, R., Diraddo, R.W., Laroche, D., Experimental study and numerical simulation of preform or sheet exposed to infrared radiative heating. *J. Mater. Process. Technol.*, 119, 1–3, 90–97, 2001.
32. Crawford, R.J. and Lui, S.K.L., Prediction of wall thickness distribution in thermoformed mouldings. *Eur. Polym. J.*, 18, 699–705, 1982.
33. Ayhan, Z., Zhang, Q.H., Wall thickness distribution in thermoformed food containers produced by a Benco aseptic packaging machine. *Polym. Eng. Sci.*, 40, 1, 1–10, 2000.
34. Aroujalian, A., Ngadi, M.O., Emond, J.-P., Wall thickness distribution in plug-assist vacuum formed strawberry containers. *Polym. Eng. Sci.*, 37, 1, 178–182, 1997.
35. O'Connor, C., Martin, P.J., Sweeney, J., Menary, G.H., Caton-Rose, P., Spencer, P., Simulation of the plug-assisted thermoforming of polypropylene using a

large strain thermally coupled constitutive model. *J. Mater. Process. Technol.*, 213, 9, 1588–1600, 2013.
36. Azdast, T., Doniavi, A., Ahmadi, S.R., Amiri, E., Numerical and experimental analysis of wall thickness variation of a hemispherical PMMA sheet in thermoforming process. *Int. J. Adv. Manuf. Technol.*, 64, 113–122, 2013.
37. McCool, R. and Martin, P.J., The role of process parameters in determining wall thickness distribution in plug-assisted thermoforming. *Polym. Eng. Sci.*, 50, 10, 1923–1934, 2010.
38. Rosenzweig, N., Narkis, M., Tadmor, Z., Wall thickness distribution in thermoforming. *Polym. Eng. Sci.*, 19, 13, 946–950, 1979.
39. Laursen, T.A. and Oancea, V.G., On the Constitutive Modeling and Finite Element Computation of Rate Dependent Frictional Sliding in Large Deformations. *Comput. Methods Mech. Eng.*, 143, 197–277, 1997.
40. de Lorenzi, H.G. and Nied, H.F., Finite element simulation of thermoforming and blow moulding, in: *Modeling of Polymer Processing*, A.I. Isayev, (Ed.), pp. 117–171, Hanser Publishers, Munich, 1999.
41. Marathe, D., Shelar, S., Mahajan, S., Ahmad, Z., Gupta, S.J., Kulkarni, S., Juvekar, V.A., Lele, A.K., Study of Rheology and Plug Assist Thermoforming of Linear and Branched PP Homopolymer and Impact Copolymer. *Int. Polym. Process.*, 34, 3, 339–355, 2019.
42. Shen, Y.-d., Li, Z.-Z., Xuan, D., Heo, K.-S., Seol, S.Y., *Int. J. Precis. Eng. Manuf.*, 11, 77–81, 2010.
43. Liu, S.-J., Process optimization of thermoforming PP/CaCO3 composites. *Int. Polym. Process.*, 14, 1, 98–102, 1999.
44. Hsiao, S.W. and Kikuchi, N., Numerical analysis and optimal design of composite thermoforming process. *Comput. Methods Appl. Mech. Eng.*, 177, 1–2, 1–34, 1999.
45. Morye, S.S., Comparison of the thermoformability of a PPE/PP blend with thermoformable ABS. Part II: large deformation methods. *Polym. Eng. Sci.*, 45, 10, 1377–1384, 2005.
46. Parisi, M., Ryan, M.E., Charrier, J.M., Experimental study on the dynamics of thermoforming of polystyrene. *Polym. Eng. Sci.*, 34, 2, 102–108, 1994.
47. Lee, J.K., Scott, C.E., Virkler, T.L., Effects of ABS rubber particles on rheology, melt failure, and thermoforming. *Polym. Eng. Sci.*, 42, 7, 1541–1557, 2002.
48. Lee, J.K., Solovyov, S.E., Virkler, T.L., Scott, C.E., On modes and criteria of ABS melt failure in extension. *Rheol. Acta*, 41, 6, 567–576, 2002.
49. White, J.L., Theoretical considerations of biaxial stretching of viscoelastic fluid sheets with application to plastic sheet forming. *Rheol. Acta*, 14, 7, 600–611, 1975.
50. Nam, G.J., Ahn, K.H., Lee, J.W., Three-dimensional simulation of thermoforming process and its comparison with experiments. *Polym. Eng. Sci.*, 40, 10, 2232–2240, 2000.

51. Sukiman, M.S., Erchiqui, F., Kanit, T., Imad, A., Design and numerical modeling of the thermoforming process of a WPC based formwork structure. *Mater. Today Commun.*, 22, 100805, 2020.
52. Harron, G., Harkin-Jones, E.M.A., Martin, P.J., Influence of thermoforming parameters on final part properties. *Proc. ANTEC 2000 Int. Conf., The Society of Plastics Engineers*, Orlando, Florida, USA, pp. 3723–3727, May 2000.
53. Shafi, M.A., Lee, J.G., Flumerfelt, R.W., Prediction of cellular structure in free expansion polymer foam processing. *Polym. Eng. Sci.*, 36, 1950, 1996.
54. DiRaddo, R.W. and Meddad, A., Sensitivity of operating conditions and material properties for thermoforming process. *Plast. Rubber Compos.*, 29, 4, 163–167, 2000.
55. Takaffoli, M., Hangalur, G., Bakker, R., Chandrashekar, N., Thermo-visco-hyperelastic behavior of polycarbonate in forming of a complex geometry. *J. Manuf. Process.*, 57, 105–113, 2020.
56. Atmani, O., Abbès, F., Li, Y., Batkam, S., Abbès, B., Experimental and numerical investigation of the effects of sheet material, plug-assist tool material, and process conditions on the mechanical pre-stretching stage of plug-assist thermoforming. *Int. J. Adv. Manuf. Technol.*, 122, 7–8, 3217–3234, 2022.
57. Tsai, J.T., Some rheological aspects of fracture in thermoforming. *Polym. Eng. Sci.*, 22, 4, 265–268, 1982.
58. Chy, M.M. and Boulet, B., A conjugate gradient method for the solution of the inverse heating problem in thermoforming. *IEEE Industry Applications Society Annual Meeting*, Houston, TX, USA, pp. 1–8, 2010.
59. Pignon, B., Boyard, N., Sobotka, V., Delaunay, D., Heat transfer analysis at high cooling rate on the surface of thermoplastic parts. *Int. J. Heat Mass Transf.*, 106, 253, 2017.
60. Winkel, E., Gross, H.G., Masberg, U., Wortberg, J., Extrusion, thermoforming and stretch-blow-molding of polypropylene. *Adv. Polym. Technol.*, 2, 2, 107–140, 1982.
61. Eder, G., Janeschitz-Kriegl, H., Liedauer, S., Crystallization processes in quiescent and moving polymer melts under heat transfer conditions. *Prog. Polym. Sci.*, 15, 4, 629–714, 1990.
62. Amado, A., Schmid, M., Wegener, K., Simulation of warpage induced by non-isothermal crystallization of co-polypropylene during the SLS process. *AIP Conf. Proc.*, 1664, 160002, 2015.
63. Erchiqui, F., Hamani, I., Charette, A., Modélisation par éléments finis du chauffage infrarouge des membranes thermoplastiques semi-transparentes. *Int. J. Therm. Sci.*, 48, 1, 73, 2009.
64. Nakamura, K., Katayama, K., Amano, T., Some aspects of nonisothermal crystallization of polymers. II. Consideration of the isokinetic condition. *J. Appl. Polym. Sci.*, 17, 4, 1031–1041, 1973.
65. Erchiqui, F., Souli, M., Yedder, R.B., Nonisothermal finite-element analysis of thermoforming of polyethylene terephthalate sheet: Incomplete effect of the forming stage. *Polym. Eng. Sci.*, 47, 12, 2129, 2007.

66. Pignon, B., Boyard, N., Sobotka, V., Delaunay, D., Heat transfer analysis at high cooling rate on the surface of thermoplastic parts. *Int. J. Heat Mass Transf.*, 106, 253–262, 2017.
67. Puehringer, J.F., Zitzenbacher, G., Spreitzer, C., Study of heat absorption in thermoforming for transparent and filled polystyrene. *Int. Polym. Process.*, 28, 1, 14–23, 2013.
68. Atmani, O., Abbès, F., Li, Y., Batkam, S., Abbès, B., Experimental and numerical investigation of the effects of sheet material, plug-assist tool material, and process conditions on the mechanical pre-stretching stage of plug-assist thermoforming. *Int. J. Adv. Manuf. Technol.*, 122, 3217–3234, 2022.
69. Bernard, C.A., Correia, J.P.M., Bahlouli, N., Ahzi, S., Numerical simulation of plug-assisted thermoforming process: application to polystyrene. *Key Eng. Mater.*, 554-557, 1602–1610, 2013.
70. Kleespies III, H.S. and Crawford, R.H., Vacuum forming of compound curved surfaces with a variable geometry mold. 17, 5, 325–337, 1998.
71. Su, S., Li, M.Z., Liu, C.G., Ji, C., Setchi, R., Larkiola, J., Panteleev, I.O., Stead, I.M.N., Lopez, R.A., Flexible tooling system using reconfigurable multi-point thermoforming technology for manufacturing freeform panels. *Key Eng. Mater.*, 504-506, 839–844, 2012.
72. Baek, H.M., Giacomin, A.J., Wurz, M., Sag in commercial thermoforming. *Aiche J.*, 60, 1529–1535, 2014.
73. Michaud, R.A. and Giacomin, A.J., Sheet temperature in thermoforming, *J. Plast. Film Sheet.*, 27, 293–330, 2011.
74. Kommoji, S., Banerjee, R., Bhatnagar, N., Ghosh, A.K., Studies on the stretching behaviour of medium gauge high impact polystyrene sheets during positive thermoforming. *J. Plast. Film Sheet.*, 31, 112–96, 2015.
75. Gilham, E., Lean manufacturing. *SPE-European Thermoforming Division Conference*, Barcelona (Sitges), 2016.
76. Koziey, B.L., Ghafur, M.O., Vlachopoulos, J., Mirza, F.A,. Chapter 3 Computer simulation of thermoforming. *Compos. Mater. Ser.*, 11, 75–89, 1997.
77. Throne, J.L., Parametric study of extrusion and thermoforming conditions for CPET. *J. Plast. Film Sheet.*, 5, 222–242, 1989.
78. DiRaddo, R.W. and Meddad., A., Sensitivity of operating conditions and material properties for thermoforming process. *Plast. Rubber Compos.*, 29, 4, 163–167, 2000.
79. Duarte, F.M. and Covas, J.A., On the use of the heating stage to control the thickness distribution in thermoformed parts. *Int. Polym. Process.*, 19, 186–198, 2004.
80. Peng, H., Li, M., Liu, C., Cao, J.H., Study of multi-point forming for polycarbonate sheet. *Int. J. Adv. Manuf. Technol.*, 67, 2811–2817, 2013.
81. Progrelhof, R.C., Quintiere, J.G., Throne, J.L., Temperature distribution in semitransparent plastic sheets exposed to symmetric, unsymmetric, and pulsed radiant heating and surface cooling. *J. Appl. Polym. Sci.*, 17, 1227–1252, 1973.

82. Duarte, F.M. and Covas, J.A., Heating thermoplastic sheets for thermoforming solution to direct and inverse problems. *Plast. Rubber Compos. Process. Appl.*, 26, 213–221, 1997.
83. Progelhof, R.C., Franey, J., Haas, T.W., Absorption Coefficient of Unpigmented Poly(Methyl Methacrylate), Polystyrene, Polycarbonate, and Poly(4-Methylpentene-1) Sheets. *Appl. Polym. Sci.*, 15, 1803, 1971.
84. Buffel, B., Leeman, K., Desplentere, F., Pulsed IR Heating of Thermoplastic Sheets for Thermoforming Applications. *Int. Polym. Process.*, 36, 4, 388-397, 2021. 10.1515/ipp-2020-3981.
85. Gauthier, G., Ajersch, M., Boulet, B., Haurani, A., Girard, P., DiRaddo, R., A new absorption based model for sheet reheat in thermoforming. *ANTEC Conference Proceedings*, pp. 353–357, 2005.
86. Puehringer, J.F., Zitzenbacher, G., Spreitzer, C., Study of heat absorption in thermoforming for transparent and filled polystyrene. *Int. Polym. Process.*, 28, 1, 14–23, 2013.
87. Progelhof, R.C. and Throne, J.L., Thermoforming conditions in transparent/opaque coextruded sheet. *Polym. Eng. Sci.*, 14, 810–816, 1974.
88. Chang, Y.-W. and Cheng, J.-H., Numerical and experimental investigation of polycarbonate vacuum-forming process. *J. Chin. Inst. Eng.*, 36, 831–841, 2013.
89. Macauley, N.J., Harkin-Jones, E., Murphy, W.R., The influence of nucleating agents on the extrusion and thermoforming of polypropylene. *Polym. Eng. Sci.*, 38, 516–523, 1998.
90. Sameer, T.S., Panigrahi, P., Anirudh, P.V., Smrithin, S., Raghavendra, U.P., Development of thermal vacuum forming machine. *Int. J. Sci. Technol. Res.*, 9, 1673–1677, 2020.
91. Chy, M.M., II, Boulet, B., Haidar, A., A model predictive controller of plastic sheet temperature for a thermoforming process. *Proceedings of the 2011 American Control Conference*, pp. 4410–4415, 2011.
92. Morales, R.A., Candal, M.V., Santana, O., Gordillo, A., Salazar, R., Effect of the thermoforming process variables on the sheet friction coefficient. *Mater. Des.*, 53, 1097–1103, 2014.
93. Hernández-Ambato, J., Rodríguez-Flores, J., Cortes-Llanganate, J., Cabrera-Aguayo, F., Classic controllers design applied to temperature control for a plastic thermoforming machine, in: *2018 IEEE 23rd International Conference on Emerging Technologies and Factory Automation (ETFA)*, vol. 1, pp. 551–558, 2018.
94. Mohammadian-Gezaz, S., Ghasemi, I., Karrabi, M., Azizi, H., Investigation on the thermoformability of polyolefin blends by hot tensile and rheological tests. *Polym. Test.*, 25, 504–511, 2006.
95. Sasimowski, E., ., A pressure-bubble vacuum forming process for polystyrene sheet. *Adv. Sci. Technol. Res. J.*, 11, 2, 180–186, 2017.

96. Gotsis, A.D., Zeevenhoven, B.L., Hogt, A.H., The effect of long chain branching on the processability of polypropylene in thermoforming. *Polym. Eng. Sci.*, 44, 973–981, 2004.
97. Hosseini, H., Berdyshev, B.V., Mehrabani-Zeinabad, A., Modeling of Deformation Processes in Vacuum Thermoforming of a Pre-stretched Sheet. *Polym.-Plast. Technol. Eng.*, 45, 1357–1362, 2006.
98. Yang, S. and Boulet, B., Cycle-to-Cycle Control of the Thermoforming Reheat Process. *Proceedings of the World Congress on Engineering and Computer Science*, San Francisco, CA, USA, 2008.
99. Torres, F.G. and Bush, S.F., Sheet extrusion and thermoforming of discrete long glass fibre reinforced polypropylene. *Compos. Part A*, 31, 1293, 2000.
100. Schmidt, F.M., Le Maoult, Y., Monteix, S., Modelling of infrared heating of thermoplastic sheet used in thermoforming process. *J. Mater. Process. Technol.*, 143–144, 226, 2003.
101. Lau, H.C., Bhattacharya, S.N., Field, G.J., Melt strength of polypropylene: Its relevance to thermoforming, Polymer Engineering and Science. *Polym. Eng. Sci.*, 38, 11, 1915, 2004.
102. Walczyka, D.F. and Yoo, S., Design and fabrication of a laminated thermoforming tool with enhanced features. *J. Manuf. Process.*, 11, 9, 2009.
103. Christopherson, R., Debbaut, B., Rubin, Y., Simulation of pharmaceutical blister pack thermoforming using a non-isothermal integral model. *Proceedings of the SPE ANTEC 2000 Conference*, Orlando, FL, USA, vol. 46, p. 773, 2000.
104. Dooling, P.J., Buckley, C.P., Rostami, S., Zahlan, N., Hot-drawing of poly (methyl methacrylate) and simulation using a glass–rubber constitutive model. *Polymer*, 43, 8, 2451–2465, 2002.
105. Arruda, E.M. and Boyce, M.C., A three-dimensional constitutive model for the large stretch behavior of rubber elastic materials. *J. Mech. Phys. Solids*, 41, 2, 389–412, 1993.
106. Johansson, K.T. and Reber, D.H., In-line extrusion/thermoforming techniques for deep draw polyolefin containers. *J. Plast. Film Sheet.*, 5, 2, 140–153, 1989.
107. Chang, Y.-W. and Cheng, J.-H., Numerical and experimental investigation of polycarbonate vacuum-forming process. *J. Chin. Inst. Eng.*, 36, 7, 831–841, 2013.
108. Chen, S.-C., Huang, S.-T., Lin, M.-C., Chien, R.-D., Study on the thermoforming of PC films used for in-mold decoration. *Int. Commun. Heat Mass Transf.*, 35, 8, 967–973, 2008.
109. Lee, R.E., Guo, Y., Tamber, H., Planeta, M., Leung, S.N.S., Thermoforming of polylactic acid foam sheets: Crystallization behaviors and thermal stability. *Ind. Eng. Chem. Res.*, 55, 3, 560–567, 2016.
110. Ghobadnam, M., Mosaddegh, P., Rejani, M.R., Amirabadi, H., Ghaei, A., Numerical and experimental analysis of HIPS sheets in thermoforming process. *Int. J. Adv. Manuf. Technol.*, 76, 1079–1089, 2015.

111. Nied, H.F., Taylor, C.A., Delorenzi, H.G., Three-dimensional finite element simulation of thermoforming. *Polym. Eng. Sci.*, 30, 20, 1314–1322, 1990.
112. Sala, G., Di Landro, L., Cassago, D., A numerical and experimental approach to optimise sheet stamping technologies: Polymers thermoforming. *Mater. Des.*, 23, 1, 21–39, 2002.
113. Ayhan, Z. and Zhang, Q.H., Wall thickness distribution in thermoformed food containers produced by a Benco aseptic packaging machine. *Polym. Eng. Sci.*, 40, 1, 1–10, 2000.
114. Wineman, A., On axisymmetric deformations of nonlinear viscoelastic membranes. *J. Non-Newton. Fluid Mech.*, 4, 3, 249–260, 1978.
115. Lee, J.K., Virkler, T.L., Scott, C.E., Influence of initial sheet temperature on ABS thermoforming. *Polym. Eng. Sci.*, 41, 1830, 2001.
116. Warby, M.K., Whiteman, J.R., Jiang, W.-G., Warwick, P., Wright, T., Finite element simulation of thermoforming processes for polymer sheets. *Math. Comput. Simul.*, 61, 3-6, 209–218, 2003.
117. Lee, J.K., Scott, C.E., Virkler, T.L., Effects of rheological properties and processing parameters on ABS thermoforming. *Polym. Eng. Sci.*, 41, 2, 240–261, 2001.
118. Sorrentino, L. and Bellini, C., Potentiality of hot drape forming to produce complex shape parts in composite material. *Int. J. Adv. Manuf. Technol.*, 85, 5-8, 945–954, 2016.
119. *Polypropylene Sheet Extrusion and Thermoforming for Packaging Applications*, Basell Polyolefins, Rotterdam, The Netherlands, 054 PP e, 2005.
120. aus der Wiesche, S., Industrial thermoforming simulation of automotive fuel tanks. *Appl. Therm. Eng.*, 24, 2391–2409, 2004.
121. Martin, P.J., Keaney, T., McCool, R., Development of a multivariable online monitoring system for the thermoforming process. *Polym. Eng. Sci.*, 54, 12, 2815–2823, 2014.
122. Nam, G.J., Lee, J.W., Ahn, K.H., Three-dimensional simulation of thermoforming process and its comparison with experiments. *Polym. Eng. Sci.*, 40, 10, 2232–2240, 2000.
123. Throne, J.L., Chapter 1, in: *Understanding Thermoforming*, Hanser, Munich, 1999.
124. Yamaguchi, M. and Suzuki, K.-I., Enhanced strain hardening in elongational viscosity for HDPE/crosslinked HDPE blend. II. Processability of thermoforming. *J. Appl. Polym. Sci.*, 86, 1, 79–83, 2002.
125. Rosenzweig, N., Narkis, M., Tadmor, Z., Wall thickness distribution in thermoforming. *Polym. Eng. Sci.*, 19, 946–951, 1979.
126. Baumard, T., Menary, G., De Almeida, O., Schmidt, F., Martin, P., Bikard, J., Towards a coupled heating-forming simulation of the thermoforming of thermoplastic composites, in: *AIP Conference Proceedings*, vol. 1896, AIP Publishing, p. 030035, 2017.

Chapter 6

1. Bertsekas, D.P., *Dynamic Programming and Optimal Control*, 3rd ed, pp. 12–17, Athena Scientific, Belmont, Massachusetts, 2000.
2. Schiffers, R., Kruppa, S., Moser, S., The right changeover point for each shot. *J. Kunststoffe*, 11, 26–29, 2014.
3. Jalham, I.S., *Process Control and Quality Improvement of Plug-Assist Thermoforming Process: A Case Study. JKAU Eng. Sci.*, 16, 1, 17–33, 2005.
4. Beine, J., Varied requirements: Innovations in thermoforming. *Kunststoffe-Plast Europe*, 93, 10, 155, 2003.
5. Berdyshev, B. and Hosseini, H., Rheological modeling of warpage in polymeric products under high temperature. *Proceedings of the SPE Conference-ANTEC 2006*, Charlotte, NC, USA, p. 2610, 2006.
6. Marshall, R.A., Effect of crystallinity on PVC physical properties. *J. Vinyl Addit. Technol.*, 16, 35–38, 1994.
7. Graessley, W.W., Glasscock, S.D., Crawley, R.L., Die swell in molten polymers. *Trans. Soc Rheol.*, 14, 4, 519–544, 1970.
8. White, J.L. and Roman, J.F., Extrudate swell during the melt spinning of fibers—influence of rheological properties and take-up force. *J. Appl. Polym. Sci.*, 20, 4, 1005–1023, 1976.
9. Utracki, L.A., Bakerdjian, Z., Kamal, M.R., A method for the measurement of the true die swell of polymer melts. *J. Appl. Polym. Sci.*, 19, 2, 481–501, 1975.
10. Gourisaria, M.K., Agrawal, R., Harshvardhan, G., Pandey, M., Rautaray, S.S., Application of machine learning in industry 4.0, in: *Studies in Big Data*, vol. 4, pp. 57–87, Springer Singapore, Singapore, 2021.
11. Zhang, Y., Jia, Z., Dai, Y., Real-time performance analysis of industrial serial production systems with flexible manufacturing, in: *2018 Joint 10th International Conference on Soft Computing and Intelligent Systems (SCIS) and 19th International Symposium on Advanced Intelligent Systems (ISIS)*, pp. 360–365, 2018.
12. Pearson, G.H. and Garfield, L.J., The effect of molecular weight and weight distribution upon polymer melt rheology. *Polym. Eng. Sci.*, 18, 7, 583–589, 1978.
13. Hosseini, H. and Berdyshev, B.V., A rheological model for mechanical pre-stretching vacuum-forming and solving of thermoforming defects. *Structural Plastics 2005*, Society of Plastics Industry, Orlando, FL, USA, March 21–24, 2005.
14. Hosseini, H., Berdyshev, B.V., Mehrabani-Zeinabad, A., Rheological modeling of fracture in plug-assisted vacuum thermoforming. *J. Appl. Polym. Sci., 117,* 1, 171–177, 2010.
15. Hosseini, H., Berdyshev, B.V., Mehrabani Zeinabad, A., A solution for warpage in polymeric products by plug-assisted thermoforming. *Eur. Polym. J.*, 42, 8, 1836–1843, 2006.

16. Atmani, O., Abbès, F., Li, Y., Batkam, S., Abbès, B., Experimental and numerical investigation of the effects of sheet material, plug-assist tool material, and process conditions on the mechanical pre-stretching stage of plug-assist thermoforming. *Int. J. Adv. Manuf. Technol.*, 122, 7-8, 3217–3234, 2022.
17. Nam, G.J., Lee, J.W., Ahn, K.H., Three-dimensional simulation of thermoforming process and its comparison with experiments. *Polym. Eng. Sci.*, 40, 10, 2232–2240, 2000.
18. Giacomin, A.J., Mix, A.W., Mahmood, O., Sag in thermoforming. *Polym. Eng. Sci.*, 50, 10, 2060–2068, 2010.
19. Berdyshev, B. and Hosseini, H., A solution for rupture of polymeric sheets during plug-assist thermoforming. *Proceedings of the SPE Conference-ANTEC 2006*, Charlotte, NC, USA, p. 2610, 2006.
20. Baek, H.M., Giacomin, A.J., Wurz, M.J., Sag in commercial thermoforming. *AIChE J.*, 60, 4, 1529–1535, 2014.
21. Min, S.C., Zhang, H.Q., Yang, H.-J., Thermoformed container wall thickness effects on orange juice quality. *J. Food Process. Preserv.*, 35, 6, 758–766, 2011.
22. Cha, J., Song, H.Y., Hyun, K., Go, J.S., Rheological measurement of the nonlinear viscoelasticity of the ABS polymer and numerical simulation of thermoforming process. *Int. J. Adv. Manuf. Technol.*, 107, 2449–2464, 2020.
23. Bird, R.B., Curtiss, C.F., Armstrong, R.C., Hassager, O., *Dynamics of Polymer Liquids, Vol. 2*, 2nd Ed, J. Wiley & Sons, New York, 1987.
24. Keddie, J.L., Jones, R.A.L., Cory, R.A., Size-dependent depression of the glass transition temperature in polymer films. *Europhys. Lett.*, 27, 1, 59, 1994.
25. Fetters, L.J., Lohse, D.J., Richter, D., Witten, T.A., Zirkel, A., Connection between polymer molecular weight, density, chain dimensions, and melt viscoelastic properties. *Macromolecules*, 27, 17, 4639–4647, 1994.
26. Aharoni, S.M., On entanglements of flexible and rodlike polymers. *Macromolecules*, 16, 11, 1722–1728, 1983.
27. Ronca, G., Frequency spectrum and dynamic correlations of concentrated polymer liquids. *J. Chem. Phys.*, 79, 2, 1031–1043, 1983.
28. Ljung, L., *System Identification: Theory for the User*, 2nd ed, Prentice-Hall, Upper Saddle River, NJ, 1999.
29. Song, Y., Zhang, K.F., Wang, Z.R., Diao, F.X., Yan, Y.N., Zhang, R.J., Coupled thermo-mechanical analysis for plastic thermoforming. *Polym. Eng. Sci.*, 40, 8, 1736–1746, 2000.
30. Hegemann, B., Tessier, N., Bush, T., Various plug assist materials and their effect on the thermoforming characteristics of polymer sheet. *Thermoforming Q., Fourth Quarter*, 21, 4, 12–16, 2002.
31. Aroujalian, A., Ngadi, M.O., Emond, J.-P., Wall thickness distribution in plug-assist vacuum formed strawberry containers. *Polym. Eng. Sci.*, 37, 1, 178–182, 1997.
32. Throne, J.L., Guidelines for thermoforming part wall thickness. *Polym.-Plast. Technol. Eng.*, 30, 7, 685–700, 1991.

33. Lieg, K.L. and Giacomin, A.J., Thermoforming triangular troughs. *Polym. Eng. Sci.*, 49, 1, 189–199, 2009.
34. Rosenzweig, N., Narkis, M., Tadmor, Z., Wall thickness distribution in thermoforming. *Polym. Eng. Sci.*, 19, 13, 946–951, 1979.
35. Harron, G.W., Harkin-Jones, E.M.A., Martin, P.J., Influence of thermoforming parameters on final part properties. *The Annual Technical Conference ANTEC 2000 Conference Proceedings*, pp. 3723–3727, 2000.
36. Dong, Y., Lin, R.J.T., Bhattacharyya, D., Finite element simulation on thermoforming acrylic sheets using dynamic explicit method. *Polym. Polym. Compos.*, 14, 3, 307–328, 2006.
37. Azdast, T., Doniavi, A., Ahmadi, S.R., Amiri, E., Numerical and experimental analysis of wall thickness variation of a hemispherical PMMA sheet in thermoforming process. *Int. J. Adv. Manuf. Technol.*, 64, 113–122, 2013.

Chapter 7

1. Klein, P.W., Fundamentals of plastic thermoforming, in: *Synthesis Lectures on Materials Engineering*, pp. 1–79, Morgan & Claypool Publishers, Kentfield, CA, 2009.
2. Throne, J.L., *Understanding thermoforming*, Hanser Publishers, Munich, 2008; Schwarzmann, P., *Thermoforming – A practical guide*, Hanser Publishers, Munich, 2001.
3. Wang, P., Hamila, N., Boisse, P., Thermoforming simulation of multilayer composites with continuous fibres and thermoplastic matrix. *Compos. B*, 52, 127–136, 2013.
4. Chen, S.-C., Huang, S.-T., Lin, M.-C., Chien, R.-D., Study on the thermoforming of PC films used for in-mold decoration. *Int. Commun. Heat Mass Transfer*, 35, 967–973, 2008.
5. Morales, R.A., Candal, M.V., Santana, O.O., Gordillo, A., Salazar, R., Effect of the thermoforming process variables on the sheet friction coefficient. *Mater. Des.*, 53, 1097–1103, 2014.
6. Pignon, B., Boyard, N., Sobotka, V., Delaunay, D., Heat transfer analysis at high cooling rate on the surface of thermoplastic parts. *Int. J. Heat Mass Transfer*, 106, 253, 2017.
7. Baumard, T., Menary, G., De Almeida, O., Schmidt, F., Martin, P., Bikard, J., Towards a coupled heating-forming simulation of the thermoforming of thermoplastic composites, in: *AIP Conference Proceedings*, vol. 1896, no. 1, AIP Publishing, 2017.
8. Christopherson, R., Debbaut, B., Rubin, Y.B., Simulation of pharmaceutical blister pack thermoforming using a non-isothermal integral model. *J. Plast. Film Sheeting*, 17, 239–251, 2001.

9. Dong, Y., Lin, R., Bhattacharyya, D., Finite element simulation on thermoforming acrylic sheets using dynamic explicit method. *Polym. Polym. Compos.*, 14, 307–328, 2006.
10. Adams, A.M., Paul Buckley, C., Jones, D.P., Biaxial hot drawing of poly (ethylene terephthalate): measurements and modelling of strain-stiffening. *Polymer*, 41, 771–786, 2000.
11. Capt, L., Rettenberger, S., Münstedt, H., Kamal, M.R., Simultaneous biaxial deformation behavior of isotactic polypropylene films. *Polym. Eng. Sci.*, 43, 1428–1441, 2003.
12. Harron, G.W., Harkin-Jones, E., Martin, P.J., Plug force monitoring for the control and optimization of the thermoforming process. *Proc. Inst. Mech. Eng., Part E: J. Process Mech. Eng.*, 217, 181–188, 2003.
13. O'Connor, C.P.J., Martin, P.J., Sweeney, J., Menary, G., Caton-Rose, P., Spencer, P.E., Simulation of the plug-assisted thermoforming of polypropylene using a large strain thermally coupled constitutive model. *J. Mater. Process. Technol.*, 213, 9, 1588–1600, 2013.
14. Khoun, L., Centea, T., Hubert, P., Characterization methodology of thermoset resins for the processing of composite materials—Case study: CYCOM 890RTM epoxy resin. *J. Compos. Mater.*, 44, 1397–1415, 2010.
15. D'Oria, F., Bourgin, P., Coincenot., L., Progress in numerical modeling of the thermoforming process. *Adv. Polym. Technol.*, 14, 291–301, 1995.
16. Nam, G.J. and Lee, J.W., Numerical and experimental studies of 3-dimensional thermoforming process. *J. Reinf. Plast. Compos.*, 20, 14, 1182–1190, 2001.
17. Ayhan, Z. and Zhang, H., Wall thickness distribution in thermoformed food containers produced by a Benco aseptic packaging machine. *Polym. Eng. Sci.*, 40, 1, 1–10, 2000.
18. Harron, G.W., Harkin-Jones, E.M.A., Martin, P.J., Influence of thermoforming parameters on final part properties. *The Annual Technical Conference ANTEC 2000 Conference Proceedings*, pp. 3723–3727, 2000.
19. Myers, R.H., Montgomery, D.C., Anderson-Cook, Ch.M., *Response surface methodology: Process and product optimization using designed experiments*, John Wiley and Sons, New York, 2009.
20. Montgomery, D.C., *Design and analysis of experiments*, 8th. Ed., pp. 1–23, John Wiley & Sons, USA, 2009.
21. Rosenzweig, N., Narkis, M., Tadmor, Z., Wall thickness distribution in thermoforming. *Polym. Eng. Sci.*, 19, 13, 946–951, 1979.
22. Throne, J.L. and Mooney, P.J., Thermoforming: growth and evolution I. *Thermoforming Q.*, 24, 1, 18–20, 2005.
23. Lieg, K.L. and Giacomin, A.J., Thermoforming triangular troughs. *Polym. Eng. Sci.*, 49, 1, 189–199, 2009.
24. Maekawa, Y. and Yanagi, H., Examination of modeling for computer simulation of thermoforming process. *Seikei-Kakou*, 6, 788–796, 1994.

25. Azdast, T., Doniavi, A., Ahmadi, S.R., Amiri, E., Numerical and experimental analysis of wall thickness variation of a hemispherical PMMA sheet in thermoforming process. *Int. J. Adv. Manuf. Technol.*, 64, 1, 113–122, 2013.
26. Bénard, A. and Advani, S.G., Energy equation and the crystallization kinetics of semi-crystalline polymers: regimes of coupling. *Int. J. Heat Mass Transfer*, 38, 819–832, 1995.
27. Erchiqui, F., Annasabi, Z., Kaddami, H., Ben Hamou, K., Ngoma, G.-D., Effect of the cooling temperature of a PET sheet on the crystallinity and mould removal time for thermoforming applications. *Can. J. Chem. Eng.*, 99, 695–707, 2021.
28. Eder, G., Janeschitz-Kriegl, H., Liedauer, S., Crystallization processes in quiescent and moving polymer melts under heat transfer conditions. *Prog. Polym. Sci.*, 15, 629–714, 1990.
29. Patel, R.M., Crystallization kinetics modeling of high density and linear low density polyethylene resins. *J. Appl. Polym. Sci.*, 124, 1542–1552, 2012.
30. Kouba, K., Bartos, O., Vlachopoulos, J., Computer simulation of thermoforming in complex shapes. *Polym. Eng. Sci.*, 32, 699–704, 1992.
31. Karamanou, M., Warby, M.K., Whiteman, J.R., Computational modelling of thermoforming processes in the case of finite viscoelastic materials. *Comput. Methods Appl. Mech. Eng.*, 195, 5220–5238, 2006.
32. O'Connor, C.P.J., Martina, P.J., Sweeney, J., Menary, G., Caton-Rose, P., Spencer, P.E., Simulation of the plug-assisted thermoforming of polypropylene using a large strain thermally coupled constitutive model. *J. Mater. Process. Technol.*, 213, 1588–1600, 2013.
33. Wang, P., Hamila, N., Boisse, P., Thermoforming simulation of multilayer composites with continuous fibres and thermoplastic matrix. *Compos. B*, 52, 127–136, 2013.
34. R. J. Crawford and S. K. L. Lui, *Eur. Polym. J.*, 18, 699, 1982.
35. DeLorenzi, H.G. and Nied, H.F., Finite element simulation of thermoforming and blow molding, in: *Modeling of Polymer Processing*, ch. 5, A.I. Isayev (Ed.), Hanser Verlag, Munich, 1991.
36. Song, W.N., Mirza, F.A., Vlachopoulos, J., Finite element analysis of inflation of an axisymmetric sheet of finite thickness. *J. Rheol.*, 35, 1, 93–111, 1991.
37. Marchal, T.M., Clemeur, N.P., Agarwal, A.K., Optimisation of the thermoforming process: A few industrial examples. *The Annual Technical Conference ANTEC 1998 Conference Proceedings*, pp. 701–705, 1998.
38. Dong, Y., Lin, R.J.T., Bhattacharyya, D., Finite element simulation on thermoforming acrylic sheets using dynamic explicit method. *Polym. Polym. Compos.*, 14, 3, 307–328, 2006.
39. Throne, J.L., *Technology of Thermoforming*, Hanser/Gardner Publications, Cincinnati, 1996.
40. Atmani, O., Abbès, B., Abbès, F., Li, Y.M., Batkam, S., Identification of a thermo-elasto-viscoplastic behavior law for the simulation of thermoforming of high impact polystyrene. *AIP Conf. Proc.*, 1960, 1–7, 2018.

41. Jain, R., *The art of computer systems performance analysis - techniques for experimental design, measurement, simulation, and modeling*, Wiley, New York, Professional Computing, 1991.

Chapter 8

1. Martin, P.J., Tan, C.W., Tshai, K.Y., McCool, R., Menary, G., Armstrong, C.G., Harkin-Jones, E.M.A., Biaxial characterisation of materials for thermoforming and blow moulding. *Plast. Rubber Compos.*, 34, 5/6, 276–282, 2005.
2. Collins, P., Harkin-Jones, E.M.A., Martin, P.J., The role of tool/sheet contact in plug-assisted thermoforming. *Int. Polym. Process.*, 17, 4, 361–369, 2002.
3. McCool, R., Martin, P.J., Harkin-Jones, E.M.A., Control of product wall thickness distribution in plug-assisted thermoforming. *Proceedings of PPS-2005 Americas Regional Meeting*, Polymer Processing Society, Quebec, Canada, 2005.
4. Sweeney, J., Collins, T.L.D., Coates, P.D., Duckett, R.A., High-temperature large strain viscoelastic behavior of polypropylene modeled using an inhomogeneously strained network. *J. Appl. Polym. Sci.*, 72, 4, 563–575, 1998.
5. Walton, K.L., Weaver, L.B., Waszeciak, D.P., The effect of formulation and process on thermoformable TPO gloss. *Proceedings of the SPE Conference - ANTEC 2006*, Charlotte, NC, USA, p. 2636, 2006.
6. Reid, C. and Mahan, M., *Proceedings of the SPE Automotive TPO Global Conference 2001*, Dearborn, MI, USA, p. 213, 2001.
7. Walton, K.L., Laughner, M.K., Gisler, E.S., *Proceedings of the SPE Automotive TPO Global Conference 2004*, Dearborn, MI, USA, p. 12, 2004.
8. Maltby, A. and Marquis, R.E., Slip additives for film extrusion. *J. Plast. Film Sheeting*, 14, 4, 111–120, 1998, (2004), p.5.
9. Reid, C., Mahan, M., Tavakoli, H., *Proceedings of the SPE Automotive TPO Global Conference 2002*, SPE, Dearborn, MI, USA, p. 193, 2002.
10. Tolinski, M., Fuel tanks face stricter standards. *Plast. Eng.*, 61, 6, 18–19, 2005.

Chapter 9

1. Throne, J.L. and Mooney, P.J., Thermoforming: Growth and evolution I. *Thermoforming Q.*, 24, 2, 19, 2005.
2. Lau, H.C.P., Bhattacharya, S.N., Field, G., Influence of rheological properties on the sagging of polypropylene and abs sheet for thermoforming applications. *Polym. Eng. Sci.*, 40, 1564–15705, 2000.
3. Yamaguchi, M., Rheological properties of linear and crosslinked polymer blends: Relation between crosslink density and enhancement of elongational viscosity. *J. Polym. Sci. Part B*, 39, 228–2355, 2001.

4. Rosenzweig, N., Narkis, M., Tadmor, Z., Wall thickness distribution in thermoforming. *Polym. Eng. Sci.*, 19, 946–951, 1979.
5. Aroujalian, A., Ngadi, M.O., Emond, J.-P., Wall thickness distribution in plug-assist vacuum formed strawberry containers. *Polym. Eng. Sci.*, 37, 1, 178–182, 1997.
6. American Meat Science Association, *Meat color measurement guidelines*, Am. Meat Sci. Assoc., Chicago, IL, 2012.
7. Gahleitner, M., Tranninger, C., Doshev, P., Polypropylene copolymers, in: *Polypropylene Handbook*, J. Karger-Kocsis, and T. Bárány, (Eds.), pp. 295–355, Springer International Publishing, Cham, Switzerland, 2019.
8. R.C. Christensen, R.P. Genske, D.E. Kester, W.F. Ossian, United States Patent, 4,405,667, 1983.
9. Roerink, A.M. and Beebe, S., Cryovac® brand 2021 national meat case study—A snapshot of the meat case today, in: *Sealed Air*, North American Meat Institute, Charlotte, NC, USA, 2021.
10. Rojas, M.C. and Brewer, M.S., Effect of natural antioxidants on oxidative stability of frozen, vacuum-packed beef and pork. *J. Food Qual.*, 31, 173–1885, 2008.
11. Faustman, C. and Cassens, R.G., The biochemical basis for discoloration in fresh meat: A review. *J. Muscle Foods*, 1, 217–2435, 1990.
12. Kaiser, K., Schmid, M., Schlummer, M., Recycling of polymer-based multilayer packaging: A review. *Recycling*, 3, 1, 2017.
13. Boone, J.S., Lox, F., Pottie, S., Deficiencies of polypropylene in its use as a food-packaging material — A review. *Packag. Technol. Sci.*, 6, 277–2815, 1993.
14. Dobrucka, R. and Przekop, R.E., New perspectives in active and intelligent food packaging. *J. Food Process. Preserv.*, 43, e14194, 2019.
15. Abdel-Ghany, A.M., Picuno, P., Al-helal, I.M., Alsadon, A.A., Ibrahim, A.A., Shady, M., Radiometric characterization, solar and thermal radiation in a greenhouse as affected by shading configuration in an arid climate. *Energies*, 8, 13928–13937, 2015.
16. Antunes, D.T., da Silva Ponce, F., Pataraico Junior, V., Fernandes Junior, F., Campos, R.A.S., de Carvalho Campos Botelho, S., do Nascimento, M.S.P., Seabra Júnior, S., Zanuzo, M.R., Effect of polycarbonate and agricultural film on production and biochemical compounds of tomato fruits. *Res. Soc. Dev.*, 11, 13, e513111335848, 2022.
17. Desimone, J., Branching out into new polymer markets. *Science*, 269, 1060–10615, 1995.
18. Mai, Y. and Eisenberg, A., Self-assembly of block copolymers. *Chem. Soc. Rev.*, 41, 18, 5969–85, 2012.
19. Egli, S., Schlaad, H., Bruns, N., Meier, W.P., Functionalization of block copolymer vesicle surfaces. *Polymers*, 3, 252–280, 2011.

20. Kunori, T. and Geil, P.H., Morphology-property relationships in polycarbonate-based blends. I. Modulus. *J. Macromol. Sci., Part B*, 18, 93–134, 1980.
21. Calin, L., Cătinean, A., Bilici, M., Dăscălescu, L., Samuilă, A., Electrostatic separation of HIPS/ABS and HIPS/ABS-PC plastic mixtures from IT equipment using fluidized bed tribocharging. *Part. Sci. Technol.*, 40, 1, 113–122, 2021.
22. Kurauchi, T. and Ohta, T., Energy absorption in blends of polycarbonate with ABS and SAN. *J. Mater. Sci.*, 19, 1699–1709, 1984.
23. Pilati, F., Marianucci, E., Berti, C., Study of the reactions occurring during melt mixing of poly(ethylene terephthalate) and polycarbonate. *J. Appl. Polym. Sci.*, 30, 1267–12755, 1985.
24. Inoue, N., Nakayama, T., Ariyama, T., Hydrostatic extrusion of amorphous polymers and properties of extrudates. *J. Macromol. Sci. Part B*, 19, 543–563, 1981.
25. Dharmarajan, N. and Kaufman, L.G., High flow TPO compounds containing branched EPDM modifiers. *Rubber Chem. Technol.*, 71, 778–7945, 1998.
26. Zeyana binti Taib, N., Rahman, Md.R., Huda, Kuok, K.K., Bin Hamdan, S., Bin Bakri, M.K., Mirza bin Julaihi, M.R., Khan, A., A review on poly lactic acid (PLA) as a biodegradable polymer. *Polym. Bull.*, 80, 1179–12135, 2022.
27. Bauwens-Crowet, C., Bauwens, J.C., Homes, G., The temperature dependence of yield of polycarbonate in uniaxial compression and tensile tests. *J. Mater. Sci.*, 7, 176–1835, 1972.
28. Anoshkin, A.N., Shipunov, G.S., Baranov, M.A., Golovin, D.V., Features of the technological process of thermoforming of products made of thermoplastic composite materials. *AIP Conf. Proc.*, 2167, 1, 020019, 19 November 2019.
29. Saleem, M., Khan, A.N., Abbas, S.R., Saboor, A., Super toughening, strengthening, and antimicrobial behaviors of cyclic olefinic copolymer/few layer graphene nanocomposites. *Polym. Compos.*, 40, 536–5435, 2019.
30. Matuana, L.M., WoodhamsRaymond, T., Balatinecz, J.J., Park, C.B., Influence of interfacial interactions on the properties of PVC/cellulosic fiber composites. *Polym. Compos.*, 19, 446–4555, 1998.
31. Cooper-White, J.J. and Mackay, M.E., Rheological properties of poly(lactides). Effect of molecular weight and temperature on the viscoelasticity of poly(l-lactic acid). *J. Polym. Sci. B Polym. Phys.*, 37, 1803–1814, 1999.
32. Agari, Y. and Ueda, A., Thermal conductivity of poly(vinyl chloride)/polycaprolactone blends. *J. Polym. Sci. Part B*, 32, 59–62, 1994.
33. De Gaspari, J., Mutual attraction. *Plast. Technol.*, 44, 6, 44–48, 1998.
34. Mahmood, H., Dorigato, A., Pegoretti, A., Thermal mending in novel epoxy/cyclic olefin copolymer blends. *eXPRESS Polym. Lett.*, 14, 4, 368–383, 2020.
35. Argon, Ali, S., Bessonov, M.I., Plastic flow in glassy polymers. *Polym. Eng. Sci.*, 17, 3, 174–182, 1977.

36. Kausch, H.H., *Advanced thermoplastic composites*, Hanser Publishers, New York, 1993.
37. Riga, A.T., Heat distortion and mechanical properties of polymers by thermal-mechanical analysis. *Polym. Eng. Sci.*, 14, 11, 764–767, 1974.
38. Palade, L.-I., Lehermeier, H.J., Dorgan, J.R., Melt rheology of high L-content poly (lactic acid). *Macromolecules*, 34, 5, 1384–1390, 2001.
39. Woodhams, R.T., Thomas, G., Rodgers, D.K., Wood fibers as reinforcing fillers for polyolefins. *Polym. Eng. Sci.*, 24, 15, 1166–1171, 1984.
40. Rosenzweig, N., Narkis, M., Tadmor, Z., Wall thickness distribution in thermoforming. *Polym. Eng. Sci.*, 19, 13, 946–951, 1979.
41. Niknejad, A.S., Bazgir, S., Kargari, A., Mechanically improved superhydrophobic nanofibrous polystyrene/high-impact polystyrene membranes for promising membrane distillation application. *J. Appl. Polym. Sci.*, 139, 36, 50917, 2021.
42. Hosseini, H., Vasilivich, B.B., Mehrabani-Zeinabad, A., Rheological modeling of plug-assist thermoforming. *J. Appl. Polym. Sci.*, 101, 6, 4148–4152, 2006.
43. Shih, W.K., Thermoforming of amorphous PET. *SPE ANTEC Technical Papers, Montreal, Quebec, Canada*, pp. 584-587, 1991.
44. Liu, S.C., Jack Hu, S., Woo, T.C., Tolerance analysis for sheet metal assemblies. *J. Mech. Des.*, 118, 62–675, 1996.
45. Fathailall, A.K. and Dixon, J.R., A method for early manufacturability evaluation of proposed tolerance plans for thin-walled parts, in: *International Design Engineering Technical Conferences and Computers and Information in Engineering Conference*, vol. 12822, American Society of Mechanical Engineers, pp. 9–17, 1994.
46. Kyotani, M., Kaito, A., Nakayama, K., Structural, mechanical, and thermal properties of extruded sheets of a liquid crystalline copolyester. *J. Appl. Polym. Sci.*, 47, 11, 2053–2063, 1993.
47. Choy, C.-L., Luk, W.H., Chen, F.C., Thermal conductivity of highly oriented polyethylene. *Polymer*, 19, 155–1625, 1978.
48. Hansen, D. and Bernier, G.A., Thermal conductivity of polyethylene: The effects of crystal size, density and orientation on the thermal conductivity. *Polym. Eng. Sci.*, 12, 3, 204–208, 1972.
49. Huh, M.-W., Kang, I.K., Lee, D.-H., Kim, W., Lee, D.-H., Park, L.S., Min, K.-E., Seo, K.-H., Surface characterization and antibacterial activity of chitosan-grafted poly(ethylene terephthalate) prepared by plasma glow discharge. *J. Appl. Polym. Sci.*, 81, 2769–27785, 2001.
50. Yang, C.J. and Hung, S.-W., Optimising the thermoforming process of polymeric foams: An approach by using the Taguchi method and the utility concept. *Int. J. Adv. Manuf. Technol.*, 24, 353–3605, 2004.
51. Milicevic, D., Micic, M., Stamboliev, G., Leskovac, A.R., Mitrić, M., Suljovrujić, E., Microstructure and crystallinity of polyolefins oriented via

solid-state stretching at an elevated temperature. *Fibers Polym.*, 13, 466–470, 2012.

52. Wolf, C.J., Bornmann, J.A., Grayson, M.A., The absorption of organic liquids in poly(aryl-ether-ether-ketone) [PEEK]. *J. Polym. Sci. Part B*, 29, 1533–15395, 1991.
53. Eller, R., New resin technology and global growth prospects for automotive interior skins and foams. *SPE Automotive TPO Global Conference '99*, Troy, MI, September 20-22, 1999.
54. Dorigato, A., Mahmood, H., Pegoretti, A., Optimization of the thermal mending process in epoxy/cyclic olefin copolymer blends. *J. Appl. Polym. Sci.*, 138, 49937, 2021.
55. Benabdallah, S.M.H. and Yelle, H., Static and quasi-dynamic coefficient of friction of three engineering thermoplastics: UHMWPE, PA 66, POM. *J. Mater. Sci.*, 26, 2445–24505, 1991.
56. Dröscher, M., Hertwig, K., Reimann, H., Wegner, G., Extended-chain crystals of POM-copolymers: Synthesis, structure and properties. *Macromol. Chem. Phys.*, 177, 1695–1706, 1976.
57. Xiao, H., Li, P., Ren, X., Jiang, T., Yeh, J.-T., Isothermal crystallization kinetics and crystal structure of poly(lactic acid): Effect of triphenyl phosphate and talc. *J. Appl. Polym. Sci.*, 118, 3558–3569, 2010.
58. He, D., Wang, Y., Shao, C., Zheng, G., Li, Q., Shen, C., Effect of phthalimide as an efficient nucleating agent on the crystallization kinetics of poly(lactic acid). *Polym. Test.*, 32, 1088–1093, 2013.
59. Kuram, E., Ozcelik, B., Yilmaz, F., The influence of recycling number on the mechanical, chemical, thermal and rheologi-calproperties of poly(butylene terephthalate)/polycarbonate binary blend and glass-fibre-reinforced composite. *J. Thermoplast. Compos. Mater.*, 29, 1443–1457, 2016.
60. Okaya, T. and Ikari, K., *Polyvinyl alcohol-developments*, C.A. Finch (Ed.), Wiley, London, 1992, Chap 8.
61. Pawar, S. and Dolas, D.R., Effect of process parameters on flexural strength and surface roughness in fused deposition modeling of PC-ABS material. *J. Micromanuf.*, 5, 164–170, 2021.
62. Singh, R., Kumar, S., Singh, A.P., Wei, Y., On comparison of recycled LDPE and LDPE–bakelite composite based 3D printed patch antenna. *Proc. Inst. Mech. Eng., Part L: J. Mater.: Des. Appl.*, 236, 842–856, 2021.
63. Abbas-Abadi, M.S., The effect of process and structural parameters on the stability, thermo-mechanical and thermal degradation of polymers with hydrocarbon skeleton containing PE, PP, PS, PVC, NR, PBR and SBR. *J. Therm. Anal. Calorim.*, 143, 1–16, 2020.
64. Vries, A.J.D. and Bonnebat, C., Uni-and biaxial stretching of chlorinated PVC sheets. *Polym. Eng. Sci.*, 16, 2, 93–100, 1976.
65. Vocke, R., Stempin, J.L., Schiebel, P., Herrmann, A.S., Fischer, A., *Key Eng. Mater.*, 809, 598–603, 2019.

66. Chou, R., Chang, C.-C., Leon Yu, T., Tseng, Y.H., Wu, M., Crystallization kinetics of poly(1,4-butylene-co-ethylene terephthalate). *Polym. Int.*, 50, 213–2215, 2001.
67. Aboamer, M.A., Aboamer, A.A., Elgohary, D.H., Alqahtani, T.M., Abdel-Hadi, A., Al-Mutairi, S.M., EL-Bagory, T.M.A.A., Alshareef, K.M., Rahman Mohamed, N.A., Comparative study of mechanical behavior of low- and high-density polyethylene based on UVB sterilization for medical usage. *J. Mech. Sci. Technol.*, 36, 5977–59935, 2022.
68. Shieh, Y.-T. and Liu, C.-M., Silane grafting reactions of LDPE, HDPE, and LLDPE. *J. Appl. Polym. Sci.*, 74, 3404–3411, 1999.
69. Nabirqudri, S.A.M., Roy, A.S., Prasad, M.A., Mechanical and thermal properties of free standing polystyrene – ABS blends. *Int. J. Sci. Res.*, 1, 1–75, 2015.
70. Truc, T., Thi, N., Le, H.A., Lee, B.-K., Sono-oxidation treatment of hazardous ABS/PC surface for its selective separation from ESR styrene plastics. *Environ. Sci. Pollut. Res.*, 28, 24771–247845, 2020.
71. Waeger, P., Film pouch for enteral feeding of patients. *J. Plast. Film Sheeting*, 4, 189–192, 1988.
72. Hussain, S. and Al-Sarraf, A.H.R.M., Influence of bioactive and bio inert ceramic powders on tribology properties of PMMA composite denture base. *J. Biomimetics Biomater. Biomed. Eng.*, 57, 1–8, 2022.
73. Radley, H., Thermoplastic elastomers in medical applications. *Med. Device Technol.*, 14, 4, 38–40, 2003.
74. Mahmood, H., Dorigato, A., Pegoretti, A., Healable carbon fiber-reinforced epoxy/cyclic olefin copolymer composites. *Materials*, 1313, 2165, 2020.
75. Ping, H., Chou, H., Hua, Z., Zhan'ao, T., Zhang-Chan, H., The friction wear properties and application of thermoplastic polyester elastomer and polyoxymethylene. *J. Wuhan Univ. Technol.- Mater. Sci. Ed.*, 19, 33–35, 2004.
76. Zhao, X.-Y., Hu, X., Yue, C.Y., Du, X., Zhou, Q., Synthesis, characterization, and application of semi-interpenetrating liquid crystalline polymer networks LCP/PS. *J. Appl. Polym. Sci.*, 76, 1141–1150, 2000.
77. Köckritz, T., Schiefer, T., Jansen, I., Beyer, E., Improving the bond strength at hybrid-yarn textile thermoplastic composites for high-technology applications by laser radiation. *Int. J. Adhes. Adhes.*, 46, 85–94, 2013.
78. Lagaron, J.M., Giménez, E., Saura, J.J., Gavara, R., Phase morphology, crystallinity and mechanical properties of binary blends of high barrier ethylene-vinyl alcohol copolymer and amorphous polyamide and a polyamide-containing ionomer. *Polymer*, 42, 7381–73945, 2001.
79. Lombardo, B.S., Keskkula, H., Paul, D.R., Influence of ABS type on morphology and mechanical properties of PC/ABS blends. *J. Appl. Polym. Sci.*, 54, 11, 1697–1720, 1994.
80. Cescato, R., Rigotti, D., Mahmood, H., Dorigato, A., Pegoretti, A., Thermal mending of electroactive carbon/epoxy laminates using a porous poly(ε-caprolactone) electrospun mesh. *Polymers*, 1313, 2723, 2021.

81. Ahmad, D., Wang, P., Hamila, N., Boisse, P., Numerical analysis of non-isothermal viscoelastic materials for thermoforming of polymer films. *Key Eng. Mater.*, 554-557, 1692–1698, 2013.
82. Lagaron, J.M., Giménez, E., Gavara, R., Saura, J.J., Study of the influence of water sorption in pure components and binary blends of high barrier ethylene–vinyl alcohol copolymer and amorphous polyamide and nylon-containing ionomer. *Polymer*, 42, 9531–95405, 2001.
83. Cormont, J.J., Differences between amorphous and crystalline plastics with respect to thermoforming. *Adv. Polym. Technol.*, 5, 209–218, 1985.
84. Seidler, S. and Grellmann, W., Fracture behaviour and morphology of PC/ ABS blends. *J. Mater. Sci.*, 28, 4078–4084, 1993.
85. Riaz, Z., Khan, A.N., Abbas, S.R., Hussain, Z., Structure-properties relationships of novel cyclic olefinic copolymer/ poly(L-lactic acid) polymer blends. *J. Mater. Res. Technol.*, 9, 7172–7179, 2020.
86. Chtourou, H., Riedl, B., Ait-Kadi, A., Reinforcement of recycled polyolefins with wood fibers. *J. Reinf. Plast. Compos.*, 11, 372–3945, 1992.
87. Greco, R. and Dong, L.-S., PC/ABS blends: Compatibilization and mechanical behaviour. *Macromol. Symp.*, 78, 141–153, 1994.
88. Lamontagne, N.D., The future looks bright for in-mold labeling: IML decoration expands into new materials, markets and processes, including thermoforming. *Plast. Eng.*, 73, 28–305, 2017.
89. Harron, G.W., Harkin-Jones, E., Martin, P.J., *Proc. Inst. Mech. Eng., Part E: J. Process Mech. Eng.*, 217, 181–188, 2003.
90. Ide, Y. and Ophir, Z.H., Orientation development in thermotropic liquid crystal polymers. *Polym. Eng. Sci.*, 23, 261–265, 1983.
91. Jackson, W.J. and Kuhfuss, H.F., Liquid crystal polymers. I. Preparation and properties of p-hydroxybenzoic acid copolyesters†. *J. Polym. Sci. Part A*, 14, 2043–2058, 1976.
92. Lamontagne, N.D., Forming the future of thermoforming: Processors urged to get creative & push the envelope. *Plast. Eng.*, 72, 34–37, 2016.
93. Wong, C.-P. and Bollampally, R.S., Thermal conductivity, elastic modulus, and coefficient of thermal expansion of polymer composites filled with ceramic particles for electronic packaging. *J. Appl. Polym. Sci.*, 74, 3396–34035, 1999.
94. Ozdemir, M., Yurteri, C.U., Sadıkoğlu, H., Physical polymer surface modification methods and applications in food packaging polymers. *Crit. Rev. Food Sci. Nutr.*, 39, 5, 457–77, 1999.
95. Kato, K., Recent Advances in thermoplastic elastomers in Japan - application, market and materials. *Int. Polym. Sci. Technol.*, 34, 226–232, 2007.
96. Keating, M.Y., High glass transitions of high-performance thermoplastics. *Thermochim. Acta*, 319, 201–2125, 1998.

Index

Also of Interest

Check out these published titles from the same author

Available at www.scrivenerpublishing.com

Polymer Drugs
By Muralisrinivasan Subramanian
Forthcoming 2024. ISBN 978-1-119-48000-6

Thermoforming
Processing and Technology
By Muralisrinivasan Subramanian
Published 2024. ISBN 978-1-119-55586-5

Plastics Waste Management, 2nd Edition
Processing and Disposal
By Muralisrinivasan Subramanian
Published 2019. ISBN 978-1-119-55587-2

Polymer Blends and Composites
By Muralisrinivasan Subramanian
Published 2017. ISBN 978-1-118-11889-4

Plastics Additives and Testing
By Muralisrinivasan Subramanian
Published 2013. ISBN 978-1-118-11890-0

The Basics of Troubleshooting in Plastics Processing
By Muralisrinivasan Subramanian
Published 2011. ISBN 978-0-470-62606-1

Printed and bound by CPI Group (UK) Ltd, Croydon, CR0 4YY
17/09/2026
14964245-0003